Serge Cosnier (Ed.)
Bioelectrochemistry

Also of interest

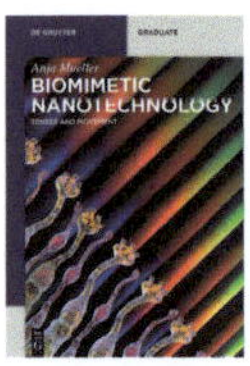

Biomimetic Nanotechnology
Senses and Movement
Mueller 2017
ISBN 978-3-11-037914-3, e-ISBN 978-3-11-037916-7

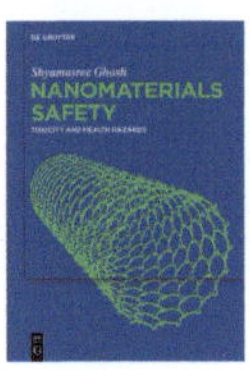

Nanomaterials Safety
Toxicity And Health Hazards
Ghosh 2018
ISBN 978-3-11-057808-9, e-ISBN 978-3-11-057909-3

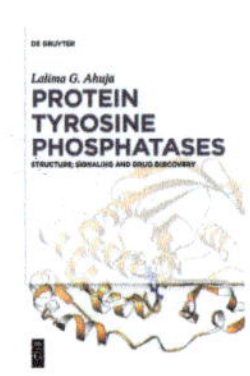

Protein Tyrosine Phosphatases
Structure, Signaling and Drug Discovery
Ahuja, 2018
ISBN 978-3-11-042643-4, e-ISBN 978-3-11-042177-4

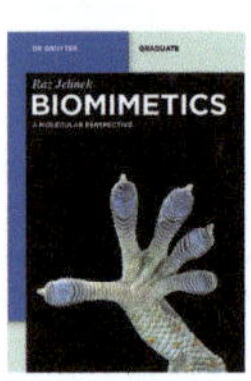

Biomimetics
A Molecular Perspective
Jelinek, 2013
ISBN 978-3-11-028117-0, e-ISBN 978-3-11-028119-4

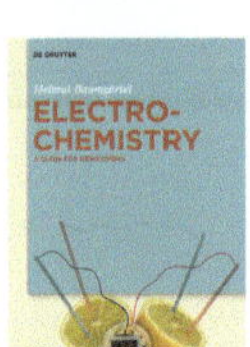

Electrochemistry
A Guide for Newcomers
Baumgärtel, 2019
ISBN 978-3-11-044340-0, e-ISBN 978-3-11-043739-3

Bioelectrochemistry

Design and Applications of Biomaterials

Edited by
Serge Cosnier

DE GRUYTER

Editor
Dr. Serge Cosnier
Department of Molecular Chemistry UMR CNRS 5250
570 rue de la Chimie
Bâtiment Nanobio
Université Grenoble-Alpes
CS 40700
38058 Grenoble cedex 9
France
Serge.Cosnier@univ-grenoble-alpes.fr

ISBN 978-3-11-056898-1
e-ISBN (PDF) 978-3-11-057052-6
e-ISBN (EPUB) 978-3-11-056926-1

Library of Congress Control Number: 2018963422

Bibliographic information published by the Deutsche Nationalbibliothek
The Deutsche Nationalbibliothek lists this publication in the Deutsche Nationalbibliografie; detailed bibliographic data are available on the Internet at http://dnb.dnb.de.

Typesetting: Integra Software Services Pvt. Ltd.
Printing and binding: CPI books GmbH, Leck
Cover image: Michael Holzinger

www.degruyter.com

MIX
Papier aus verantwortungsvollen Quellen
FSC
www.fsc.org
FSC® C083411

Contents

List of contributors

Olga Aleksejeva
Department of Biomedical Science
Faculty of Health and Society
Malmö University
20506 Malmö, Sweden
olga.aleksejeva@mau.se

Frédéric Barrière
Université Rennes
CNRS, ISCR Institut des Sciences Chimiques de Rennes
UMR 6226
F-35000 Rennes, France
frederic.barriere@univ-rennes1.fr

Zoltan Blum
Department of Biomedical Science
Faculty of Health and Society
Malmö University
20506 Malmö, Sweden
zoltan.blum@mau.se

Redouane Borsali
Centre de Recherches sur les Macromolécules Végétales (CERMAV)
Univ. Grenoble Alpes
CNRS, UPR 5301
38000 Grenoble, France
borsali@cermav.cnrs.fr

Laurent Bouffier
Univ. Bordeaux
CNRS, Bordeaux INP
ISM, UMR 5255
F-33400 Talence, France
Laurent.Bouffier@enscbp.fr

Wen-Rong Cai
School of Environmental and Biological Engineering
Nanjing University of Science and Technology
Xiaollingwei 200
Nanjing, China
315102002432@njust.edu.cn

Stefan Cirovic
Department of Biomedical Science
Faculty of Health and Society
Malmö University
20506 Malmö, Sweden
stefan.cirovic@mau.se

Serge Cosnier
Department of Molecular Chemistry, DCM
Univ. Grenoble Alpes
CNRS
38000 Grenoble, France
serge.cosnier@univ-grenoble-alpes.fr

Nazua L. Costa
Université Rennes
CNRS, ISCR Institut des Sciences Chimiques de Rennes
UMR 6226
F-35000 Rennes, France
nazua.costa@univ-rennes1.fr

Shaojun Dong
State Key Laboratory of Electroanalytical Chemistry
Changchun Institute of Applied Chemistry
Chinese Academy of Sciences
Changchun, Jilin 130022, China
dongsj@ciac.ac.cn

Mathieu Etienne
Université de Lorraine
CNRS, Laboratoire de Chimie Physique et Microbiologie pour les Matériaux et l'Environnement (LCPME)
405 rue de Vandoeuvre
F-54600 Villers-les-Nancy, France
mathieu.etienne@lcpme.cnrs-nancy.fr

Magnus Falk
Department of Biomedical Science
Faculty of Health and Society
Malmö University
20506 Malmö, Sweden
magnus.falk@mau.se

https://doi.org/10.1515/9783110570526-201

Thomas Flinois
Université Rennes
CNRS, ISCR Institut des Sciences Chimiques de Rennes
UMR 6226
F-35000 Rennes, France
thomas.flinois@univ-rennes1.fr

Fabien Giroud
Department of Molecular Chemistry, DCM
Univ. Grenoble Alpes
CNRS
38000 Grenoble, France
fabien.giroud@univ-grenoble-alpes.fr

Andrew J. Gross
Department of Molecular Chemistry, DCM
Univ. Grenoble Alpes
CNRS
38000 Grenoble, France
andrew.gross@univ-grenoble-alpes.fr

Michael Holzinger
Department of Molecular Chemistry, DCM
Univ. Grenoble Alpes
CNRS
38000 Grenoble, France
michael.holzinger@univ-grenoble-alpes.fr

Ioannis Ieropoulos
Bristol BioEnergy Centre
Bristol Robotics Laboratory
University of the West of England
Bristol, UK
ioannis2.Ieropoulos@uwe.ac.uk

Huangxian Ju
State Key Laboratory of Analytical Chemistry for Life Science
Nanjing University
Nanjing 210093, China
hxju@nju.edu.cn

Evgeny Katz
Department of Chemistry and Biomolecular Science
Clarkson University, Potsdam
NY 13699-5810, USA
ekatz@clarkson.edu

Alexander Kuhn
Univ. Bordeaux
CNRS, Bordeaux INP
ISM, UMR 5255
F-33400 Talence, France
Alexander.Kuhn@enscbp.fr

Estelle Lebègue
Université Rennes
CNRS, ISCR Institut des Sciences Chimiques de Rennes
UMR 6226
F-35000 Rennes, France
estelle.lebegue.86@gmail.com

Alan Le Goff
Department of Molecular Chemistry, DCM
Univ. Grenoble Alpes
CNRS
38000 Grenoble, France
alan.le-goff@univ-grenoble-alpes.fr

Issei Otsuka
Centre de Recherches sur les Macromolécules Végétales (CERMAV)
Univ. Grenoble Alpes
CNRS, UPR 5301,
38000 Grenoble, France
issei.otsuka@cermav.cnrs.fr

Dmitry Pankratov
Department of Chemistry
Technical University of Denmark
Kongens Lyngby, Denmark
dmpp@kemi.dtu.dk

Timothé Philippon
Université Rennes,
CNRS, ISCR Institut des Sciences Chimiques de Rennes
UMR 6226,
F-35000 Rennes, France
timothe.philippon@etudiant.univ-rennes1.fr

Joanna Rogińska
Université de Lorraine
CNRS, Laboratoire de Chimie Physique et Microbiologie pour les Matériaux et l'Environnement (LCPME)
405 rue de Vandoeuvre
F-54600 Villers-les-Nancy, France
joanna.roginska@univ-lorraine.fr

Carlo Santoro
Bristol BioEnergy Centre
Bristol Robotics Laboratory
University of the West of England
Bristol, UK
carlo.santoro830@gmail.com

Dan Shan
School of Environmental and Biological Engineering
Nanjing University of Science and Technology
Xiaollingwei 200
Nanjing, China
danshan@njust.edu.cn

Sergey Shleev
Department of Biomedical Science
Faculty of Health and Society
Malmö University
20506 Malmö, Sweden
sergey.shleev@mah.se

Francesca Soavi
Department of Chemistry "G. Ciamician"
Alma Mater Studiorum – University of Bologna
Italy
francesca.soavi@unibo.it

Neso Sojic
Univ. Bordeaux
CNRS, Bordeaux INP
ISM, UMR 5255
F-33400 Talence, France
Neso.Sojic@enscbp.fr

Seiya Tsujimura
Division of Materials Science
Faculty of Pure and Applied Sciences
University of Tsukuba
1-1-1 Tennodai
Tsukuba, Ibaraki 305-8573, Japan
seiya@ims.tsukuba.ac.jp

You Yu
State Key Laboratory of Electroanalytical Chemistry
Changchun Institute of Applied Chemistry
Chinese Academy of Sciences
yuyoufree@gmail.com

Andrew J. Gross, Michael Holzinger, Serge Cosnier

1 Buckypapers for bioelectrochemical applications

1.1 Introduction to buckypapers for bioelectrochemistry

Buckypapers are self-supporting nanostructured thin films of entangled carbon nanotubes (CNTs), resembling an electronic paper, which are held together by π–π stacking and interweaving interactions. In 1998, Richard Smalley's group was the first to report the formation of CNT sheets (coining the term "buckypaper") as a method for characterizing the production quality of CNTs [1]. CNT buckypapers and their composites have excellent prospects for a wide range of applications from aerospace materials [2] to sensors [3, 4] and fuel cells [5, 6], and are thus of interest to a vast research and industrial community. Over the last 5–7 years, buckypaper materials have emerged for the construction of implantable, wearable, and portable bioelectronic devices owing to their properties including high conductivity and porosity, flexibility, lightweight, biocompatibility, and the ability for electron transfer with enzymes and microbes [7]. In addition to being used for enzymatic and microbial biofuel cell construction [7–9], buckypapers are also being used for hybrid supercapacitors [10], photoelectric biofuel cells [11], and other bioelectronic systems including biosensors [4, 12], bioreactors [13], and biologic devices [14].

Buckypaper is an accepted term for CNT sheets, disordered or aligned, formed by vacuum filtration of aqueous (e.g., in the presence of a nonionic surfactant such as Triton X-100) and nonaqueous dispersions (e.g., in *N, N*-dimethylformamide, DMF) of single-walled, double-walled, and multiwalled carbon nanotubes (SWCNTs, DWCNTs, and MWCNTs) [1, 6, 15]. Sonication, centrifugation, and additional filtration steps are also commonly used to improve the quality and purity of the CNT dispersion prior to filtration through a porous membrane [6, 15]. Finally, free-standing "lab-made" buckypaper sheets are obtained after washing, drying, and peeling from the underlying membrane. "Commercial" buckypaper prepared by continuous manufacturing is a popular type of buckypaper. The MWCNT buckypaper from Buckeye Composites (a division of NanoTechLabs, USA) is the most widely reported commercialized buckypaper for bioelectrochemical applications [16–19]. In addition to vacuum filtration, lab-made buckypaper can also be prepared by methods including domino-pushing [20], CNT winding [21], and by using a lab-scale hand sheet former [22]. The most common types of buckypapers used in bioelectrochemistry are illustrated in Fig. 1.1.

Andrew J. Gross, Michael Holzinger, Serge Cosnier, Department of Molecular Chemistry, Université Grenoble Alpes, Grenoble, France

https://doi.org/10.1515/9783110570526-001

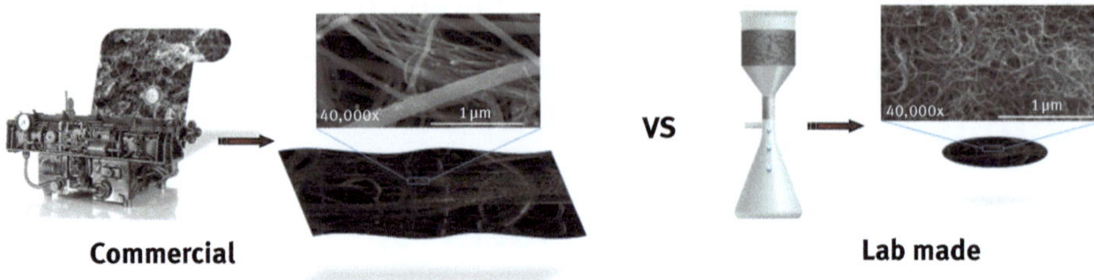

Fig. 1.1: Illustration of fabrication procedures for commercial and lab-made buckypaper via mass production and lab-scale vacuum filtration, respectively.

Fabrication of buckypaper is conceptually straightforward, but factors such as dispersion homogeneity, CNT type and chemical functionality, membrane porosity, and the presence of additives in the dispersion all create differences with respect to material reproducibility and functionality. Several studies have focused on tuning the physical and mechanical properties of buckypaper such as porosity, Young's modulus, hardness, and electrical conductivity [15, 22–30]. For example, Shen et al. investigated the length of CNTs and showed how this parameter strongly governed the viscoelasticity and permeability of buckypaper [30]. Whitby et al. demonstrated that the porosity of buckypaper could be tuned using different casting solvents [29]. Oh et al. revealed the crucial roles of CNT suspension concentration and filtration velocity for self-assembled alignment of buckypaper to enhance the mechanical properties [15]. In our recent electrochemical study, we compared the physical, chemical, electrochemical, and bioelectrocatalytic properties of lab-made and commercial buckypapers [16].

For bioelectrochemical applications, surfactant-free methods of producing buckypaper are highly desirable owing to the undesirable impact of residual surfactant on conductivity and biocompatibility risks such as cell lysis [6, 16, 29]. Buckypaper prepared from nonaqueous solvent DMF or alcohols without surfactant has been reported with success, for example, owing to improved CNT-solvent interactions and the possibility to dissolve a wide range of chemical modifiers [6, 22, 31–33]. However, the removal of nonaqueous solvents is crucial to minimize enzyme denaturation, microbe deactivation, and material toxicity, in particular for in vivo applications.

Porous 3D-nanostructured carbon structures based on CNTs [34], mesoporous carbons [35, 36], and carbon black [37] have become privileged conducting supports for bioelectrode design. Many of the best performing enzymatic bioelectrodes to date have been fabricated using CNTs due to their high specific surface area and exceptional electronic and mechanical properties, for example, that allow effective direct and mediated electron transfer with enzymes for bioelectrocatalysis [8, 34]. Porous 3D-nanostructured carbon electrodes are particularly attractive due to the high surface/volume ratio, which increases the loading of catalyst per geometric area. The size and curvature of pore and nanotube structures (e.g., nano-, micro-, and mesopores) and their surface chemistry play a crucial role on the bioelectrocatalytic parameters of the electrode, such as mass transport, enzyme orientation and activity, and electron transfer kinetics [16, 35, 38, 39].

An important aspect that is frequently overlooked in bioelectrode design is the development of porous electrodes with practical physical properties for their target application. For example, for wearable biofuel cell and biosensor applications, "skin-like" electrodes that are soft, bendable, and even stretchable are required [40, 41]. For implantation, compact and lightweight electrodes are required with flexible geometries and superior stability and biocompatibility [7, 42]. Many bioelectrodes, including 3D-nanostructured electrodes, are prepared via the simple adsorption of nanocarbons (e.g., MgO-templated carbon [35], mesoporous nanoparticles [43], and Ketjen black [37]) onto robust but bulky and nonflexible glassy carbon supports. The resulting thin films can also be fragile and prone to delamination from the electrode. Bulk CNT "pellet" electrodes formed by compression in the presence of enzyme emerged as a more practical solution to glassy carbon-based electrodes for implantable biofuel cells; however, the pellet electrodes were still quite cumbersome, fragile, and brittle [44, 45]. For wearable applications, carbon electrodes based on printed CNT inks and pellets supported on a flexible substrate have been developed [46, 47].

Buckypaper is an alternative form of 3D-nanostructured electrode, which is the electrode itself (no support is required) and has attractive qualities for interfacing biological and electrochemical systems for bioelectrochemical applications [7]. Compared to the existing CNT pellet bioelectrodes, for example, buckypapers offer a higher density of CNTs per surface area and improved stability owing to their more compact structure. A comparison of the quantity of enzymes used per electrode also reveals that buckypapers are more economical than CNT pellet electrodes [6, 48]. Nevertheless, buckypapers do suffer from being brittle and fragile in several cases. In addition, lab-made fabrication of buckypapers can also require the use of significant amounts of organic solvent.

1.2 Biological fuel cell devices

One of the important applications of buckypaper is in the field of biofuel cells [7]. Buckypaper electrodes have been used in implanted bioelectronic devices since 2012. Notably, enzymatic biofuel cells based on commercial buckypaper have been implanted in different animals and used for in vivo energy harvesting and device powering via the organism [17, 49]. Biofuel cells are emerging power sources that can generate clean electrical energy from chemical substrates present in various media such as biofluids and environmental waters using enzymes (enzymatic biofuel cells) or microorganisms (microbial biofuel cells) as the catalysts [8]. Power is generated from bioelectrochemical reactions via the oxidation of fuels such as hydrogen and sugars (at the anode) and the reduction of oxidants such as oxygen (at the cathode). Biological fuel cells offer advantages compared to conventional batteries and fuel cells including the use of renewable and nontoxic catalysts and fuels. Enzymatic biofuel cells are considered as promising power sources for low-power electronic devices for short periods (e.g., 3 months to 1 year), whereas microbial biofuel cells are better suited for large-scale powering and industrial applications over longer periods (e.g., years). A short discussion highlighting some of the latest innovations in biofuel cell research, as well as some of the hype, can be found in our recent opinion article [50].

1.2.1 Implantable enzymatic biofuel cells

Katz and coworkers were the first to report the use of buckypaper for the construction of enzymatic biofuel cells. In its pioneering research, the buckypaper-based biofuel cells were implanted in snails [17], clams [51], rats [52], and lobsters [49] and operated for durations ranging from a few hours to several weeks. The open-circuit voltages (OCVs) and maximum power outputs of the biofuel cell devices ranged from 0.14 to 0.54 V (up to 1.2 V for anode–cathode pairs connected in series) [49] and ca. 0.5–160 μW, respectively. The implanted biofuel cells exploited commercial MWCNT buckypaper (Buckeye Composites) modified with adsorbed pyrroloquinoline quinone-dependent glucose dehydrogenase (PQQ-GDH) at the anode for glucose oxidation, and laccase from *Trametes versicolor* (*Tv*Lc) at the cathode for oxygen reduction. The glucose/O_2 biofuel implanted in the shell of the snail that operates from the hemolymph is depicted in Fig. 1.2. Buckypaper proved to be an effective choice due to its bioelectrocatalytic performance as well as its physical properties such as submillimeter (typically 5–200 μm) thickness, flexibility, and shapeability, which facilitated their insertion via small incisions into the organisms (e.g., in lobsters) [49] or their conformal contact with internal muscle tissues (e.g., in rats) [52].

In an early fascinating experiment, Katz and coworkers demonstrated that feeding the snail, or allowing it to rest for up to 1 h, allowed the power output of the biofuel cell to be restored via substrate diffusion and metabolism [17]. The stable operation of the implanted biofuel cell over 2 weeks highlighted practical stability and limited inhibition and fuel cell deactivation. Katz and coworkers later demonstrated the connection of implanted biofuel cells in series and the possibility to produce sufficient energy to turn an electric motor or an electronic watch [49, 51]. In another breakthrough, the connection of five buckypaper biofuel cells in series generated an OCV of 2.8 V in human serum, highlighting the potential of buckypaper

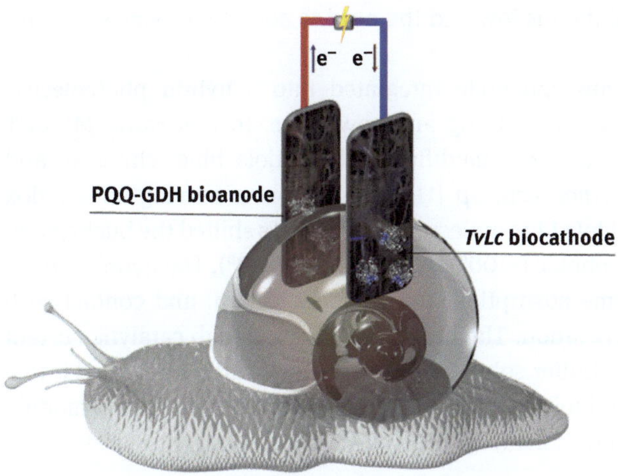

Fig. 1.2: Cartoon of an implanted glucose/oxygen biofuel cell inserted via incision of two buckypaper bioelectrodes in the snail's shell.

biofuel cells to meet the ca. 1.4 V requirement to power a range of low-power microelectronic devices without a voltage boost convertor [49].

1.2.2 Wearable enzymatic biofuel cells

Enzymatic buckypaper electrodes have great potential in the emerging field of wearable biofuel cells owing to their attractive properties, depending on the fabrication methods used, such as their flexibility and skin-conformability under bending [7, 53]. Compared to implantable biofuel cells that require surgical procedures, wearable biofuel cells can be mounted on (i.e., noninvasive) or inserted in (i.e., minimally invasive) the body without the need for surgical intervention. Wearable devices are therefore slightly easier to develop and more user friendly compared to implantable devices. Skin-based wearable biofuel cells that exploit human sweat are popular as sweat is easily accessible, can be produced in high quantities, and contains attractive fuels such as lactate and glucose. For example, lactate is particularly attractive owing to its high concentrations of ca. 10–100 mmol L^{-1} compared to ≤1 mmol L^{-1} for glucose [54, 55].

The first lactate-oxidizing buckypaper bioanode was developed by Atanassov and coworkers. Commercial MWCNT buckypaper (Buckeye Composites) was modified with methylene green as an electrocatalyst to regenerate dissolved NAD^+ cofactor, chitosan as a biocompatible stabilizing matrix, and adsorbed lactic dehydrogenase from *Lactobacillus leichmannii* as the selective catalyst [56]. The bioanode was not tested in a fuel cell setup but could potentially be exploited. However, the catalytic current of

$53.4 \pm 5.1\,\mu A\ cm^{-2}$ in buffer solution is low and the need to add the enzyme's cofactor are major limitations.

A practical lactate-oxidizing bioanode integrated into a hybrid photoelectric biofuel cell was recently reported by Dong and coworkers. In this work, MWCNT buckypaper (Buckeye Composites) was modified with Meldola blue, chitosan, and lactate oxidase (LOx) *from Pediococcus sp* [11]. In addition to serving as a redox mediator, the π–π stacked Meldola blue molecules on the CNTs shifted the buckypaper surface wettability from hydrophobic (<100°) to hydrophilic (<30°). The hydrophilicity is attractive to promote enzyme adsorption, during modification, and contact with electrolytic solution, during operation. The bioanode delivered high catalytic current densities of ca. $2.1\,mA\ cm^{-2}$ in buffer solution and was successfully integrated into a wearable epidermic lactate/O_2 biofuel cell capable of powering a bluetooth module and transmitting data to a smartphone (Fig. 1.3).

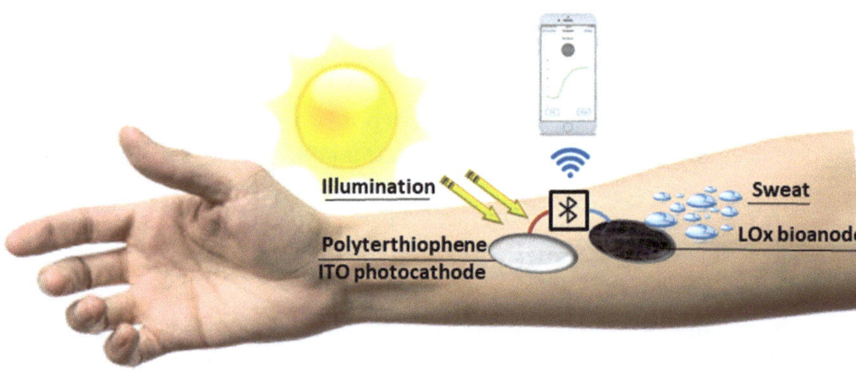

Fig. 1.3: Principle of a wearable hybrid biofuel cell with a polyterthiophene-coated indium tin oxide photocathode and a lactate-oxidizing buckypaper bioanode.

Buckypaper bioelectrodes have also been developed for energy harvesting from tear fluid to power "smart" electronic contact lenses for applications in human vision and sensing [18]. It is envisioned that such biofuel cells could be used to power contact lenses worn by type I diabetics for noninvasive glucose monitoring [57]. Minteer and coworkers exploited commercial MWCNT buckypaper (Buckeye Composites) immobilized on an elastomeric contact lens [18]. Buckypaper was used to benefit from its high surface area, laser cuttability, durability, and mechanical properties such as flexibility. For example, the authors reported that the buckypaper exhibited better mechanical properties compared to other conductive papers such as Toray paper and evaporated metal electrodes. The bioanode was constructed with NAD-dependent lactate dehydrogenase *from Escherichia coli* modified with methylene green electrocatalyst and a cross-linked polymeric hydrogel matrix to entrap the NAD^+ cofactor and enzyme on the electrode surface. The use of immobilized NAD^+ is in principal more practical than the classical approach that requires cofactor to be added to the solution. A further advantage is that this strategy could be applied to a wide range of NAD-dependent enzymes. In addition to lactate oxidation, electrocatalytic ascorbic acid oxidation was also observed, although this had parasitic effects on the fuel cell. Lisdat and coworkers also observed parasitic effects due to ascorbic acid oxidation at buckypaper electrodes in human biofluids [19]. For the biofuel cell developed by Minteer and coworkers, the lactate-oxidizing bioanode was combined with an oxygen-reducing biocathode prepared via oriented adsorption of bilirubin oxidase (*Mv*BOx) on pyrenemethyl anthracene modified buckypaper. The biofuel cell delivered an OCV of 0.41 V and a small power output up to 8 µW cm^{-2} at 0.2 V in tear solution with stability for several hours [18]. The low-power output and stability revealed that there is plenty of room for improvement. Improved performance could be achieved by removing the ascorbic acid interference, increasing the amount and stability of the immobilized enzymes and cofactor, and by reducing biocatalyst leaching.

The mechanical and chemical properties of buckypaper can also be improved for wearable and implantable applications. The strength and Young's modulus of buckypaper electrodes can be enhanced by incorporating functional polynorbornenes [58] and molecular cross-linkers such as pyrene derivatives [22, 33]. The introduction of polynorbornene copolymers, as illustrated in Fig. 1.4, revealed the possibility to introduce multiple functionalities to the buckypaper, such as pyrene groups, for cross-linking, and a reactive *N*-hydroxysuccinimide ester, for covalent tethering of molecules for enzyme wiring or the enzymes themselves [58]. Alternatively, the strength, properties, and hardness of buckypaper can be significantly enhanced by aligning the CNTs in the buckypaper with a high packing density [15]. Lab-made rather than commercial buckypapers may hold the key to improved buckypaper biofuel cell performance, for example, due to (i) the ability to better tune the chemical and physical properties of the bulk materials, (ii) their superior larger surface areas, and (iii) their apparently improved enzyme wiring for bioelectrocatalysis [6, 16].

1.2.3 Portable enzymatic biofuel cells

Enzymatic glucose/O_2 biofuel cells have been designed for powering of portable electronics such as music players and toy cars [59]. A potentially exciting application is the development of eco-friendly paper-based power sources. Atanassov and coworkers initiated the development of MWCNT buckypaper-based paper biofuel cells that delivered up to 1.8 V (3 biofuel cells in series) and power outputs up to 13.1 mW under quasisteady flow [60, 61]. For example, a digital clock was powered using the commercial soft drink, Gatorade, which was supplied via cellulose filter paper. In addition to demonstrating the portable paper-based biofuel cell concept (using

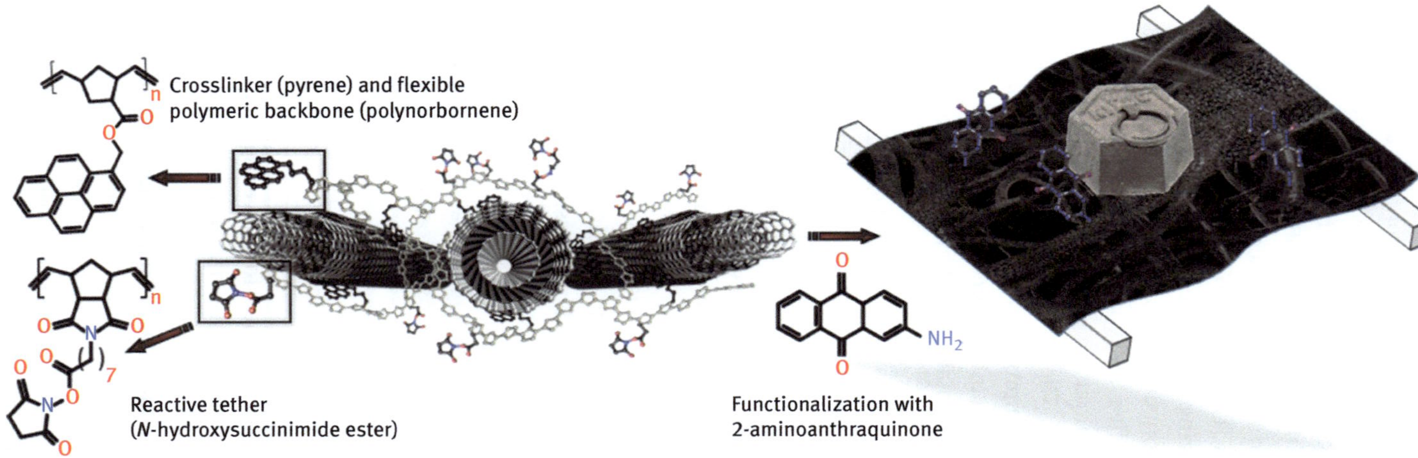

Fig. 1.4: Incorporation of "precision" polynorbornenes in buckypaper to improve the mechanical properties, such as flexibility, tensile modulus, and strength, and the chemical properties, for covalent attachment of molecules to improve enzyme orientation and wiring.

buckypaper from Buckeye Composites), this work nicely demonstrated the possibility to use commercially available liquids rather than biofluids or environmental matrices to generate power. The need to add NAD^+ cofactor for glucose oxidation at the NAD-dependent glucose dehydrogenase anode is nevertheless a significant inconvenience [60]. In later work by the authors, the same bioanode was combined with an air-breathing MWCNT buckypaper/Toray paper composite modified with *Mv*BOx as the oxygen-reducing biocathode [61]. This paper biofuel cell also benefited from a supercapacitive element that facilitated the production of mW power outputs, under pulse operation, and 4,200 discharge/recharge cycles over a period of 3 days [61].

The elegant concept of using cellulose paper as a transport medium to induce passive flow via wicking has also been explored for the construction of a paper-based "enzyme cascade"-based biofuel cell. This novel type of biofuel device was developed to improve bioanode performance by exploiting sequential catalytic reactions and multiple fuels (Fig. 1.5). Multiple NAD(H)-dependent enzymes were used at the MWCNT buckypaper (Buckeye Composites) anode for the multistep oxidation of ethanol to acetate, and complete oxidation of methanol to CO_2 [62].

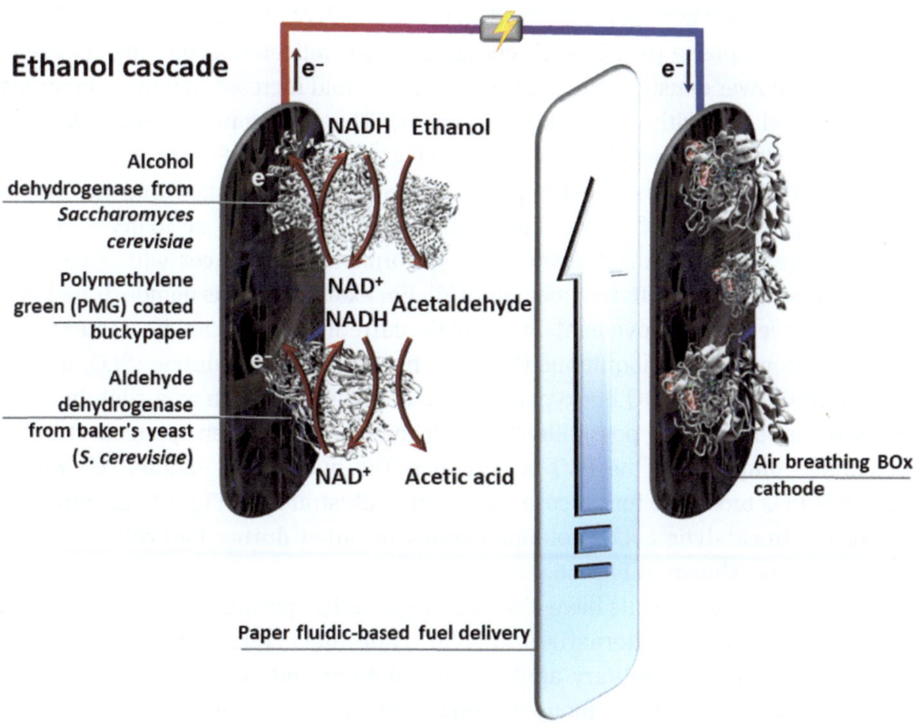

Fig. 1.5: Schematic of a biofuel cell with a multiple NAD-dependent enzyme bioanode, for the cascade oxidation of ethanol to acetic acid, coupled with an air-breathing BOx biocathode, for oxygen reduction. The fuel ethanol is delivered to the cell via paper fluidics.

The ability to use air-breathing cathodes is a major advantage for portable biofuel cells compared to in vivo biofuel cells due to the possibility to enhance oxygen delivery at the electrode interface. The testing and optimization of MWCNT buckypaper-based air-breathing biocathodes with immobilized *Mv*BOx has been reported [63, 64]. In combination with air-breathing cathodes, the use of glucose-oxidizing anodes that exploit the emerging enzyme, FAD-dependent GDH (FAD-GDH), is considered to be one of the most practical biofuel cell setups currently available. The use of FAD-GDH, which is oxygen insensitive (unlike glucose oxidase, GOx) and can now be obtained commercially, has only very recently been reported in combination with quinone-modified buckypapers and, promisingly, gave high catalytic currents up to 1.97 mA cm^{-2} [65] and 5.4 mA cm^{-2} [6]. The integration of an FAD-GDH-based buckypaper anode in biofuel cells was reported but not yet in combination with an air-breathing cathode.

The integration of energy storage elements such as supercapacitors together with biofuel cell stacks, assembled in series or parallel, is expected to become increasingly important for the powering of portable miniaturized electronic devices [10]. Hou and Liu reported a flexible lab-made buckypaper biofuel cell device connected in series based on a FAD-GDH-modified anode and a laccase-modified cathode. In addition, a supercapacitor was incorporated based on polyvinyl alcohol–H_3PO_4 and a MWCNTs–polyaniline composite [10]. The self-charging device delivered a charging voltage of 0.8 V and a power density up to 326 μW cm^{-2} (a twofold increase in power compared to the biofuel cell without the supercapacitor). The capacitance of the MWCNTs/polyaniline electrode was observed to be 329 F g^{-1} at 10 mVs^{-1}. For comparison, Tran et al. reported a potentially better-performing supercapacitor based on the in situ electropolymerization of polyaniline at buckypaper (397 Fg^{-1} at 10 mV s^{-1}) [66]. Furthermore, Gross et al. observed clear performance differences with FAD-GDH-based bioanodes with different mediators [6]. For example, it was demonstrated that around a fivefold improvement in catalytic current density can be achieved by changing from a naphthoquinone (NQ) to a phenanthroline quinone (PLQ) derivative. Nevertheless, the NQ buckypaper showed advantages such as a slightly lower onset potential and the possibility to reach the catalytic current plateau at lower potentials compared to the PLQ buckypaper. The buckypaper biofuel cell with a powerful PLQ bioanode for glucose oxidation is illustrated in Fig. 1.6. Examples of the bioelectrocatalytic cyclic voltammograms recorded during half-cell characterization are also shown in Fig. 1.6.

Enzymatic biofuel cells have also been explored as portable, low-temperature, and sulfur-insensitive alternatives to solid oxide fuels cells that run on JP-8 jet fuel, for example, for military applications. Minteer and coworkers exploited an enzyme cascade of alkane monooxygenase and alcohol oxidase at an MWCNT buckypaper (Buckeye Composites) anode for mediated bioelectrocatalytic oxidation of hexane, octane, and jet fuel [67]. This study demonstrated the possibility to exploit alkanes as fuels for biofuel cells for the first time, and, additionally,

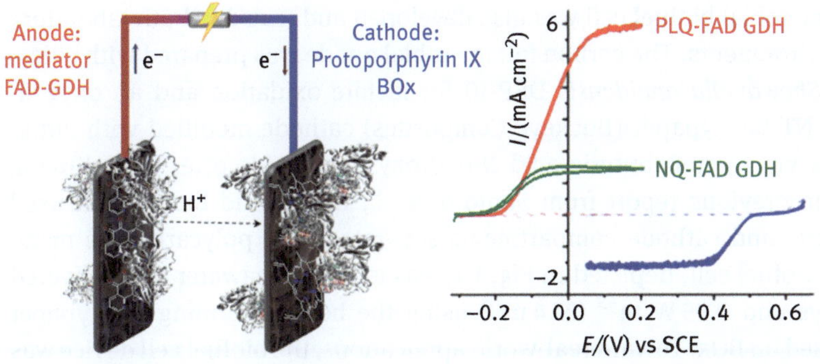

Fig. 1.6: Cartoon of the buckypaper-based glucose/oxygen biofuel cell using FAD-GDH and an artificial quinone mediator at the bioanode, and protoporphyrin IX-oriented BOx at the biocathode. Right: cyclic voltammograms of bioelectrocatalysis at FAD-GDH bioanodes, with different mediators, PLQ and NQ, and the BOx biocathode.

showed the possibility to produce power densities up to 3 mW cm^{-2} without preprocessing of alkane-based fuels.

1.2.4 Microbial biofuel cells

Microbial biofuel cells that exploit electroactive microbes such as bacteria are an eco-friendly technology for applications such as energy-generating wastewater treatment, self-powered sensors, and large-scale electricity production from biomass [68, 69]. Compared to enzymatic biofuel cells, microbial biofuels typically give lower catalytic efficiencies but are better suited to large-scale energy generation and therefore larger raw power output. Bioelectrodes based on bacteria are also generally more durable under harsh and polluted conditions compared to enzymatic electrodes. A typical microbial biofuel cell consists of an anode with acidophilic bacteria on the electrode surface, for oxidation via metabolic processes, and a cathode, typically Pt or activated carbon, for oxygen reduction. The anode and cathode are classically placed in individual chambers and separated by an ion-selective membrane.

In 2013, Santoro et al. reported for the first time the use of a microbe-based carbon cloth anode together with an MWCNT buckypaper (Buckeye Composites)–carbon cloth enzymatic air-breathing cathode in a compartmentless hybrid biofuel cell [9]. The biofuel cell operated in buffer with additional sodium acetate and wastewater solutions. A key observation was that the biofuel cell exploiting the oxygen-reducing BOx modified cathode exhibited a ca. 200 mV improvement in OCV compared to the biofuel cell with a Pt cathode under equivalent conditions. A maximum power density of 2 W m^{-2} (200 µW cm^{-2}) was achieved in 125 mL of buffer solution.

A hybrid microbial biofuel cell was also developed and tested in both laboratory and marine environments. The carbon felt microbial anode was prepared with silica-encapsulated *Shewanella oneidensis* DSP-10 for lactate oxidation and an oxygen-reducing MWCNT buckypaper (Buckeye Composites) cathode modified with either laccase or BOx via 1-pyrenebutyric acid *N*-hydroxysuccinimide ester cross-linking [70]. Unlike the previous report from Santoro et al., this hybrid biofuel cell used individual anode and cathode compartments separated by a polycarbonate membrane [9]. The biofuel cell, depicted in Fig. 1.7, was tested in seawater and delivered 0.7 V for 9 days and 0.95 W m^{-3} (22.4 μW) using the best performing buckypaper biocathode based on BOx. Toward real-world applications, the biofuel cell device was also tested in a marine environment using a floating watercraft.

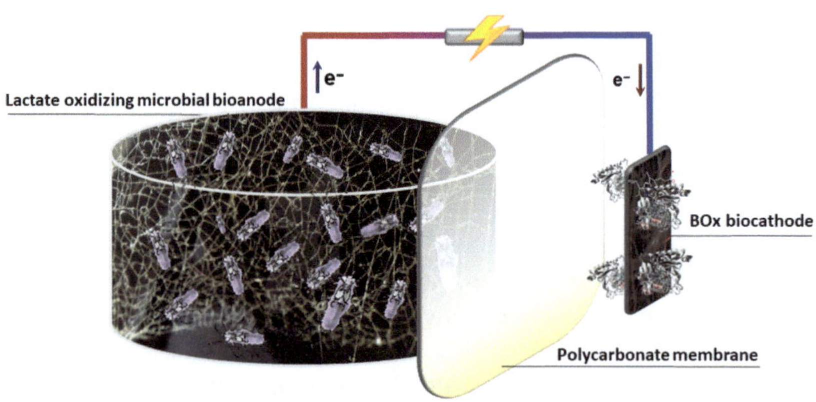

Fig. 1.7: Representation of a hybrid microbial biofuel cell for harvesting energy from sea water with an enzymatic cathode and a microbial bioanode in individual compartments.

To improve power performance for portable wastewater applications, Santoro et al. reported a "smart" supercapacitive microbial biofuel cell [71]. The fuel cell was designed with an air-breathing MWCNT buckypaper (Buckeye Composites) cathode modified with BOx, and an activated carbon/carbon black microbial anode with a mixed cultures bacteria. In addition, a carbon brush supercapacitor electrode was short-circuited to the microbial anode. The supercapacitive fuel cell delivered up to 19 mW (84.4 W m^{-2}, 152 W m^{-3}) at pulse currents as high as 45 mA, which is one of the highest power values reported for a microbial fuel cell. The authors highlighted that the integrated capacitor exhibited shorter recharging times (seconds to minutes) compared to the use of external supercapacitors (hours). Having shorter recharge times is attractive due to the possibility to increase the power output frequency. A further advantage is that the supercapacitor size can be modulated to decouple energy and power depending on the target application.

Buckypaper has also been explored as a high surface area conductive support for the preparation of microbial anodes in biofuel cells [72, 73]. Kerzenmacher and coworkers reported a comparison of bioanodes modified with *Shewanella oneidensis* MR-1 and prepared using different carbon materials, namely, activated carbon felt and cloths, graphite felt and foil, Toray paper, and lab-made MWCNT buckypaper. The high surface area activated carbon cloth with a Brunauer–Emmett–Teller (BET) specific area of ca. 800 m^2 g^{-1} outperformed the other carbon materials in terms of catalysis per geometric area and also OCP due to factors such as improved mass transport, electron transfer (e.g., mediated electron transfer via self-secreted flavins), and bacterial migration into the porous structure (e.g., enhanced catalyst loading). Nevertheless, the buckypaper exhibited the largest catalytic current in terms of the volumetric density. Volumetric density is arguably a more accurate representation of current density due to the 3D nature of the electrodes. Lower ohmic losses were also observed using buckypaper, attributed to the enhanced conductivity of CNTs. The lab-made MWCNT buckypaper reported in this work had a BET surface area of 70 m^2 g^{-1}, which compares to values of ca. 30 and 260 m^2 g^{-1} reported recently by us for commercial (Buckeye Composites) and lab-made MWCNT buckypaper electrodes [16]. In a 2018 study, Kerzenmacher and coworkers reported a systematic comparison of bioanodes with the prominent model organism, *Geobacter sulfurreducens* [73]. Similar catalytic current densities were observed at different carbon- and metal-based anodes, which included lab-made MWCNT buckypaper. The highest limiting current density of up to 756 μA cm^{-2} was observed with graphite foil, which compares to the 613 μA cm^{-2} observed when buckypaper was used. In terms of the time until the limiting current is reached, buckypaper anodes took ca. 319 h, whereas graphite foil took ca. 1,134 h (including the one week initial growth phase) using step-wise galvanostatic characterization.

1.3 Biosensors

Biosensors that employ glucose-oxidizing enzyme electrodes are devices that have already shown success in the form of electrochemical glucose biosensors for diabetes management [74]. Enzymes are attractive for biosensing owing to their exceptional substrate selectivity, high specific activity per active site, and inherent biocompatibility. The development of implantable and wearable biosensors offers the ability to continuously monitor physiological analytes with minimal patient intervention and to provide real-time health information [75, 76]. Future wearable biosensors will be noninvasive or minimally invasive devices and therefore (i) avoid the need for surgery and (ii) reduce complications associated with biocompatibility. The realization of wearable sensors is therefore more tangible than implantables. To date, wearable sensors have been almost exclusively developed for monitoring physical physiological parameters such as skin temperature, heart rate, and pressure [75]. For the development of wearable electrochemical sensors, electrode materials are required that are soft, flexible, and stretchable (e.g., mechanical properties similar to human skin), as well as being lightweight and having large surface areas. Buckypapers and their composites can offer these types of properties.

Minteer and coworkers developed a self-powered amperometric lactate biosensor based on lactate oxidase from *Pediococcus sp* cross-linked onto a MWCNT buckypaper modified with a Fc-based polyethylenimine polymer hydrogel [4]. The authors reported that the flexibility of buckypaper and its ability to adhere to nonplanar surfaces during repeated bending was attractive for the envisaged lactate skin patch and contact lens sensors. The amperometric lactate biosensor based on mediated electron transfer exhibited a detection limit of 1 µmol L^{-1}, a calibration range up to 40 mmol L^{-1}, and storage stability up to 21 days. In addition, the biosensor was stable in the presence of ascorbate, a common interference present in biofluids.

Ahmadalinezhad et al. developed a buckypaper-based biosensor platform using a gold-coated titanium substrate as a mechanical support [77]. Buckypaper was electrochemically functionalized with carboxylic acid groups then modified by adsorption of GOx *from Aspergillus niger*, horseradish peroxidase (HRP), and chitosan to complete the sensor. Amperometric glucose detection was indirectly monitored at 0.1 V versus AgAgCl based on the reduction of H_2O_2 produced by the enzymatic oxidation of glucose. The mediatorless glucose biosensor demonstrated high selectivity in the presence of common interferents (ascorbic acid, uric acid, and acetamidophenol), a dynamic range of 0.01–9 mmol L^{-1}, sensitivity of 20 µA mmol L^{-1} cm^{-2}, and a long lifetime of over 80 days. The excellent stability was attributed to the physical and chemical properties of the buckypaper combined with the use of chitosan for enzyme stabilization.

Chatterjee and Chen subsequently reported a buckypaper-based biosensor for the determination of H_2O_2 in urine for monitoring of oxidative stress in vivo [12]. The same buckypaper-titanium substrate was used as reported previously [77], followed by electrochemical surface oxidation, and immobilization of HRP, chitosan, and the mediator methylene blue. A low detection limit of 7.5×10^{-8} M, a dynamic range of 0.1–500 µmol, and a sensitivity of 54 µA mmol L^{-1} cm^{-2} were reported. Spike recovery analysis revealed recoveries of 95–105% in human urine, demonstrating the validity of the method as an in vivo biosensor.

Decoration of CNT buckypapers with nanoparticles can be used to improve conductivity [78], mechanical [79] and catalytic performance [80, 81], and enzyme wiring via electronic bridging or electron mediation [82, 83]. Papa et al. developed a glucose biosensor based on lab-made acid-treated SWCNT and MWCNT buckypapers prepared in the presence of citrate-capped gold nanoparticles [84]. A schematic showing the construction of the sensor mechanism is shown in Fig. 1.8. GOx enzyme, HRP, and chitosan were subsequently cast onto the buckypaper to generate the bioelectrode. The SWCNT BP with embedded gold nanoparticles exhibited the best performance with a dynamic range of 0.02–7 mmol L^{-1}, sensitivity of 21.5 µA mmol L^{-1} cm^{-2}, and no

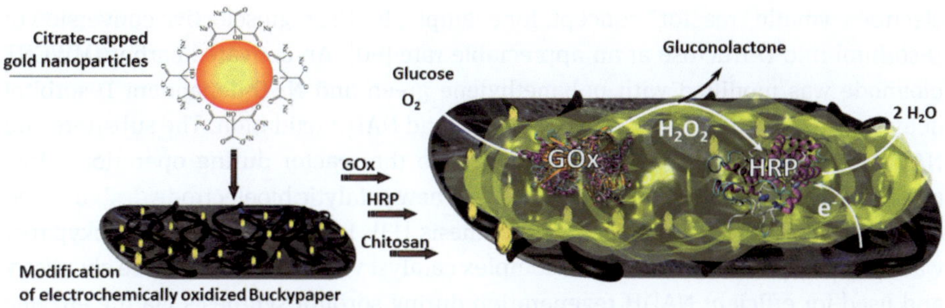

Fig. 1.8: A scheme showing the fabrication of a buckypaper-based biosensor for the indirect detection of glucose. The electrochemical signal is generated by the electrocatalytic reduction of H_2O_2, produced by GOx after oxidation of glucose via HRP.

significant response from physiological interferents (ascorbic acid and uric acid). More recently, the same authors expanded the development of the gold nanoparticle-modified SWCNT and MWCNT buckypapers for the electrochemical detection of biologically important molecules such as tryptophan, L-carnitine, tyrosine, and myoglobin with a 1,000-fold increase in the signal in buffer compared with classical glassy carbon electrode sensors [85, 86]. For the myoglobin sensor, methylene blue was integrated into the buckypaper to facilitate electron transfer between the electrode and the protein's active site. The reported patent covers methods for the construction of functionalized buckypaper biosensors as an alternative to the use of screen-printed and glassy carbon-based electrodes [86]. It is noted that the specificity of the GOx-gold nanoparticle-based biosensors in the presence of interferences was not demonstrated.

As an alternative to enzyme-based biosensing, the work of Desai et al. demonstrates the successful application of a lab-made MWCNT buckypaper as an electrical nanosensor on a printable circuit. The nonenzymatic bioelectrode sensor was constructed by immobilization of single stranded DNA and used to monitor DNA hybridization of pathogen DNAs based on device resistance changes resulting from the wrapping and unwrapping of DNA [87, 88].

1.4 Biosynthesis

An exciting and largely unexplored application of buckypaper electrodes is in the field of electroenzymatic synthesis. Enzymes offer attractions compared to chemical catalysts such as high selectivity and reaction rates under eco-friendly conditions, and are therefore recognized as powerful catalysts in industrial biotechnology, for example, for the generation of pharmaceuticals and fine chemicals including enantiomerically pure compounds [89]. Etienne and coworkers have made progress on an electroenzymatic "reactor" concept, for example, for the regioselective conversion of D-sorbitol into D-fructose at an appreciable rate [90]. An activated carbon/MWCNT bioanode was modified with polymethylene green and NAD-dependent D-sorbitol dehydrogenase (DSDH) for catalytic sorbitol and NADH oxidation. The substrate and NAD^+ cofactor were flowed continuously into the reactor during operation. More recently, Etienne and coworkers developed a new catalytic bioelectrode design based on buckypaper for electroenzymatic synthesis [13]. Lab-made MWCNT buckypaper was elegantly modified with an Rh complex catalyst via diazonium surface chemistry and used for efficient NADH regeneration during sorbitol synthesis via the enzyme DSDH (Fig. 1.9). The work constitutes significant advances in terms of reusability, good catalytic activity for NADH regeneration (turnover frequency of 1.3 s^{-1}), and efficient conversion over several days (conversion rate of 87% after 95 h).

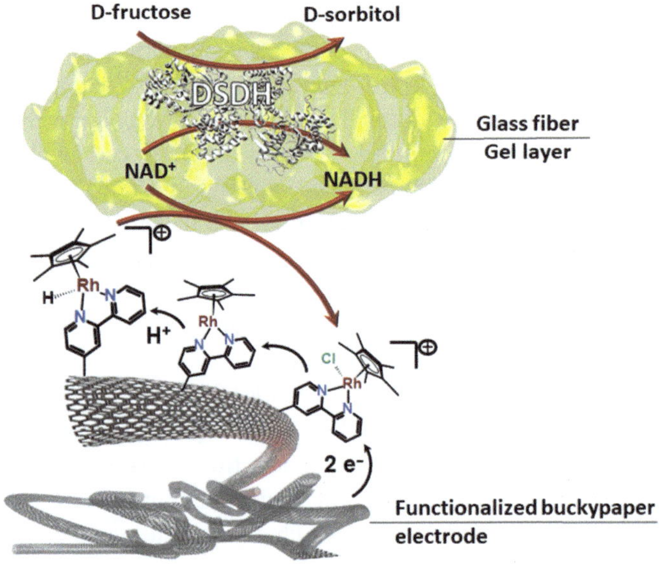

Fig. 1.9: A schematic representation of the bioelectrocatalytic synthesis of sorbitol from fructose using an Rh complex for the regeneration of NAD and the enzyme catalyst, sorbitol dehydrogenase (DSDH).

1.5 Biologic

Buckypaper bioelectrodes also offer interesting possibilities in which their biocatalytic reactions may be used for biocomputing, for example, for novel multisignal sensor applications. The review of Katz and Privman serves as an excellent introduction to this topic [91]. Strack et al. demonstrated for the first time that buckypaper bioelectrodes could be used for sustained built-in information processing [14]. A direct electron transfer biocathode configuration was used in which a laccase-based MWCNT buckypaper (Buckeye Composites) responded directly to dissolved oxygen, as a continuous input signal, and could generate an amperometric response, as the output signal. Nitrogen and oxygen were pumped into a flow system to give stable and reproducible current signals at an applied potential of 0.4 V over more than 100 cycles and a 20 day period. The current output with nitrogen served as "on" or "1" and with oxygen served as "off" or "0". The device was used to generate a binary-based code sequence (ASCII 12-bit Code 39) that could be read by a conventional barcode scanner. This proof-of-concept study ultimately demonstrated the prospect of using biocatalytic buckypapers for responding to incoming signals, processing them, and relaying them ex vivo. The use of signals from substrates present in biological fluids such as sugars and oxygen would open up the prospects for in vivo data processing.

1.6 Conclusion

Buckypaper materials based on CNTs have witnessed great success for the construction of bioelectrodes based on enzymes and microbes owing to a unique combination of properties such as their high conductivity, mechanical strength and flexibility, lightweight, and their porous high surface area structures. Both lab-made and commercial buckypapers have proved to be practical supporting frameworks for enzymes and catalytic components including nanoparticles and nonmetal catalysts, as well as functional building blocks including polymers and cross-linking molecules. A wide range of functional catalytic biointerfaces have been reported over the last 5–7 years for applications including bioenergy conversion and storage (enzymatic and microbial biofuel cells, hybrid biofuel-capacitor systems, and photoelectric biofuel cells), chemical and biochemical detection (enzymatic and nonenzymatic biosensors), and information-processing devices (biologic systems). The greatest achievement of buckypaper to date has been its use in *actual* implanted biofuel cells for energy harvesting from organisms such as snails and lobsters; no other electrode can boast as much success in the field of biological fuel cells. There is still a huge amount of work to be done to improve the performance of implantable and wearable biofuel cells, and we note in particular the importance of evaluating the stability, biocompatibility, and toxicity of buckypaper bioelectrodes in future studies. Until now, the majority of reported buckypaper bioelectrodes have been prepared using commercial buckypaper from Buckeye Composites, and this type of readily available buckypaper has proved to be an excellent electrode platform. We nevertheless envisage a bright future for lab-made buckypaper electrodes that can be prepared using relatively straightforward procedures in the laboratory, and which permit a higher level of control over the physical, chemical, and catalytic properties of the electrode. We also recognize that a wide variety of parameters affect the performance of lab-made buckypaper and thus we take this opportunity to emphasize the importance of providing accurate details with respect to the fabrication methods and materials used for buckypaper preparation. With all of this in mind, there should be little doubt that buckypaper will continue to play an important role not only for the construction of future bioelectronic devices but also for other applications in the fields of electrochemistry, energy, and materials science.

References

[1] Rinzler, A. G., Liu, J., Dai, H., et al. Large-scale purification of single-wall carbon nanotubes: process, product, and characterization. Appl Phys A. 1998, 67, 29–37.

[2] Wang, S., Downes, R., Young, C., et al. Carbon fiber/carbon nanotube buckypaper interply hybrid composites: manufacturing process and tensile properties. Adv Eng Mater. 2015, 17, 1442–1453.

[3] Li, Z., Dharap, P., Nagarajaiah, S., Barrera, E. V., & Kim, J. D. Carbon nanotube film sensors. Adv Mater. 2004, 16, 640–643.

[4] Hickey, D. P., Reid, R. C., Milton, R. D., & Minteer, S. D. A self-powered amperometric lactate biosensor based on lactate oxidase immobilized in dimethylferrocene-modified LPEI. Biosens Bioelectron. 2016, 77, 26–31.

[5] Zhu, S., Zheng, J., Huang, J., Dai, N., Li, P., & Zheng, J. P. Fabrication of three-dimensional buckypaper catalyst layer with Pt nanoparticles supported on polyelectrolyte functionalized carbon nanotubes for proton exchange membrane fuel cells. J Power Sources. 2018, 393, 19–31.

[6] Gross, A. J., Chen, X., Giroud, F., et al. A high power buckypaper biofuel cell: exploiting 1,10-phenanthroline-5,6-dione with FAD-dependent dehydrogenase for catalytically-powerful glucose oxidation. ACS Catal. 2017, 7, 4408–4416.

[7] Gross, A. J., Holzinger, M., & Cosnier, S. Buckypaper bioelectrodes: emerging materials for implantable and wearable biofuel cells. Energy Environ Sci. 2018, 11, 1670–1687.

[8] Cosnier, S., Gross A, J., Goff, A. L., & Holzinger, M. Recent advances on enzymatic glucose/oxygen and hydrogen/oxygen biofuel cells: achievements and limitations. J Power Sources. 2016, 325, 252–263.

[9] Santoro, C., Babanova, S., Atanassov, P., Li, B., Ieropoulos, I., & Cristiani, P. High power generation by a membraneless single chamber microbial fuel cell (SCMFC) using enzymatic bilirubin oxidase (BOx) air-breathing cathode. J Electrochem Soc. 2013, 160, H720–H726.

[10] Hou, C., & Liu, A. An integrated device of enzymatic biofuel cells and supercapacitor for both efficient electric energy conversion and storage. Electrochim Acta. 2017, 245, 303–308.

[11] Yu, Y., Zhai, J., Xia, Y., & Dong, S. Single wearable sensing energy device based on photoelectric biofuel cells for simultaneous analysis of perspiration and illuminance. Nanoscale. 2017, 9, 11846–11850.

[12] Chatterjee, S., & Chen, A. Functionalization of carbon buckypaper for the sensitive determination of hydrogen peroxide in human urine. Biosens Bioelectron. 2012, 35, 302–307.

[13] Zhang, L.; Etienne, M.; Vilà, N.; Le, T. X. H.; Kohring, G.-W.; Walcarius, A., Electrocatalytic Biosynthesis using a Bucky Paper Functionalized by [Cp*Rh(bpy)Cl]+ and a Renewable Enzymatic Layer. ChemCatChem 2018, 10 (18), 4067–4073.

[14] Strack, G. Bioelectrocatalytic generation of directly readable code: harnessing cathodic current for long-term information relay. Chem Commun. 2011, 47, 7662–7664.

[15] Oh, J. Y., Yang, S. J., Park, J. Y., et al. Easy preparation of self-assembled high-density buckypaper with enhanced mechanical properties. Nano Lett. 2015, 15, 190–197.
[16] Chen, X., Gross, A. J., Giroud, F., Holzinger, M., & Cosnier, S. Comparison of commercial and lab-made MWCNT buckypaper: physicochemical properties and bioelectrocatalytic O_2 reduction. Electroanalysis. 2018, 30, 1511–1520.
[17] Halámková, L., Halámek, J., Bocharova, V., Szczupak, A., Alfonta, L., & Katz, E. Implanted biofuel cell operating in a living snail. J Am Chem Soc. 2012, 134, 5040–5043.
[18] Reid, R. C., Minteer, S. D., & Gale, B. K. Contact lens biofuel cell tested in a synthetic tear solution. Biosens Bioelectron. 2015, 68, 142–148.
[19] Göbel, G., Beltran, M. L., Mundhenk, J., Heinlein, T., Schneider, J., & Lisdat, F. Operation of a carbon nanotube-based glucose/oxygen biofuel cell in human body liquids: performance factors and characteristics. Electrochim Acta. 2016, 218, 278–284.
[20] Wang, D., Song, P., Liu, C., Wu, W., & Fan, S. Highly oriented carbon nanotube papers made of aligned carbon nanotubes. Nanotechnology. 2008, 19, 075609.
[21] Zhang, M., Fang, S., Zakhidov, A. A., et al. Strong, transparent, multifunctional, carbon nanotube sheets. Science. 2005, 309, 1215–1219.
[22] Choi, J. A study on the effect of pyrene derivatives on the noncovalent sidewall functionalisation of carbon nanotube buckypapers. Thin Solid Films. 2018, 651, 77–84.
[23] Chen, I.-W. P., Liang, R., Zhao, H., Wang, B., & Zhang, C. Highly conductive carbon nanotube buckypapers with improved doping stability via conjugational cross-linking. Nanotechnology. 2011, 22, 485708.
[24] Li, Y., & Kroeger, M. A theoretical evaluation of the effects of carbon nanotube entanglement and bundling on the structural and mechanical properties of buckypaper. Carbon. 2012, 50, 1793–1806.
[25] Yang, K., He, J., Puneet, P., et al. Tuning electrical and thermal connectivity in multiwalled carbon nanotube buckypaper. J Phys-Condens Matter. 2010, 22, 334215.
[26] Young, J. P. Continuous buckypaper manufacturing process: process investigation and improvement. MSc thesis, Florida State University, 2009.
[27] Wang, S., Haldane, D., Liang, R., Smithyman, J., Zhang, C., & Wang, B. Nanoscale infiltration behaviour and through-thickness permeability of carbon nanotube buckypapers. Nanotechnology. 2013, 24, 015704.
[28] Zhang, S., E., Leonhardt, B., Nguyen, N., et al. Roll-to-roll continuous carbon nanotube sheets with high electrical conductivity. RSC Adv. 2018, 8, 12692–12700.
[29] Whitby, R. L. D., Fukuda, T., Maekawa, T., James, S. L., & Mikhalovsky, S. V. Geometric control and tuneable pore size distribution of buckypaper and buckydiscs. Carbon 2008, 46, 949–956.
[30] Shen, Z., Roding, M., Kroger, M., & Li, Y. Carbon nanotube length governs the viscoelasticity and permeability of buckypaper. Polymers. 2017, 9, 115.
[31] Roy, S., Jain, V., Bajpai, R., et al. Formation of carbon nanotube bucky paper and feasibility study for filtration at the nano and molecular scale. J Phys Chem C. 2012, 116, 19025–19031.
[32] Yang, X., Lee, J., Yuan, L., et al. Removal of natural organic matter in water using functionalised carbon nanotube buckypaper. Carbon. 2013;59:160–166.
[33] Bourourou, M., Elouarzaki, K., Holzinger, M., et al. Freestanding redox buckypaper electrodes from multi-wall carbon nanotubes for bioelectrocatalytic oxygen reduction via mediated electron transfer. Chem Sci. 2014, 5, 2885–2888.
[34] Cosnier, S., Holzinger, M., & Goff, A. L. Recent advances in carbon nanotube-based enzymatic fuel cells. Front Bioeng Biotechnol. 2014, 2, 45.
[35] Funabashi, H., Takeuchi, S., & Tsujimura, S. Hierarchical meso/macro-porous carbon fabricated from dual MgO templates for direct electron transfer enzymatic electrodes. Sci Rep. 2017, 7, 45147.

[36] Tsujimura, S., Murata, K., & Akatsuka, W. Exceptionally high glucose current on a hierarchically structured porous carbon electrode with "wired" flavin adenine dinucleotide-dependent glucose dehydrogenase. J Am Chem Soc. 2014, 136, 14432–14437.
[37] Sugimoto, Y., Kitazumi, Y., Shirai, O., & Kano, K. Effects of mesoporous structures on direct electron transfer-type bioelectrocatalysis: facts and simulation on a three-dimensional model of random orientation of enzymes. Electrochemistry. 2017, 85, 82–87.
[38] So, K., Kitazumi, Y., Shirai, O., & Kano, K. Analysis of factors governing direct electron transfer-type bioelectrocatalysis of bilirubin oxidase at modified electrodes. J Electroanal Chem. 2016, 783, 316–323.
[39] So, K., Onizuka, M., Komukai, T., Kitazumi, Y., Shirai, O., & Kano, K. Significance of the length of carbon nanotubes on the bioelectrocatalytic activity of bilirubin oxidase for dioxygen reduction. Electrochim Acta. 2016, 192, 133–138.
[40] Bandodkar, A. J., & Wang, J. Wearable biofuel cells: a review. Electroanalysis. 2016, 28, 1188–1200.
[41] Bandodkar, A. J., & Wang, J. Non-invasive wearable electrochemical sensors: a review. Trends Biotechnol. 2014, 32, 363–371.
[42] Katz, E. Implanted biofuel cells operating in vivo. In: Implantable bioelectronics. Weinheim: Wiley-VCH; 2014. p. 363–379.
[43] Trifonov, A., Herkendell, K., Tel-Vered, R., Yehezkeli, O., Woerner, M., & Willner, I. Enzyme-capped relay-functionalized mesoporous carbon nanoparticles: effective bioelectrocatalytic matrices for sensing and biofuel cell applications. ACS Nano. 2013, 7, 11358–11368.
[44] Cinquin, P., Gondran, C., Giroud, F., et al. A glucose biofuel cell implanted in rats. PloS One. 2010, 5, e10476.
[45] Zebda, A., Cosnier, S., Alcaraz, J.-P., et al. Single glucose biofuel cells implanted in rats power electronic devices. Sci Rep. 2013, 3, 1516.
[46] Jia, W., Valdés-Ramírez, G., Bandodkar, A. J., Windmiller, J. R., & Wang, J. Epidermal biofuel cells: energy harvesting from human perspiration. Angew Chem Int Ed. 2013, 52, 7233–7236.
[47] Bandodkar, A. J., You, J.-M., Kim, N.-H., et al. Soft, stretchable, high power density electronic skin-based biofuel cells for scavenging energy from human sweat. Energy Environ Sci. 2017, 10, 1581–1589.
[48] Abreu, C., Nedellec, Y., Gross, A. J., et al. Assembly and stacking of flow-through enzymatic bioelectrodes for high power glucose fuel cells. ACS Appl Mater Interfaces. 2017, 9, 23836–23842.
[49] MacVittie, K., Halámek, J., Halámková, L., et al. From "cyborg" lobsters to a pacemaker powered by implantable biofuel cells. Energy Environ Sci. 2013, 6, 81–86.
[50] Cosnier, S., Gross, A. J., Giroud, F., & Holzinger, M. Beyond the hype surrounding biofuel cells: what's the future of enzymatic fuel cells? Curr Opin Electrochem. [Internet] just accepted 2018. DOI: 10.1016/j.coelec.2018.06.006.
[51] Szczupak, A., Halámek, J., Halámková, L., Bocharova, V., Alfonta, L., & Katz, E. Living battery: biofuel cells operating in vivo in clams. Energy Environ Sci. 2012, 5, 8891–8895.
[52] Castorena-Gonzalez, J. A., Foote, C., MacVittie, K., et al. Biofuel cell operating in vivo in rat. Electroanalysis. 2013, 25, 1579–1584.
[53] Bandodkar, A. J. Wearable biofuel cells: past, present and future. J Electrochem Soc. 2017, 164, H3007–HH3014.
[54] Lee, H., Song, C., Hong, Y. S., et al. Wearable/disposable sweat-based glucose monitoring device with multistage transdermal drug delivery module. Sci Adv. 2017, 3, e1601314.
[55] Onor, M., Gufoni, S., Lomonaco, T., et al. Potentiometric sensor for non-invasive lactate determination in human sweat. Anal Chim Acta. 2017, 989, 80–87.
[56] Villarrubia, C. W. N., Garcia, S. O., Lau, C., & Atanassov, P. Biofuel cell anodes integrating NAD (+)-dependent enzymes and multiwalled carbon nanotube papers. ECS J Solid State Sci Technol. 2013, 2, M3156–M3159.

[57] Bruen, D., Delaney, C., Florea, L., & Diamond, D. Glucose sensing for diabetes monitoring: recent developments. Sensors. 2017, 17, E1866.
[58] Gross, A. J., Robin, M. P., Nedellec, Y., O'Reilly, R. K., Shan, D., & Cosnier, S. Robust bifunctional buckypapers from carbon nanotubes and polynorbornene copolymers for flexible engineering of enzymatic bioelectrodes. Carbon. 2016, 107, 542–547.
[59] Sony Corporation. Sony develops "bio battery" generating electricity from sugar [Internet]. Available from: https://www.sony.net/SonyInfo/News/Press/200708/07-074E/index.html.
[60] Villarrubia, C. W. N., Lau, C., Ciniciato, G. P. M. K., et al. Practical electricity generation from a paper based biofuel cell powered by glucose in ubiquitous liquids. Electrochem Commun. 2014, 45, 44–47.
[61] Villarrubia, C. W. N., Soavi, F., Santoro, C., et al. Self-feeding paper based biofuel cell/self-powered hybrid μ-supercapacitor integrated system. Biosens Bioelectron. 2016, 86, 459–465.
[62] Lau, C., Moehlenbrock, M. J., Arechederra, R. L., et al. Paper based biofuel cells: incorporating enzymatic cascades for ethanol and methanol oxidation. Int J Hydrog Energy. 2015, 40, 14661–14666.
[63] Babanova, S., Artyushkova, K., Ulyanova, Y., Singhal, S., & Atanassov, P. Design of experiments and principal component analysis as approaches for enhancing performance of gas-diffusional air-breathing bilirubin oxidase cathode. J Power Sources. 2014, 245, 389–397.
[64] Santoro, C., Babanova, S., Erable, B., Schuler, A., & Atanassov, P. Bilirubin oxidase based enzymatic air-breathing cathode: operation under pristine and contaminated conditions. Bioelectrochemistry. 2016, 108, 1–7.
[65] Hou, C., Lang, Q., & Liu, A. T. 1,4-naphthoquinone with electron-withdrawing group: toward developing redox polymer and FAD-GDH based hydrogel bioanode for efficient electrocatalytic glucose oxidation. Electrochim Acta. 2016, 211, 663–670.
[66] Tran, T. P.; Do, Q. H., High-Performance Supercapacitor Electrode Based on Buckypaper/Polyaniline Composite. Journal of Electronic Materials 2017, 46 (10), 6056–6062.
[67] Ulyanova, Y., Arugula, M. A., Rasmussen, M., et al. Bioelectrocatalytic oxidation of alkanes in a JP-8 enzymatic biofuel cell. ACS Catal. 2014, 4, 4289–4294.
[68] Santoro, C., Arbizzani, C., Erable, B., & Ieropoulos, I. A. Microbial fuel cells: from fundamentals to applications. A review. J Power Sources. 2017, 356, 225–244.
[69] Pasternak, G., Greenman, J., & Ieropoulos, I. A. Self-powered, autonomous biological oxygen demand biosensor for online water quality monitoring. Sens Actuators B Chem. 2017, 244, 815–822.
[70] Strack, G., Luckarift, H. R., Sizemore, S. R., et al. Power generation from a hybrid biological fuel cell in seawater. Bioresour Technol. 2013, 128, 222–228.
[71] Santoro, C., Soavi, F., Serov, A., Arbizzani, C., & Atanassov, P. Self-powered supercapacitive microbial fuel cell: the ultimate way of boosting and harvesting power. Biosens Bioelectron. 2016, 78, 229–235.
[72] Kipf, E., Koch, J., Geiger, B., et al. Systematic screening of carbon-based anode materials for microbial fuel cells with Shewanella oneidensis MR-1. Bioresour Technol. 2013, 146, 386–392.
[73] Kipf, E., Erben, J., Zengerle, R., Gescher, J., & Kerzenmacher, S. Systematic investigation of anode materials for microbial fuel cells with the model organism G. sulfurreducens. Bioresour Technol Rep. 2018, 2, 29–37.
[74] Wang, J. Glucose biosensors: 40 years of advances and challenges. Electroanalysis. 2001, 13, 983–988.
[75] Wang, J. Special issue for wearable electrochemical sensors. Electroanalysis. 2016, 28, 1148–1148.
[76] Kotanen, C. N., Moussy, F. G., Carrara, S., & Guiseppi-Elie, A. Implantable enzyme amperometric biosensors. Biosens Bioelectron. 2012, 35, 14–26.

[77] Ahmadalinezhad, A., Wu, G., & Chen, A. Mediator-free electrochemical biosensor based on buckypaper with enhanced stability and sensitivity for glucose detection. Biosens Bioelectron. 2011, 30, 287–293.
[78] Byrne, M. T., Hanley, C. A., & Gun'ko, Y. K. Preparation and properties of buckypaper–gold nanoparticle composites. J Mater Chem. 2010, 20, 2949–2951.
[79] Byrne, M. T., Hernandez, Y. R., Conaty, T., Blighe, F. M., Coleman, J. N., & Gun'ko, Y. K. Preparation of buckypaper–copper composites and investigation of their conductivity and mechanical properties. ChemPhysChem. 2009, 10, 774–777.
[80] Tan, P., Shyy, W., Zhao, T. S., Zhu, X. B., & Wei, Z. H. A RuO_2 nanoparticle-decorated buckypaper cathode for non-aqueous lithium–oxygen batteries. J Mater Chem A. 2015, 3, 19042–19049.
[81] Holade, Y., MacVittie, K., Conlon, T., et al. Pacemaker activated by an abiotic biofuel cell operated in human serum solution. Electroanalysis. 2014, 26, 2445–2457.
[82] Gutiérrez-Sánchez, C., Pita, M., Vaz-Domínguez, C., Shleev, S., & De Lacey, A. L. Gold nanoparticles as electronic bridges for laccase-based biocathodes. J Am Chem Soc. 2012, 134, 17212–17220.
[83] Gross, A. J., Chen, X., Giroud, F., Travelet, C., Borsali, R., & Cosnier, S. Redox-active glyconanoparticles as electron shuttles for mediated electron transfer with bilirubin oxidase in solution. J Am Chem Soc. 2017, 139, 16076–16079.
[84] Papa, H., Gaillard, M., Gonzalez, L., & Chatterjee, J. Fabrication of functionalized carbon nanotube buckypaper electrodes for application in glucose biosensors. Biosensors. 2014, 4, 449–460.
[85] Chatterjee, J., Cardenal, J., & Shellikeri, A. Engineered carbon nanotube buckypaper: a platform for electrochemical biosensors. J Biomed Nanotechnol. 2015, 1, 150–156.
[86] Chatterjee, J. Nanomaterial based electrodes and methods [Internet]. Available from: http://www.google.com/patents/WO2015168435A1
[87] Desai, V., Aliane, B., Tsai, P.-J., Chan, S. L. I., Miao H.-Y., & Sinha, S. Frequency analysis and application of a buckypaper-based bionanosensor. J Nanoparticle Res. 2013, 15, 1503.
[88] Desai, V., Sanisetty, S., Steber, B., et al. Intermediate frequency AC signal analysis for bionanosensor. J Nanotechnol 2011, 1–9.
[89] Choi, J.-M., Han, S.-S., & Kim, H.-S. Industrial applications of enzyme biocatalysis: current status and future aspects. Biotechnol Adv. 2015, 33, 1443–1454.
[90] Mazurenko, I., Etienne, M., Kohring, G.-W., Lapicque, F., & Walcarius, A. Enzymatic bioreactor for simultaneous electrosynthesis and energy production. Electrochim Acta. 2016, 199, 342–348.
[91] Katz, E., & Privman, V. Enzyme-based logic systems for information processing. Chem Soc Rev. 2010, 39, 1835–1857.

Alan Le Goff, Fabien Giroud

2 Molecular electrocatalysts for carbon-based biofuels, H_2 and O_2 activation: an alternative to precious metals and enzymes in fuel cells

2.1 Introduction

In future, power production will be based on solar panels for both the production of electricity and green production of renewable fuels. This system will be coupled to fuel cells in order to produce electrical power from renewable fuels. This type of strategy is aimed at being independent from fossil fuels or nuclear energy. Fuel cells are composed of two electrodes: the anode achieves the oxidation of the fuel and the cathode achieves the reduction of the oxidant, mostly oxygen, or hydrogen peroxide in some cases. The majority of conventional fuel cells rely on precious metal or metal alloys in order to perform long-term electrocatalytic oxidation of the fuel or reduction of the oxidant. A promising alternative has been the use of enzymes within enzymatic fuel cells (EFCs). In EFCs, purified redox enzymes, instead of metal catalysts, achieve the electrocatalytic reactions. EFCs are particularly attractive since enzymes are catalysts with high substrate specificity, high catalytic turnover frequency, and low overpotentials at low temperature. Furthermore, these biomacromolecules are constituted of organic amino acids and cofactors and/or base metal active sites. However, enzymes are complicated catalysts to handle and purify, are unstable over long periods of time, and can be considered as big catalysts when compared to metal atoms. This is the reason why molecular chemists and electrochemists have envisioned the use of molecular catalysts that are inspired from enzymatic mechanisms and structures to perform electrocatalytic reactions. However, in order to come close to matching the efficiency of molecular catalysts with enzymes in the electrocatalytic reactions performed at fuel cells, synthetic molecular catalysts have to be designed and integrated at electrodes by taking care of the structure of the catalyst, its electrocatalytic mechanisms, and its performance. The first part of this chapter describes the main advances that have been made in the design of bioinspired complexes for H_2 oxidation and O_2 reduction. The synthesis and studies of bioinspired complexes have led to highly efficient molecular catalysts based on nonnoble metals. Some of these catalysts have shown superior performances as compared to enzymes, as well as stable catalysis in functional fuel cells. The second part describes the design of molecular catalysts for the oxidation of carbon-based biofuels such as alcohols, glucose, or glycerol. Organic and inorganic molecular catalysts have both been envisioned as anodic catalysts in alcohol or glucose fuel cells or for green organic synthesis.

Alan Le Goff, Fabien Giroud, Univ. Grenoble Alpes, CNRS, Grenoble, France

https://doi.org/10.1515/9783110570526-002

2.2 Bioinspired complexes for hydrogen/O_2 fuel cells

2.2.1 Molecular catalysts for H_2 oxidation

The oxidation of H_2 by molecular catalysts relies mostly on bioinspiration from metalloenzymes. Hydrogenases catalyze the reversible oxidation of H_2 into protons [1–5]. The main families of hydrogenases are [FeFe] hydrogenases based on a FeFe dinuclear active site, which is known as the cluster H, NiFe hydrogenases based on a Ni–Fe active site, and [Fe] hydrogenases based on a mononuclear Fe center which is not active toward H_2 oxidation. In active hydrogenases, a chain of iron–sulfur clusters is responsible for the electron transfers between the active site and the redox partner or the electrode surface if the enzyme is immobilized (Fig. 2.1).

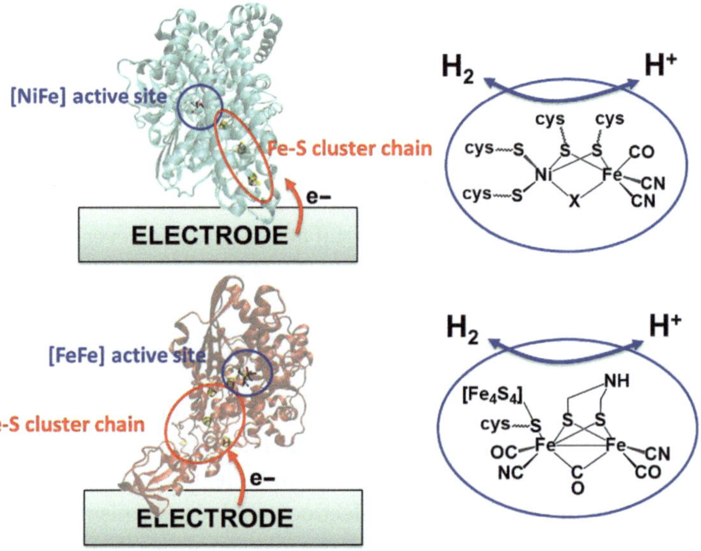

Fig. 2.1: NiFe and FeFe hydrogenases and their respective active sites.

It has been demonstrated that NiFe and FeFe hydrogenases exhibit exceptional electrocatalytic activity toward both H_2 evolution and oxidation with near-zero overpotential. For these reasons, when immobilized on electrodes, these enzymes have demonstrated to be competitive to Pt-based catalysts for the conversion of H_2 into protons [4, 6]. However, most hydrogenases are sensitive to inactivation by O_2. This is a major bottleneck for their implementation in a H_2/air fuel cell. Several strategies have been developed to circumvent this major issue: (i) site-directed mutagenesis [7], (ii) the entrapment in protective viologen polymers [8–11] or redox protein nanowires [12], (iii) the use of oxygen-tolerant NiFe hydrogenases [13–16], and (iv) the design of specific gas-diffusion electrodes for preventing direct contact with O_2 [17–19]. Furthermore, because of the complexity of enzyme structure and function, these catalysts are expensive to produce, and they are fragile and active under narrow conditions (temperature between 25 °C and 60 °C, pH between 4 and 8). A strategy to overcome the instability of enzymes while using base metal catalysts is the development of bioinspired iron- and/or nickel-based catalysts. However, in order not to rely on complex catalysts such as enzymes, the synthesis of base-metal catalysts, inspired from the nickel or iron active sites of hydrogenases, has been proposed as an alternative in order to provide stable and oxygen-resistant catalysts [20, 21].

First, a rich and fascinating chemistry has been developed for the synthesis of structural biomimetic complexes of the H cluster, which have several objectives such as isolating terminal hydride species or providing proton or electron relays in the second coordination sphere [20–22]. In particular, based on this research in 2012, Camara and Rauchfuss [23] synthesized a biomimetic iron–iron cluster bearing an azadithiolate proton relay and a decamethylferrocene group playing the role of the iron–sulfur electron relay in the protein (Fig. 2.2).

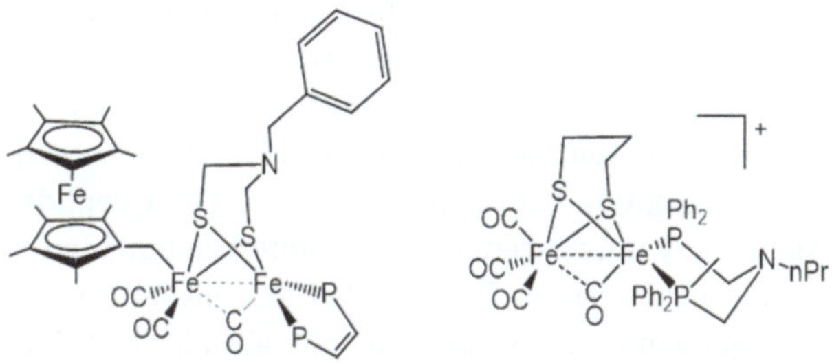

Fig. 2.2: First examples of cluster H biomimetic complexes achieving catalytic H_2 oxidation.

Furthermore, this complex has been the first example of H cluster mimic to achieve H_2 oxidation. At this time, iron–iron biomimetic complexes were only able to perform proton reduction. Then, Sun and colleagues [24] have shown that their iron–iron biomimetic clusters bearing azadiphosphine ligands acting as proton relay exhibited H_2 oxidation catalytic activity (Fig. 2.3).

For the biomimetic chemistry of NiFe active site, structures of Fe–Ni biomimetic clusters have shown only the proton reduction activity. And no true NiFe complexes have achieved H_2 oxidation yet. However, the design of nickel-based bioinspired

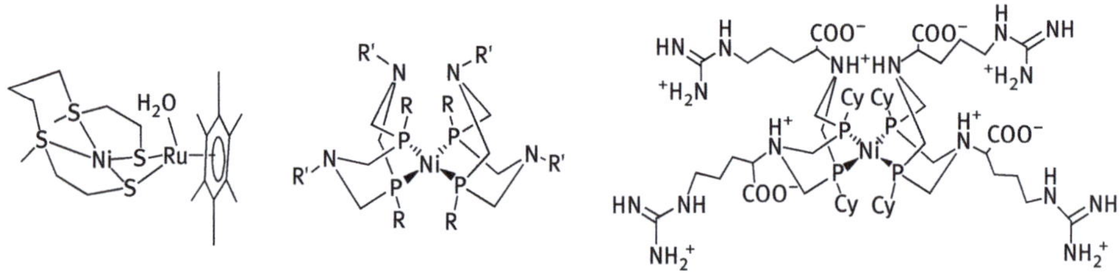

Fig. 2.3: Bioinspired nickel-based molecular catalysts for H_2 oxidation.

functional models has especially led to Ni–Ru complexes achieving H_2 oxidation. Ru(II) has been introduced in binuclear clusters in order to mimic the Fe metal center and as a well-known metal center for activation of H_2 and formation of hydride complexes. Following the work of Rauchfuss and colleagues, which demonstrated the structural mimic of the H-cluster based on Ru_2 [25] or Ir_2 [26] complexes that were able to achieve H_2 activation, Ogo et al. [27] in 2007 synthesized Ni–Ru complexes showing H_2 activation in water.

In the course of functional models of FeFe and NiFe complexes, Dubois and colleagues have designed exceptional mononuclear nickel complexes achieving highly efficient proton reduction and H_2 oxidation. The structure of these complexes is based on a mononuclear nickel bis-diphosphine complex $[Ni(P^R{}_2N^{R'}{}_2)_2]^{2+}$. A 1,5-diaza-3,7-diphosphacyclooctane $(P^R{}_2N^{R'}{}_2)$ ligand mimics the 2-azapropanedithiolate found in the H cluster. $(P^R{}_2N^{R'}{}_2)$ ligands provide an electron-rich environment and four pendant amines acting as proton relays [28, 29]. The nature of R and R′ groups influences the catalytic bias toward proton reduction or H_2 oxidation [29–31]. Furthermore, by modification of the R′ group with amino acids [32, 33] or peptides [34], Shaw and colleagues have recently introduced a second proton channel in the second coordination sphere of the nickel center. This has led to a nickel catalyst bearing arginine residues and being the most active molecular catalyst for H_2 oxidation in water. The complex exhibits a time of flight (TOF) of 210 s^{-1} (1 atm H_2, room temperature) and 106 s^{-1} (100 atm H_2, 72 °C) [35].

The immobilization of such nickel complexes on carbon nanotubes (CNTs) has especially unraveled their reversible activity in water, leading to highly performing proton-reducing or H_2–oxidizing electrodes [36, 37]. Immobilization strategies have relied on the rich surface chemistry at CNT sidewalls (Fig. 2.4). Reduction of aryldiazonium salts is an efficient means in order to introduce functional groups at the surface of CNTs via reaction of aryl radicals with the pi-extended network of CNT sidewalls [36]. Besides, this conjugated network strongly interacts with polycyclic aromatics such as modified pyrene in order to adsorb pyrene-modified species such as pyrene-modified nickel bis-diphosphine complexes [37]. In particular, the immobilization of the arginine-based nickel complex has been performed on CNTs

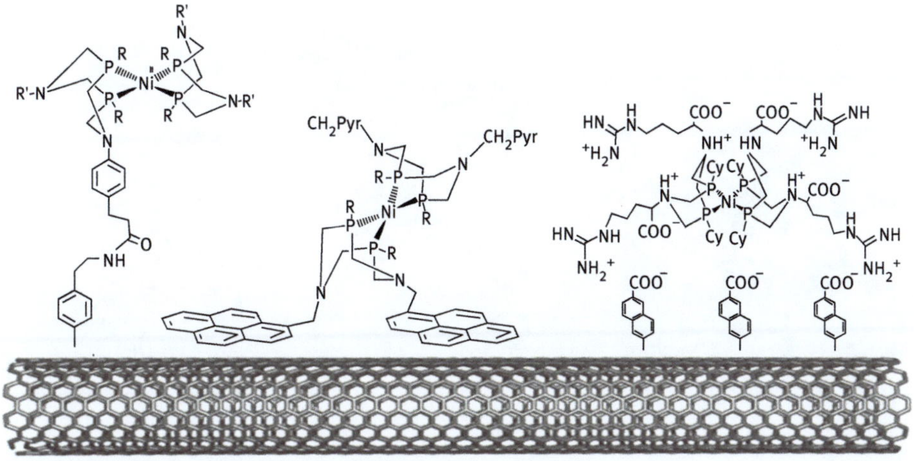

Fig. 2.4: Immobilization of bioinspired nickel complexes on CNTs.

modified by aryldiazonium salts [38]. These electrodes exhibit reversible H_2 oxidation over a wide range of pH with a near-zero overpotential and TOF of 120 s^{-1} in sulfuric acid.

2.2.2 Bioinspired complexes for O_2 reduction

Most bioinspired molecular complexes for O_2 reduction have been inspired from heme and copper enzymes. Early works have been especially devoted to the study of biomimetic iron porphyrins [39]. In particular, many studies have been devoted to iron or cobalt porphyrins mimicking the oxygen-activating hemes in metalloproteins. However, heme enzymes are only active at potentials of −0.1 V versus Normal Hydrogen Electrode (NHE) at neutral pH, which is low as compared to the thermodynamic potential of the $4H^+/4e^-$ O_2/H_2O couple (0.82 V vs. NHE). In this respect, copper-containing enzymes have been targeted for the ability of some copper family to achieve low overpotential $4H^+/4e^-$ oxygen reduction reaction (ORR). Models of cytochrome oxidases that possess more positive potentials for O_2 activation (+0.35 V) [40] have been widely investigated by Karlin and coworkers [41–46] and Collman et al. [47–51] (Fig. 2.5). These complexes have also been deeply studied after adsorption on electrodes or covalent immobilization [46, 51–54]. While efficient $4e^-/4H^+$ has been demonstrated, low overpotential has to be overcome in such complexes to be implemented in fuel cells. These studies have been particularly devoted to understand the role of the copper center in the vicinity of the oxygen-activating Fe(II) center.

Mononuclear and dinuclear bioinspired copper complexes have tried to mimic the ORR activity of enzymes such as galactose oxidases and polyphenol

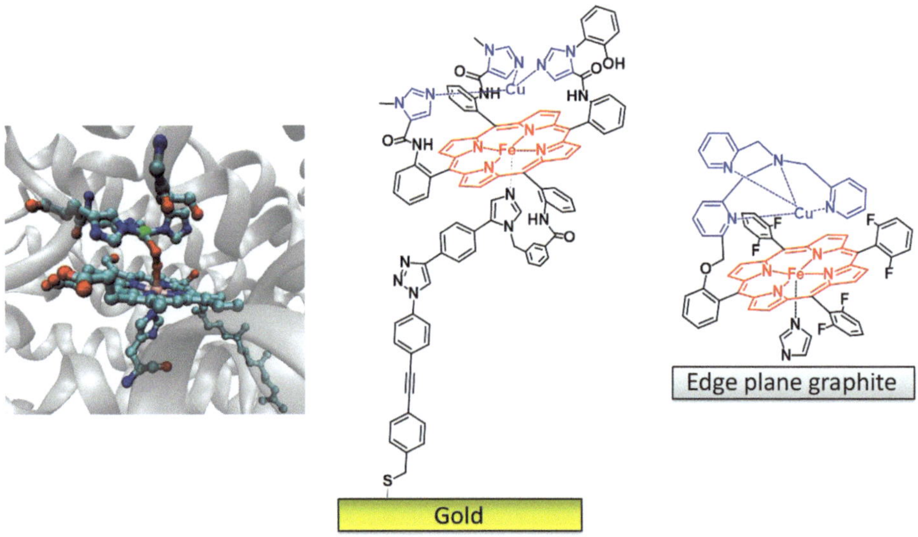

Fig. 2.5: Structure of the CcO active site and examples of immobilized CcO models for $4H^+/4e^-$ ORR. From Chatterjee et al. [46] and Collman et al. [49].

oxidases. These enzymes have shown ORR electroactivity when immobilized at nanostructured electrodes with ORR potentials of +0.7 V [55] and +0.9 V [56] versus Reversible Hydrogen Electrode (RHE), respectively. Mononuclear and dinuclear complexes have also been integrated at electrodes by adsorption, covalent binding, supramolecular pi-stacking interactions with CNTs or at functionalized polymers (Fig. 2.6). The best performances have been obtained with polyvinylimidazole-capped CNTs modified with nitrophenanthroline copper complexes. These modified nanostructured polymers reach onset potentials of 1.05 V in 0.1 M KOH [57]. Immobilized dinuclear Cu models that are active under acidic to neutral conditions exhibit ORR potential of 0.7 V versus RHE at pH 5 [58, 59] for copper complexes based on substituted phenanthrolines, 0.67 V at pH 7 for copper-hexaazamacrocyclic complexes [60], 0.73 V versus RHE at pH 7 for copper complexes based on 3,5-diamino-1,2,4-triazole ligand [61], 0.65 V at pH 5 for copper complexes based on tris(pyridin-2-ylmethyl)amine ligand [62], and 0.60 V at pH 5 for copper-phenolate complexes [63].

However, for now, the best copper enzymes for ORR belong to the multicopper oxidase (MCO) enzyme family [64, 65]. In fuel cells based on enzymes, the cathode catalyst is an MCO such as laccase or bilirubin oxidase [5, 64–67]. The active site of these enzymes is composed of a type 1 copper center located at the surface of the protein, accommodating the substrate of the enzyme and a trinuclear copper center (TNC) at which O_2 is fully reduced to H_2O (Fig. 2.7). MCOs have demonstrated electrocatalytic performances close to platinum catalysts in terms of

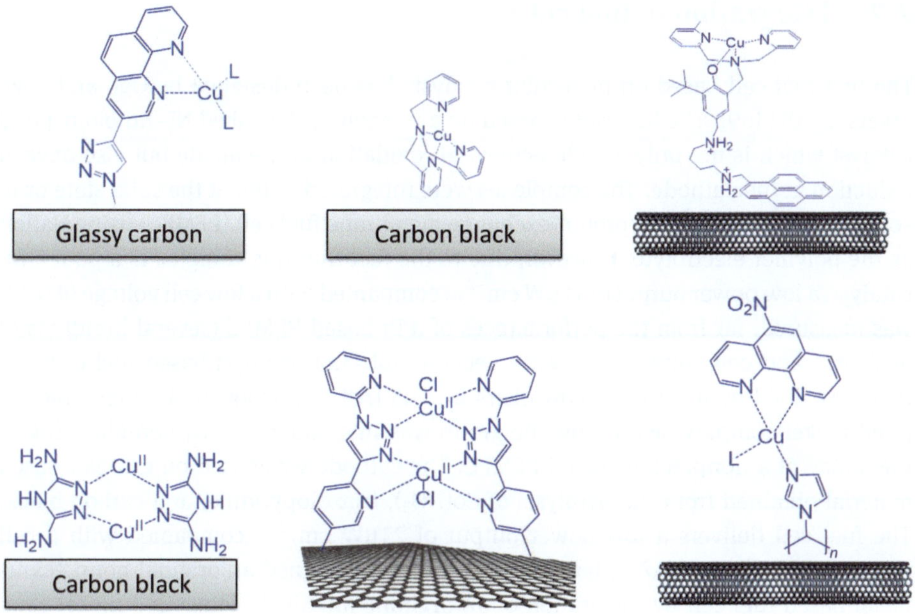

Fig. 2.6: Examples of immobilized copper complexes (up) and coordination polymers based on copper complexes (down) for $4e^-/4H^+$ ORR.

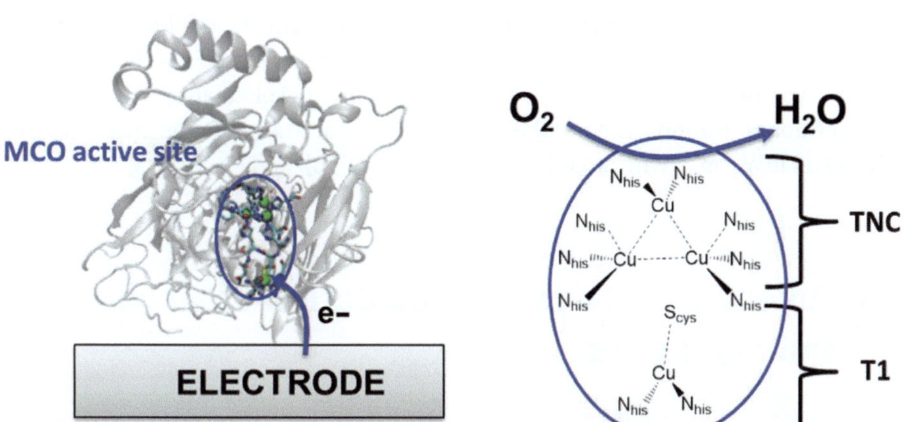

Fig. 2.7: Bilirubin oxidase from *Myrothecium verrucaria* and its active site.

overpotentials and TOF. High-potential MCOs such as laccase from *Trametes versicolor* achieves ORR at a potential as high as 1.05 V versus RHE at acidic pH. However, no efficient binuclear or trinuclear clusters mimicking the TNC have been synthesized yet [68].

2.2.3 Integration in fuel cells

The first fuel cell based on molecular catalysts has been designed by Ogo and coworkers in 2011 [69]. The fuel cell is based on a previously described Ni–Ru bioinspired catalyst which is not only able to achieve H_2 oxidation of the anode but also oxygen reduction at the cathode. The complexes were integrated either at the solid state or in solution in a conventional proton-exchange membrane fuel cell (PEMFC) using Nafion as the polymer electrolyte. However, due to the fact that this complex is a poor ORR catalyst, a low power output of 11 μW cm^{-2} accompanied with a low cell voltage of 0.3 V was measured, far from the performances of a Pt-based PEMFC (several hundreds of mW cm^{-2} and cell voltages of 1 V). A second molecular-catalyst-based fuel cell was then described by Artero and coworkers in 2015 [70]. A previously described bioinspired nickel complex bearing pyrene group was integrated as a carbon-black ink at the anode of a nonprecious metal PEMFC. The cathode was based on a cobalt-based material obtained from the pyrolysis of $Co(NO_3)$, triazolopyridine, and carbon black. The fuel cell delivers a low power output of 23 μW cm^{-2} accompanied with a cell voltage of 0.74 V. In 2017, Artero and Le Goff have designed an original nonprecious metal-based fuel cell by integration of an arginine-modified bioinspired nickel catalyst at the anode and an MCO, bilirubin oxidase from *Myrothecium verrucaria*, at the cathode [38]. The biocathode was designed by favoring the orientation of the enzyme at the surface of CNTs. A gas diffusion electrode was specifically engineered for such type of biocatalysts [71]. Owing to the low overpotential of the Ni catalyst and the enzyme for hydrogen oxidation reaction and ORR, respectively, the fuel cell delivers almost 2 mW cm^{-2} accompanied with cell voltages of 1 V, a record for a fuel cell based on such type of catalysts (Fig. 2.8). Furthermore, the nickel catalyst can operate in a conventional Nafion-based PEMFC, using Pt/C at the cathode. The fuel cell delivers 14 mW cm^{-2} with a cell voltage of 0.9 V [38].

2.3 Molecular catalysts for oxidation of alcohols and aldehydes

2.3.1 Organic electrocatalysts

Nitroxyl or nitroxide radicals are secondary amine nitrogen oxides and present unique properties as organic catalysts in different heavy metal free electrochemical oxidation reactions. *N*-Oxyl radicals have been shown to be efficient electrocatalysts for oxidation of amines, primary and secondary alcohols, ketones, aldehydes, thiols, and saccharides [72–76]. Electrooxidation of alcohols usually leads to the formation of ketones or aldehydes but deeper oxidation of the preceding products

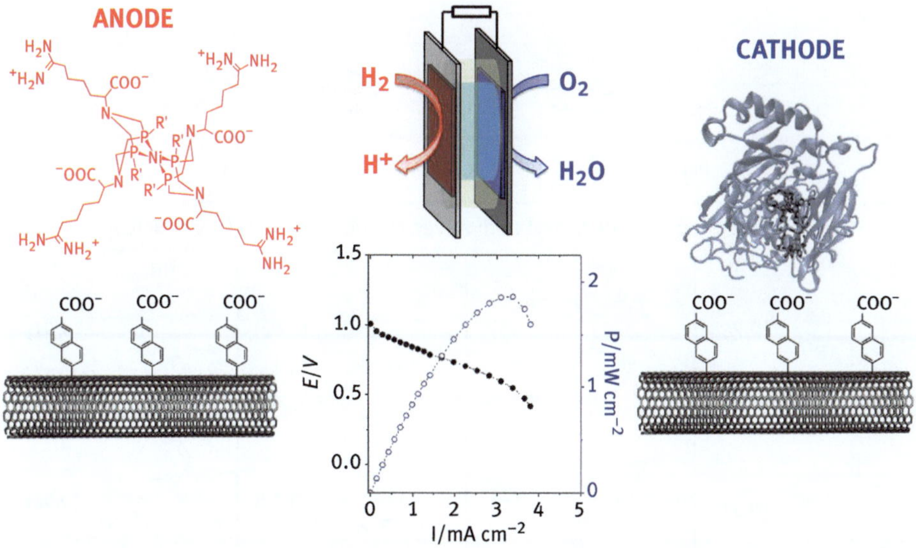

Fig. 2.8: Nonprecious metal fuel cell based on a nickel molecular catalyst and a copper enzyme. From Gentil et al. [38].

leads to the formation of carboxylates, while reducing sugars are transformed to glucoronates. Their general redox reactions, the structures of the most used *N*-oxyl radicals in electrocatalysis, and their respective redox potential leading to the conversion of the reactive form of substrate oxidation are shown in Fig. 2.9. As dealing with sustainable electrocatalysis, this section will be dedicated on probably the most widely used cyclic *N*-oxyl radical: the 2,2,6,6-tetramethylpiperidine-*N*-oxyl (TEMPO) for the oxidation of hydroxyl/carbonyl groups as the chemistry of other aminoxyl radicals is similar. However, the reactivity of imidoxyl radicals such as the phthalimide-*N*-oxyl (PINO; 8) is slightly different as the active form is the radical itself and it is obtained from the oxidation of the hydroxyimide species. For a comprehensive study of these electrocatalysts class, we direct the readers to a new review by Rafiee and coworkers [77].

Electrochemical characterization of TEMPO was initiated by Tamura and colleagues [78]. In their study they compared the electrochemical reversibility of two stable radicals: the cyclic radical TEMPO and the opened di-tert-butyl nitroxide (DTBNO). Only TEMPO showed a reversible behavior displayed by cyclic voltammetry. On the other hand, the DTBNO showed poor reversibility as its oxidized cationic form was not stable. Later, Semmelhack et al. [73] demonstrated for the first time the use of TEMPO as an electrocatalyst for alcohol oxidation and studied the mechanism to achieve such a reaction. Most of the works proposing the electrooxidation of alcohols using this *N*-oxyl radical in aqueous solvent were performed

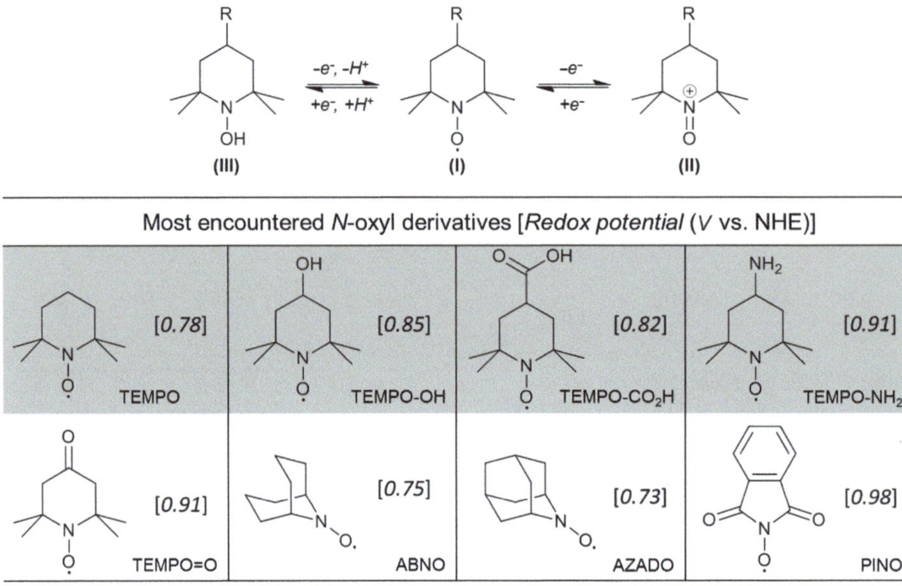

Fig. 2.9: Redox cycle underlining the E and EC mechanisms for aminoxyl radicals (I) conversion to oxoammonium ion (II) and hydroxylamine (III), respectively. Structures of TEMPO, derivatives of N-oxyl radicals with their experimental redox potentials (aminoxyl/oxoammonium (1–7) and hydroxylamine/imidoxyl (8)).

in alkaline conditions or in the presence of a stoichiometric amount of a base to facilitate the regeneration of the catalyst [79–83]. It was also demonstrated that due to lower pK_a values and smaller steric hindrances, primary alcohols are preferentially oxidized than secondary alcohols under basic conditions [84, 85]. The mechanism of TEMPO proceeds through an initial electrochemical oxidation step from the stable nitroxyl radical to the catalytically active oxoammonium salt ($TEMPO^+$). The successive oxidation of the substrate leads in the formation of a hydroxylamine (TEMPOH). From slightly acidic to highly alkaline solutions, the regeneration of the nitroxyl radical to complete the catalytic cycle is obtained through two possible pathways: (1) the direct 1 e^-/1 H^+ oxidation of TEMPOH to TEMPO at the electrode surface or (2) the comproportionation between TEMPOH and $TEMPO^+$ into TEMPO (Fig. 2.10). This is illustrated by the Pourbaix diagrams deciphered by the Stahl group [86].

Brown and coworkers reported a microfluidic electrolytic cell capable of alcohol oxidation on a laboratory scale [87]. The "microchannel" (depth × width × length: 200 µm × 1.5 mm × 70 cm) was designed to offer a large surface area to allow high conversion yield. In the best performing protocol, the cell was operated at 20 ± 5 mA between the carbon-based anode and the stainless steel cathode at controlled pH, at room temperature (to avoid loss of selectivity), and with a flow rate of 0.1 mL min^{-1}.

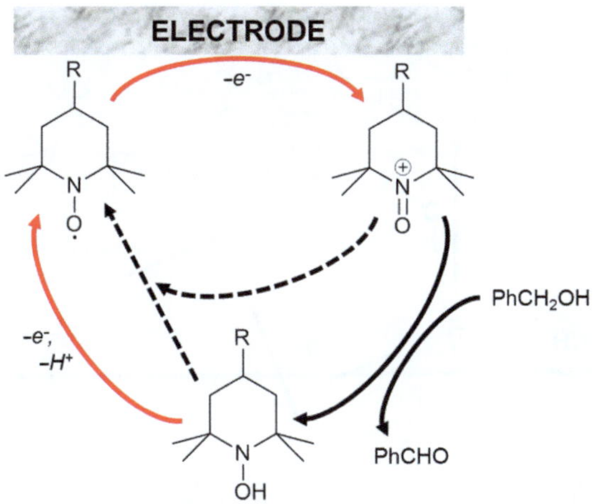

Fig. 2.10: Possible electrochemical/chemical mechanisms for the oxidation of benzyl alcohol and regeneration of the TEMPO electrocatalyst in its radical form. Red arrows represent electrochemical reactions, solid black arrows represent oxidation of the substrate, and dashed black arrows depict the comproportionation reaction in neutral to alkaline conditions. From Gerken et al. [86] to Adapted from Nutting et al. [77].

They obtained ranges of yield efficiency and selectivity from 21% to 94% and 50% to 100%, respectively, based on a survey of 15 different alcohols. The efficiency was directly linked to the steric hindrance of the alcohols. However, only 60% of the catalyst could be recovered at the end of electrolysis. In order to enhance the radical reusability, several approaches have been proposed. Tanaka and coworkers synthesized highly water-soluble *N*-oxyl compounds for the oxidation of different alcohols to aldehydes and ketones [88]. Used for green organic chemistry, the organic products were obtained by extraction from the aqueous phase with ethyl acetate while the electrocatalyst remained in the aqueous phase after the extraction procedure. Other strategies rely on the immobilization on a wide range of electrode substrates (or surfaces) utilizing various chemical properties such as the formation of self-assembled monolayers on gold or during the synthesis of gold nanoparticles [89, 90], the electropolymerization of a thiophene- or pyrrole-modified TEMPO [91, 92], the electrodeposition of a sol–gel doped with TEMPO [93], the formation of chemical bonding within hydrogels [94–96], or the noncovalent adsorption on CNTs through pyrene moieties attached on the TEMPO structure [97] (Fig. 2.11).

Very recently, Sigman and colleagues integrated TEMPO in bioelectronic systems in a hybrid organic/enzymatic electrocatalytic cascade with a single enzyme (the oxalate oxidase from barley (OO) [98], oxalate decarboxylase from *Bacillus subtilis* (OxDc) [99], or the NAD-dependent formate dehydrogenase from *Candida boidinii* (FDH)) [100] for the complete oxidation of glycerol to CO_2 (Fig. 2.12). The only downside to these nitroxyl radicals for substrate oxidation is the high electrode potentials

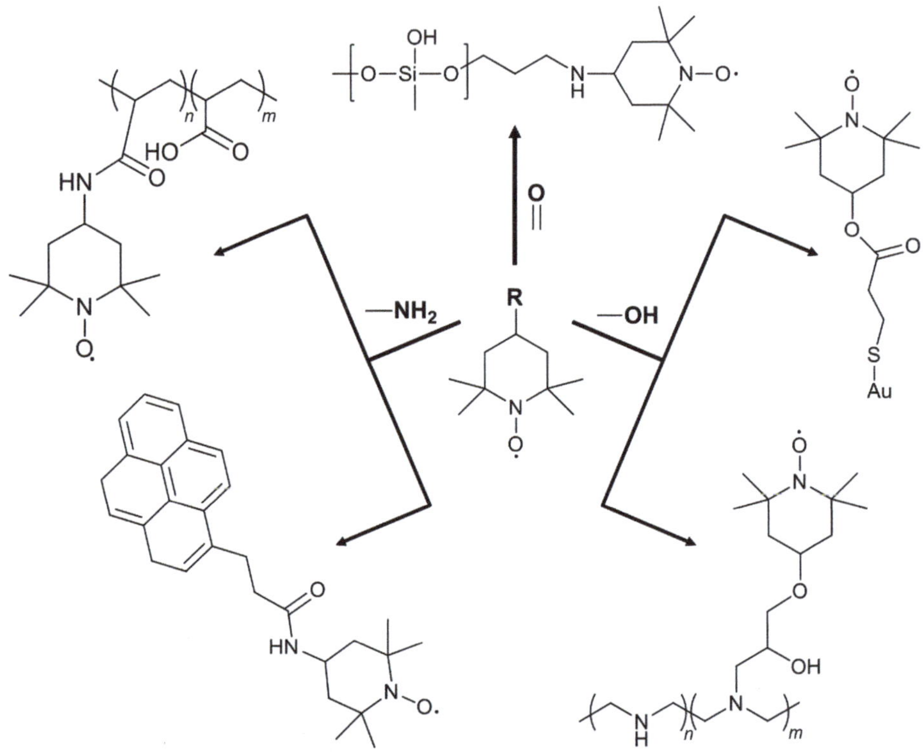

Fig. 2.11: Structures of some TEMPO derivatives for the preparation of TEMPO-modified electrodes.

(i.e., 0.45–0.70 V vs. Saturated Calomel Electrode (SCE)) required to generate the active oxoammonium species that limit their use in energy conversion such as EFCs. Nonetheless, this negative aspect was elegantly used for the continuous mesoxalate oxidation by FDH as TEMPO was also shown to promote NAD^+ electroregeneration [100]. Despite its high chemical stability, TEMPO does not present the highest catalytic rate among *N*-oxyl radicals necessarily.

Steric hindrance induced by the presence of four methyl groups near the nitroxyl radical compared to the higher substrate accessibility for other radicals such as 9-azabicyclo[3.3.1]nonane-*N*-oxyl could explain the lower activity towards alcohol oxidation [101]. Li and coworkers extended the applicability of *N*-oxyl radicals in the electrosynthesis of nitriles over the course of TEMPO-mediated electrooxidation of alcohols in dry acetonitrile containing ammonium acetate (NH_4OAc) as the nitrogen source [102]. First, they oxidized the benzyl alcohol to benzaldehyde. Subsequently, the benzaldehyde reacted with NH_4OAc to form the corresponding imine, which underwent itself a second electrooxidation by TEMPO to form the final nitrile product.

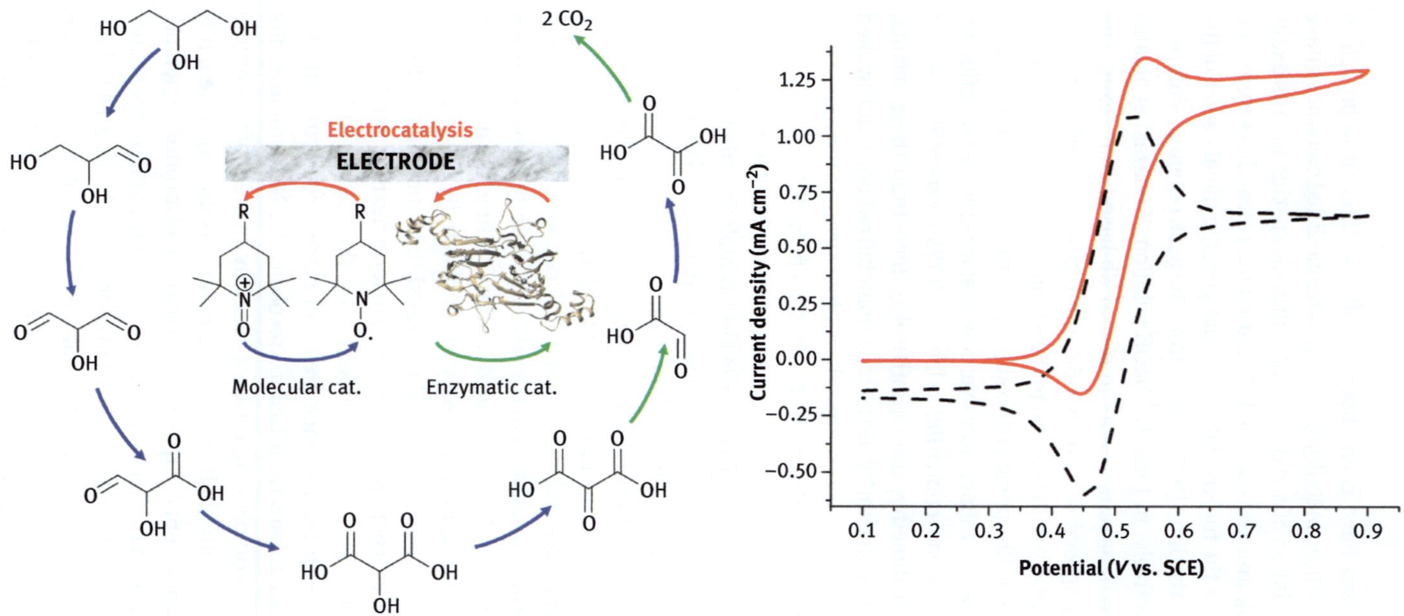

Fig. 2.12: (Left) Simplified electrocatalytic oxidation cascade of glycerol by TEMPO-R and oxalate oxidase to CO_2. (Right) Catalytic voltammograms recorded on GC electrode of 25 mmol L^{-1} TEMPO in the absence (dashed) and in the presence of 100 mmol L^{-1} glycerol with 400 mmol L^{-1} phosphate buffer (pH 6.5), at 1 mV s^{-1} and 25 °C.

2.3.2 Inorganic electrocatalysts

Some transition-metal complexes have been shown to offer a different approach to perform carbon-based fuel oxidation. Metallophthalocyanine-modified electrodes have also been employed for the oxidation of a wide variety of organic species such as oxalic acid [103–105], methanol [106], and glucose [107]. Additionally, researchers were able to decrease the overpotential for the fuel oxidation by changing the metal center of the macrocycle from cobalt [107] to rhodium [108] as seen from the peak potential shift from 450 to 5 mV versus SCE, respectively, in 1 mol L^{-1} NaOH solution. Yamazaki et al. also examined a different heterocyclic macrocycle as to obtain Rh-based complexes. The porphyrin ring structure is composed of four pyrrole subunits and is well known in biochemistry as it chelates naturally with iron ion to form heme. This group studied the electrochemical properties of the Rh–porphyrin complex for the low-potential oxidation of sugars [109]. They reported extremely low overpotential for glucose oxidation on carbon-supported electrodes with onset potential as low as −0.795 V versus SCE, thanks to a Rh (III) deuteroporphyrin dimethyl ester (Rh(DPDE)). Rich from these results, Cosnier and coworkers presented an (Rh(DPDE)) anode immobilized onto multiwalled CNTs (MWCNTs) by adsorption of the π-conjugated porphyrin core. This anode was integrated with a cobalt (II) phthalocyanine cathode to form a fully molecular catalyst-based glucose/O_2 fuel cell for the oxidation of glucose and the reduction of oxygen at the anode and cathode, respectively [110]. To increase electrocatalytic activity lifetime, they also reported a bispyrene-appended Rh (III) deuteroporphyrin (Rh(DP)pyr_2) and a tetrapyrene-appended Co (III) tetracarboxyphenyl porphyrin to enhance the π–π interactions of the porphyrinic systems to the CNT layers, hence maintaining high catalyst loading over the course of the fuel cell operation (Fig. 2.13) [111].

The overall performances of the molecular-based fuel cell were increased from 0.2 mW cm^{-2} (without pyrene anchoring site) to 0.9 mW cm^{-2} with the newly synthesized pyrene-based electrocatalysts. They evaluated for the first time the use of these catalysts in a fuel cell at neutral pH, however, only residual activity of the catalysts remained in this condition. Elouarzaki et al. [112] proposed the functionalization of porphyrinic-based electrocatalysts with pyrene moieties to form flexible, conductive, freestanding MWCNT electrodes. This strategy should allow the authors to fabricate three-dimensional electrodes with high surface area and mesoporous structure for improved catalyst loading and substrate diffusion, respectively. Nevertheless, the current density obtained for the complete fuel cell exceeded by an order of magnitude the current density of each individual half-cell. As this point was not explained, it is complicated to draw any enhancement compared to the regular adsorption of pyrene-modified electrocatalysts. Furthermore, in such type of molecular-catalyst-based fuel cells, different types of membranes have been tested in order to study the effect of the membrane on the fuel performances [110, 113, 114]. The same group also investigated gold as the metallic center in the porphyrin core. The Au (III) porphyrins were adsorbed onto MWCNTs and they displayed electrocatalytic properties for glucose

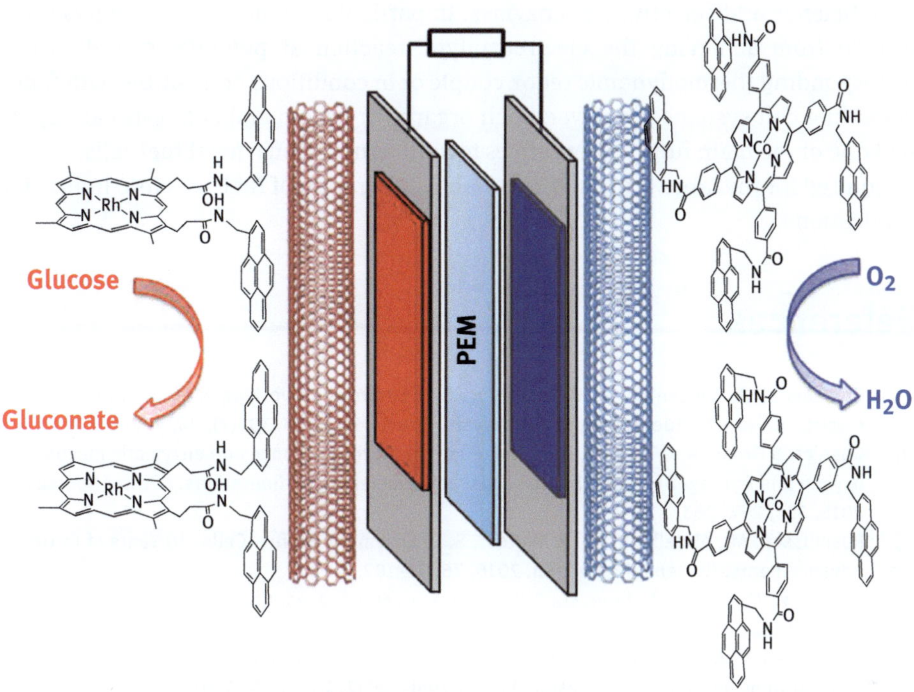

Fig. 2.13: Schematic representation of the molecular-catalyst-based glucose/O_2 fuel cell using a pyrene-modified Rh and Co porphyrin for the respective oxidation of glucose and reduction of O_2. The electrodes are separated by a proton-exchange membrane (PEM).

oxidation in alkaline media at a slightly higher overpotential than the Rh counterpart [115]. The only main drawback to such electrocatalysts is their activity in alkaline medium, which hampers their use for the analyte sensing in physiological fluids.

2.4 Conclusion

This chapter demonstrates that major challenges have been tackled in recent years in the development of molecular catalysts for fuel cell applications. Ranging from bioinspired catalysts for H_2 oxidation to organic catalysts for oxidation of alcohols, these molecular catalysts have shown to achieve low overpotential or high current electrocatalytic activity in the oxidation of fuels at the electrode of fuel cells. In particular, their combination with highly conductive nanomaterials has boosted molecular catalyst loadings and has made possible their facile integration in devices. However, improvements and novel molecular catalysts are still to be developed in order to merge the efficiency and stability of metal catalysts with the efficiency and selectivity of enzymes. In particular, some of these catalysts are still far from achieving the electrocatalytic reaction at potential closed to the corresponding thermodynamic redox couple or in conditions compatible with functional fuel cell systems. However, such organic or base-metal catalysts can represent one of the main future alternatives toward nonprecious metal fuel cells, which is required for the large-scale and renewable integration of fuel cells into industrial applications.

References

[1] Cracknell, J. A., Vincent, K. A., & Armstrong, F. A. Enzymes as working or inspirational electrocatalysts for fuel cells and electrolysis. Chem Rev. 2008, 108(7), 2439–2461.

[2] Cosnier, S Gross, A. J., Le Goff, A., & Holzinger, M. Recent advances on enzymatic glucose/oxygen and hydrogen/oxygen biofuel cells: achievements and limitations. J Power Sources. 2016, 325, 252–263.

[3] Rasmussen, M., Abdellaoui, S., & Minteer, S. D. Enzymatic Biofuel Cells: 30 Years of Critical Advancements. Biosens Bioelectron. 2016, 76, 91–102.

[4] Lubitz, W., Ogata, H., Rüdiger, O., & Reijerse, E. Hydrogenases. Chem Rev. 2014, 114(8), 4081–4148.

[5] Mazurenko, I., Wang, X., Poulpiquet, A., & de Lojou E. H_2/O_2 enzymatic fuel cells: from proof-of-concept to powerful devices. Sustain Energy Fuels. 2017, 1(7), 1475–1501.

[6] Woolerton, T W.; Sheard, S, Chaudhary, Y. S., & Armstrong, F A. Enzymes and bio-inspired electrocatalysts in solar fuel devices. Energy Environ Sci, 2012, 7470–7490.

[7] Liebgott, P. P., de Lacey, A. L., Burlat, B., Cournac, L., Richaud, P., Brugna, M., Fernandez, V M., Guigliarelli, B., Rousset, M., Léger, C., et al. Original design of an oxygen-tolerant [nife] hydrogenase: major effect of a valine-to-cysteine mutation near the active site. J Am Chem Soc. 2011, 133(4) 986–997. https://pubs.acs.org/doi/abs/10.1021/ja108787s

[8] Plumere, N., Ruediger, O., Oughli, A. A., Williams, R., Vivekananthan, J., Poeller, S., Schuhmann, W., & Lubitz, W. A redox hydrogel protects hydrogenase from high-potential deactivation and oxygen damage. Nat Chem. 2014, 6(9), 822–827.

[9] Baur, J., Le Goff, A., Dementin, S., Holzinger, M., Rousset, M., & Cosnier, S. Three-dimensional carbon nanotube–polypyrrole–[nife] hydrogenase electrodes for the efficient electrocatalytic oxidation of H_2. Int J Hydrog Energy. 2011, 36(19), 12096–12101.

[10] Ruff A., Szczesny J., Zacarias S., Pereira I A C., Plumeré N., & Schuhmann, W. Protection and reactivation of the [NiFeSe] hydrogenase from desulfovibrio vulgaris hildenborough under oxidative conditions. ACS Energy Lett. 2017, 2(5), 964–968.

[11] Oughli, A A., Conzuelo, F., Winkler, M., Happe, T., Lubitz, W. Schuhmann, W. Rüdiger, O. & Plumeré, N. A redox hydrogel protects the O_2–sensitive [FeFe]-hydrogenase from chlamydomonas reinhardtii from oxidative damage. Angew Chem Int Ed. 2015, 54(42), 12329–12333.

[12] Rengaraj, S., Haddad, R., Lojou, E., Duraffourg, N., Holzinger, M., Le Goff, A., & Forge, V. Interprotein electron transfer between FeS–protein nanowires and oxygen–tolerant NiFe hydrogenase. Angew Chem Int Ed. 2017, 56(27), 7774–7778.

[13] Ciaccafava, A., De Poulpiquet, A., Techer, V., Giudici-Orticoni, M T., Tingry, S., Innocent, C., & Lojou, E. An innovative powerful and mediatorless H_2/O_2 biofuel cell based on an outstanding bioanode. Electrochem Commun. 2012, 23, 25–28.

[14] de Poulpiquet, A., Ciaccafava, A., Gadiou, R., Gounel, S., Giudici-Orticoni, M. T., Mano, N., & Lojou, E. Design of a H_2/O_2 biofuel cell based on thermostable enzymes. Electrochem Commun. 2014, 42, 72–74.

[15] Monsalve, K., Mazurenko, I., Lalaoui, N., Le Goff, A., Holzinger, M., Infossi, P., Nitsche, S., Lojou, J. Y., Giudici-Orticoni, M. T., Cosnier, S., et al. A H_2/O_2 enzymatic fuel cell as a sustainable power for a wireless device. Electrochem Commun. 2015, 60, 216–220.

[16] Gentil, S., Che Mansor, S. M., Jamet, H., Cosnier, S., Cavazza, C., & Le Goff, A. Oriented immobilization of [NiFeSe] hydrogenases on covalently and noncovalently functionalized carbon nanotubes for H_2/Air enzymatic fuel cells. ACS Catal. 2018, 8(5), 3957–3964.

[17] Lalaoui, N., Poulpiquet, A. de., Haddad, R., Le Goff, A., Holzinger, M., Gounel, S., Mermoux, M., Infossi, P., Mano, N., Lojou, E., et al. A membraneless air-breathing hydrogen biofuel cell based on direct wiring of thermostable enzymes on carbon nanotube electrodes. Chem Commun. 2015, 51(35), 7447–7450.

[18] So, K., Kitazumi, Y., Shirai, O., Nishikawa, K., Higuchi, Y., & Kano, K. Direct electron transfer-type dual gas diffusion H_{-2}/O_{-2} biofuel cells. J Mater Chem A 2016, 4(22), 8742–8749.

[19] Xia, H., So, K., Kitazumi, Y., Shirai, O., Nishikawa, K., Higuchi, Y., & Kano, K. Dual gas-diffusion membrane- and mediatorless dihydrogen/air-breathing biofuel cell operating at room temperature. J Power Sources. 2016, 335 (Supplement C), 105–112.

[20] Coutard, N., Kaeffer, N., & Artero, V. Molecular engineered nanomaterials for catalytic hydrogen evolution and oxidation. Chem Commun. 2016, 52(95), 13728–13748.

[21] Brazzolotto, D., Gennari, M., Queyriaux, N., Simmons, T. R., Pécaut, J., Demeshko, S., Meyer, F., Orio, M., Artero, V., & Duboc, C. Nickel-centred proton reduction catalysis in a model of [NiFe] hydrogenase. Nat Chem. 2016, 8(11), 1054–1060.

[22] Capon, J. F., Gloaguen, F., Pétillon, F. Y., Schollhammer, P., & Talarmin, J. Electron and proton transfers at diiron dithiolate sites relevant to the catalysis of proton reduction by the [FeFe]-hydrogenases. Coord Chem Rev. 2009, 253(9), 1476–1494.

[23] Camara, J. M., & Rauchfuss, T. B. Combining acid–base, redox and substrate binding functionalities to give a complete model for the [FeFe]-hydrogenase. Nat Chem. 2012, 4(1), 26–30.

[24] Wang, N., Wang, M., Wang, Y., Zheng, D., Han, H., Ahlquist, M. S. G., & Sun, L. Catalytic activation of H_2 under mild conditions by an [FeFe]-hydrogenase model via an active μ-hydride species. J Am Chem Soc. 2013, 135(37), 13688–13691.

[25] Justice, A. K., Linck, R. C., Rauchfuss, T. B., & Wilson, S. R. Dihydrogen activation by a diruthenium analogue of the Fe-only hydrogenase active site. J Am Chem Soc. 2004, 126(41), 13214–13215.

[26] Linck, R. C., Pafford, R. J., & Rauchfuss, T. B. Heterolytic and homolytic activation of dihydrogen at an unusual iridium (II) sulfide. J Am Chem Soc. 2001, 123(36), 8856–8857.

[27] Ogo, S., Kabe, R., Uehara, K., Kure, B., Nishimura, T., Menon, S. C., Harada, R., Fukuzumi, S., Higuchi, Y., Ohhara, T., et al. A dinuclear Ni(μ-H)Ru complex derived from H_2. Science. 2007, 316(5824), 585–587.

[28] Curtis, C. J., Miedaner, A., Ciancanelli, R., Ellis, W. W., Noll, B. C., Rakowski DuBois, M., & DuBois, D. L. $[Ni(Et_2PCH_2NMeCH_2PEt_2)_2]^{2+}$ as a functional model for hydrogenases. Inorg Chem. 2003, 42(1), 216–227.

[29] O'Hagan, M., Shaw, W. J., Raugei, S., Chen, S., Yang, J. Y., Kilgore, U. J., DuBois, D. L., & Bullock, R. M. Moving protons with pendant amines: proton mobility in a nickel catalyst for oxidation of hydrogen. J Am Chem Soc. 2011, 133(36), 14301–14312.

[30] Wilson, A. D., Newell, R. H., McNevin, M. J., Muckerman, J. T. Rakowski DuBois, M. & DuBois, D. L. Hydrogen oxidation and production using nickel-based molecular catalysts with positioned proton relays. J Am Chem Soc. 2006, 128(1), 358–366.

[31] Shaw, W. J. Helm, M. L. & DuBois, D. L. A modular, energy-based approach to the development of nickel containing molecular electrocatalysts for hydrogen production and oxidation. Biochim Biophys Acta-Bioenerg. 2013, 1827 (8–9), 1123–1139.

[32] Dutta, A., Lense, S., Hou, J., Engelhard, M. H., Roberts, J. A. S., & Shaw, W. J. Minimal proton channel enables H_2 oxidation and production with a water-soluble nickel-based catalyst. J Am Chem Soc. 2013, 135(49), 18490–18496.

[33] Dutta, A., Roberts, J. A. S., & Shaw, W. J. Arginine-containing ligands enhance H_2 oxidation catalyst performance. Angew Chem Int Ed. 2014, 53(25), 6487–6491.

[34] Reback, M. L. Buchko, G. W. Kier, B. L. Ginovska-Pangovska, B. Xiong, Y. Lense, S. Hou, J. Roberts, J. A. S., Sorensen, C. M. Raugei, S. et al. Enzyme design from the bottom up: an active nickel electrocatalyst with a structured peptide outer coordination sphere. Chem Eur J. 2014, 20(6), 1510–1514.

[35] Dutta, A., Ginovska, B., Raugei, S., Roberts, J. A. S., & Shaw, W. J. Optimizing conditions for utilization of an H_2 oxidation catalyst with outer coordination sphere functionalities. Dalton Trans. 2016, 45, 9786–9793.

[36] Le Goff, A., Artero, V., Jousselme, B., Tran, P. D., Guillet, N., Metaye, R., Fihri, A., Palacin, S., & Fontecave, M. From hydrogenases to noble metal-free catalytic nanomaterials for H_2 production and uptake. Science. 2009, 326(5958), 1384–1387.

[37] Tran, P. D., Le Goff, A., Heidkamp, J., Jousselme, B., Guillet, N., Palacin, S., & Dau, H., Fontecave, M., Artero, V. Noncovalent modification of carbon nanotubes with pyrene-functionalized nickel complexes: carbon monoxide tolerant catalysts for hydrogen evolution and uptake. Angew Chem Int. *Ed.* 2011, 50(6), 1371–1374.

[38] Gentil, S., Lalaoui, N., Dutta, A., Nedellec, Y., Cosnier, S., Shaw, W. J., Artero, V., & Le Goff, A. Carbon-nanotube-supported bio-inspired nickel catalyst and its integration in hybrid hydrogen/air fuel cells. Angew Chem Int Ed. 2017, 56, 1845–1849.

[39] Zhang, W. Lai, W. & Cao, R. Energy-related small molecule activation reactions: oxygen reduction and hydrogen and oxygen evolution reactions catalyzed by porphyrin- and corrole-based systems. Chem Rev. 2017, 117(4), 3717–3797.

[40] Dutton, P. L., Wilson, D. F., & Lee, C. P. Oxidation-reduction potentials of cytochromes in mitochondria. Biochemistry (Mosc.). 1970, 9(26), 5077–5082.

[41] Ghiladi, R. A., Kretzer, R. M., Guzei, I., Rheingold, A. L., Neuhold, Y. M., Hatwell, K. R., Zuberbühler, A. D., & Karlin, K. D. (F8TPP)FeII/O_2 reactivity studies {F8TPP = Tetrakis(2,6-Difluorophenyl)Porphyrinate(2−)}: spectroscopic (UV–Visible and NMR) and kinetic study of solvent-dependent (Fe/O_2 = 1:1 or 2:1) reversible O_2-reduction and ferryl formation. Inorg Chem. 2001, 40(23), 5754–5767.

[42] Ghiladi, R. A., Hatwell, K. R., Karlin, K. D., Huang, H., Moënne-Loccoz, P., Krebs, C., Huynh, B. H., Marzilli, L. A., Cotter, R. J., Kaderli, S., et al. Dioxygen reactivity of mononuclear heme and copper components yielding a high-spin Heme–Peroxo–Cu complex. J Am Chem Soc. 2001, 123 (25), 6183–6184.

[43] Kopf, M. A., & Karlin, K. D. Dioxygen reactivity of reduced heme and heme–copper complexes utilizing tetraarylporphyrinates tethered with both a pyridyl axial ligand and N, N-Bis[2-(2-Pyridyl)Ethyl]Amine chelate. Inorg Chem. 1999, 38(22), 4922–4923.

[44] Ghiladi, R. A., Ju, T. D., Lee, D. H., Moënne-Loccoz, P., Kaderli, S., Neuhold, Y. M., Zuberbühler, A. D., Woods, A. S., Cotter, R. J., Karlin, K. D. Formation and characterization of a high-spin heme-copper dioxygen (peroxo) complex. J Am Chem Soc. 1999, 121(42), 9885–9886.

[45] Halime, Z., Kotani, H., Li, Y., Fukuzumi, S., & Karlin, K. D. Homogeneous catalytic O_2 reduction to water by a cytochrome c oxidase model with trapping of intermediates and mechanistic insights. Proc Natl Acad Sci. 2011, 108(34), 13990–13994.

[46] Chatterjee, S., Sengupta, K., Hematian, S., Karlin, K. D., & Dey, A. Electrocatalytic O_2-reduction by synthetic cytochrome c oxidase mimics: identification of a "Bridging Peroxo" intermediate involved in facile $4e^-/4H^+$ O_2-reduction. J Am Chem Soc. 2015, 137(40), 12897–12905.
[47] Collman, J. P., Rapta, M., Bröring, M., Raptova, L., Schwenninger, R., Boitrel, B., Fu, L., & L'Her, M. Close structural analogues of the cytochrome c oxidase Fea3/CuB center show clean 4e- electroreduction of O_2 to H_2O at physiological PH. J Am Chem Soc. 1999, 121(6), 1387–1388.
[48] Collman, J. P., Ghosh, S., Dey, A., Decréau, R. A., & Yang, Y. Catalytic reduction of O_2 by cytochrome c using a synthetic model of cytochrome c oxidase. J Am Chem Soc. 2009, 131(14), 5034–5035.
[49] Collman, J. P., Fu, L., Herrmann, P. C., & Zhang, X. A functional model related to cytochrome c oxidase and its electrocatalytic four-electron reduction of O_2. Science. 1997, 275(5302), 949–951.
[50] Collman, J. P., Fu, L., Herrmann, P. C.; Wang, Z., Rapta, M., Bröring, M., Schwenninger, R., & Boitrel, B. A functional model of cytochrome c oxidase: thermodynamic implications. Angew Chem Int Ed. 37 (24), 3397–3400.
[51] Collman, J. P., Devaraj, N. K., Decréau, R. A., Yang, Y., Yan, Y. L., Ebina, W., Eberspacher, T. A., & Chidsey, C. E. D. A cytochrome c oxidase model catalyzes oxygen to water reduction under rate-limiting electron flux. Science. 2007, 315(5818), 1565–1568.
[52] Samanta, S., Das, P. K., Chatterjee, S., Sengupta, K., Mondal, B., & Dey, A. O_2 reduction reaction by biologically relevant anionic ligand bound iron porphyrin complexes. Inorg Chem. 2013, 52(22), 12963–12971.
[53] Collman, J. P., Decréau, R. A., Lin, H., Hosseini, A., Yang, Y., Dey, A., & Eberspacher, T. A. Role of a distal pocket in the catalytic O_2 reduction by cytochrome c oxidase models immobilized on interdigitated array electrodes. Proc Natl Acad Sci. 2009, 106(18), 7320–7323.
[54] Didier, A., L'Her, M., & Boitrel, B. Substituted tren-capped porphyrins: probing the influence of copper in synthetic models of cytochrome c oxidase. Org Biomol Chem. 2003, 1(8), 1274–1276.
[55] Abad, J. M., Gass, M., Bleloch, A., & Schiffrin, D. J. Direct electron transfer to a metalloenzyme redox center coordinated to a monolayer-protected cluster. J Am Chem Soc. 2009, 131(29), 10229–10236.
[56] Reuillard, B., Le Goff, A., Agnès, C., Zebda, A., Holzinger, M., & Cosnier, S. Direct electron transfer between tyrosinase and multi-walled carbon nanotubes for bioelectrocatalytic oxygen reduction. Electrochem Commun. 2012, 20, 19–22.
[57] Wang, F. F., Zhao, Y. M., Wei, P. J., Zhang, Q. L., & Liu, J. G. Efficient electrocatalytic O_2 reduction at copper complexes grafted onto polyvinylimidazole coated carbon nanotubes. Chem Commun. 2017, 53(9), 1514–1517.
[58] McCrory, C. C. L., Devadoss, A., Ottenwaelder, X., Lowe, R. D., Stack, T. D. P., Chidsey, C. E. D. Electrocatalytic O_2 reduction by covalently immobilized mononuclear copper(I) complexes: evidence for a binuclear Cu_2O_2 intermediate. J Am Chem Soc. 2011, 133(11), 3696–3699.
[59] McCrory, C. C. L.; Ottenwaelder, X., Stack, T. D. P., & Chidsey, C. E. D. Kinetic and mechanistic studies of the electrocatalytic reduction of O_2 to H_2O with mononuclear Cu complexes of substituted 1,10-phenanthrolines. J Phys Chem A. 2007, 111(49), 12641–12650.
[60] Slowinski, K., Kublik, Z., Bilewicz, R., & Pietraszkiewicz, M. Electrocatalysis of oxygen reduction by a copper(II) hexaazamacrocyclic complex. J Chem Soc Chem Commun. 1994, 0(9), 1087–1088. https://pubs.rsc.org/en/content/articlelanding/1994/c3/c39940001087#!divAbstract
[61] Thorum, M. S., Yadav, J., & Gewirth, A. A. Oxygen reduction activity of a copper complex of 3,5-diamino-1,2,4-triazole supported on carbon black. Angew Chem Int Ed. 2009, 48(1), 165–167.

[62] Thorseth, M. A., Letko, C. S.; Rauchfuss, T. B., & Gewirth, A. A. Dioxygen and hydrogen peroxide reduction with hemocyanin model complexes. Inorg Chem. 2011, 50(13), 6158–6162.
[63] Gentil, S., Serre, D., Philouze, C., Holzinger, M., Thomas, F., & Le Goff, A. Electrocatalytic O_2 reduction at a bio-inspired mononuclear copper phenolato complex immobilized on a carbon nanotube electrode. Angew Chem Int Ed. 2016, 55(7), 2517–2520.
[64] Mano, N., & de Poulpiquet, A. O_2 reduction in enzymatic biofuel cells. Chem Rev. 2017, 118(5), 2392–2468.
[65] Le Goff, A., Holzinger, M., & Cosnier, S. Recent progress in oxygen-reducing laccase biocathodes for enzymatic biofuel cells. Cell Mol Life Sci. 2015, 72(5), 941–952.
[66] Rasmussen, M., Abdellaoui, S., & Minteer, S. D. Enzymatic biofuel cells: 30 years of critical advancements. Biosens Bioelectron. 2016, 76, 91–102.
[67] Cosnier, S., Gross, AJ., Le Goff, A., & Holzinger, M. Recent advances on enzymatic glucose/oxygen and hydrogen/oxygen biofuel cells: achievements and limitations. J Power Sources. 2016, 325, 252–263.
[68] Tse, E. C. M., Schilter, D., Gray, D. L., Rauchfuss, T. B., & Gewirth, A. A. Multicopper models for the laccase active site: effect of nuclearity on electrocatalytic oxygen reduction. Inorg Chem. 2014, 53(16), 8505–8516.
[69] Matsumoto, T., Kim, K., & Ogo, S. Molecular catalysis in a fuel cell. Angew Chem Int Ed. 2011, 50(47), 11202–11205.
[70] Tran, P. D., Morozan, A., Archambault, S., Heidkamp, J., Chenevier, P., Dau, H., Fontecave, M., Martinent, A., Jousselme, B., & Artero, V. A noble metal-free proton-exchange membrane fuel cell based on bio-inspired molecular catalysts. Chem Sci. 2015, 6(3), 2050–2053.
[71] Lalaoui, N., Holzinger, M., Le Goff, A., & Cosnier, S. Diazonium functionalisation of carbon nanotubes for specific orientation of multicopper oxidases: controlling electron entry points and oxygen diffusion to the enzyme. Chem Eur J. 2016, 22(30), 10494–10500.
[72] Semmelhack, M. F., & Schmid, C. R. Nitroxyl-mediated electro-oxidation of amines to nitriles and carbonyl compounds. J Am Chem Soc. 1983, 105(22), 6732–6734.
[73] Semmelhack, M. F., Chou, C. S., & Cortes, D. A. Nitroxyl-mediated electrooxidation of alcohols to aldehydes and ketones. J Am Chem Soc. 1983, 105(13), 4492–4494.
[74] Kashiwagi, Y., Ohsawa, A., Osa, T., Ma, Z., & Bobbitt, J. M. Electrocatalytic oxidation of thiols on a TEMPO modified electrode. Chem Lett. 1991, 20(4), 581–584.
[75] Parpot, P., Servat, K., Bettencourt, A. P., Huser, H., & Kokoh, K. B. TEMPO mediated oxidation of carbohydrates using electrochemical methods. Cellulose. 2010, 17(4), 815–824.
[76] Isogai, T., Saito, T., & Isogai, A. TEMPO electromediated oxidation of some polysaccharides including regenerated cellulose fiber. Biomacromolecules. 2010, 11(6), 1593–1599.
[77] Nutting, J. E., Rafiee, M., & Stahl, S. S. Tetramethylpiperidine *N*-Oxyl (TEMPO), Phthalimide *N*-Oxyl (PINO), and related *N*-Oxyl species: electrochemical properties and their use in electrocatalytic reactions. Chem Rev. 2018, 118(9), 4834–4885.
[78] Tsunaga, M., Iwakura, C., & Tamura, H. Electrode reactions of nitroxide radicals at platinum in acetonitrile. Electrochim Acta. 1973, 18(3), 241–245.
[79] Semmelhack, M. F., Schmid, C. R., & Cortés, D. A. Mechanism of the oxidation of alcohols by 2,2,6,6-tetramethylpiperidine nitrosonium cation. Tetrahedron Lett. 1986, 27(10), 1119–1122.
[80] Anelli, P. L., Banfi, S., Montanari, F., & Quici, S. Oxidation of diols with alkali hypochlorites catalyzed by oxammonium salts under two-phase conditions. J Org Chem. 1989, 54(12), 2970–2972.
[81] Hahn, Y., & Song, S. K. Electrochemical behavior of a TEMPO-modified electrode and its electrocatalytic oxidation of benzyl alcohol. Anal Sci. 1997, 13 (Supplement), 329–332.

[82] Zhao, M., Li, J., Mano, E., Song, Z., Tschaen, D. M., Grabowski, E. J. J., & Reider, P. J. Oxidation of primary alcohols to carboxylic acids with sodium chlorite catalyzed by TEMPO and bleach. J Org Chem. 1999, 64(7), 2564–2566.
[83] Figiel, P. J., Leskelä, M., & Repo, T. TEMPO-Copper(II) diimine-catalysed oxidation of benzylic alcohols in aqueous media. Adv Synth Catal. 2007, 349(7), 1173–1179.
[84] Bailey, W. F., Bobbitt, J. M., & Wiberg, K. B. Mechanism of the oxidation of alcohols by oxoammonium cations. J Org Chem. 2007, 72(12), 4504–4509.
[85] Bobbitt, J. M., BrüCkner, C., & Merbouh, N. Oxoammonium- and nitroxide-catalyzed oxidations of alcohols. In *Organic Reactions*; American Cancer Society. 2010; pp 103–424.
[86] Gerken, J. B., Pang, Y. Q., Lauber, M. B., & Stahl, S. S. Structural effects on the PH-dependent redox properties of organic nitroxyls: pourbaix diagrams for TEMPO, ABNO, and three TEMPO analogs. J Org Chem. 2018, 83(14), 7323–7330.
[87] Hill-Cousins, J. T., Kuleshova, J., Green, R. A., Birkin, P. R., Pletcher, D., Underwood, T. J., Leach, S. G., & Brown, R. C. D. TEMPO-mediated electrooxidation of primary and secondary alcohols in a microfluidic electrolytic cell. ChemSusChem. 2012, 5(2), 326–331.
[88] Kubota, J., Shimizu, Y., Mitsudo, K., & Tanaka, H. Water-soluble N-Oxyl compounds-mediated electrooxidation of alcohols in water: a prominent access to a totally closed system. Tetrahedron Lett. 2005, 46(52), 8975–8979.
[89] Blanchard, P. Y., Alévêque, O., Breton, T., & Levillain, E. TEMPO mixed SAMs: electrocatalytic efficiency versus surface coverage. Langmuir. 2012, 28(38), 13741–13745.
[90] Swiech, O., Bilewicz, R., & Megiel, E. TEMPO coated Au nanoparticles: synthesis and tethering to gold surfaces. RSC Adv. 2013, 3(17), 5979–5986.
[91] Iragi, A., Crayston, J. A., Walton, J. C., & Harrison, A. Aminoxyl functionalized polythiophene: synthesis and electrochemical applications. J Mater Chem. 1995, 5(9), 1291–1295.
[92] Lu, J., Ma, J. Yi, J., Shen, Z., Zhong, Y., Ma, C., & Li, M. Electrochemical polymerization of pyrrole containing TEMPO side chain on Pt electrode and its electrochemical activity. Electrochim Acta. 2014, 130, 412–417.
[93] Palmisano, G., Ciriminna, R., & Pagliaro, M. Waste-free electrochemical oxidation of alcohols in water. Adv Synth Catal. 2006, 348(15), 2033–2037.
[94] Osa, T., Akiba, U., Segawa, I., & Bobbitt, J. M. Electrocatalytic oxidation of nerol with nitroxyl radical covalently immobilized to Poly(Acrylic Acid) coated on carbon electrodes. Chem Lett. 1988, 17(8), 1423–1426.
[95] Hickey, D. P., Milton, R. D., Chen, D., Sigman, M. S., & Minteer, S. D. TEMPO-modified linear Poly(Ethylenimine) for immobilization-enhanced electrocatalytic oxidation of alcohols. ACS Catal. 2015, 5(9), 5519–5524.
[96] Schille, B., Giltzau, N. O., & Francke, R. On the use of polyelectrolytes and polymediators in organic electrosynthesis. Angew Chem Int Ed. 2018, 57(2), 422–426.
[97] Das, A., & Stahl, S. S. Noncovalent immobilization of molecular electrocatalysts for chemical synthesis: efficient electrochemical alcohol oxidation with a pyrene–TEMPO conjugate. Angew Chem. 2017, 129(30), 9018–9023.
[98] Hickey, D. P., McCammant, M. S., Giroud, F., Sigman, M. S., & Minteer, S. D. Hybrid enzymatic and organic electrocatalytic cascade for the complete oxidation of glycerol. J Am Chem Soc. 2014, 136(45), 15917–15920.
[99] Macazo, F. C., Hickey, D P., Abdellaoui, S., Sigman, M. S., & Minteer, S. D. Polymer-immobilized, hybrid multi-catalyst architecture for enhanced electrochemical oxidation of glycerol. Chem Commun. 2017, 53(74), 10310–10313.
[100] Abdellaoui, S., Chavez, M. S., Matanovic, I., Stephens, A. R., Atanassov, P., & Minteer, S. D. Hybrid molecular/enzymatic catalytic cascade for complete electro-oxidation of glycerol using

a promiscuous NAD-dependent formate dehydrogenase from candida boidinii. Chem Commun. 2017, 53(39), 5368–5371.
[101] Shibuya, M., Tomizawa, M., Sasano, Y., & Iwabuchi, Y. An expeditious entry to 9-Azabicyclo [3.3.1]Nonane N-Oxyl (ABNO): another highly active organocatalyst for oxidation of alcohols. J Org Chem. 2009, 74(12), 4619–4622.
[102] Fan, Z., Yang, X., Chen, C., Shen, Z., & Li, M. One-pot electrochemical oxidation of alcohols to nitriles mediated by TEMPO. J Electrochem Soc. 2017, 164(4), G54–G58.
[103] Santos, L. M., & Baldwin, R. P. Electrocatalytic response of 2 cobalt phthalocyanine chemically modified electrodes toward oxalic acid and alpha.-keto acids. Anal Chem. 1986, 58(4), 848–852.
[104] Ribeiro, E. S., & Gushikem, Y. Cobalt(II) Tetrasulfophthalocyanine complex adsorbed on a silica gel surface chemically modified with 3-N-Propylpyridinium chloride: oxalic acid oxidation study. Electroanalysis. 1999, 11(17), 1280–1284.
[105] Yamazaki, S., Yamada, Y., Fujiwara, N., Ioroi, T., Siroma, Z., Senoh, H., & Yasuda, K. Electrochemical oxidation of oxalic acid by Rh Octaethylporphyrin adsorbed on carbon black at low overpotential. J Electroanal Chem. 2007, 602(1), 96–102.
[106] Lu, Y., & Reddy, R. G. Electrocatalytic properties of carbon supported cobalt phthalocyanine–platinum for methanol electro-oxidation. Int J. Hydrog Energy. 2008, 33(14), 3930–3937.
[107] Barrera, C., Zhukov, I., Villagra, E., Bedioui, F., Páez, M. A., Costamagna, J., & Zagal, J. H. Trends in reactivity of unsubstituted and substituted cobalt-phthalocyanines for the electrocatalysis of glucose oxidation. J Electroanal Chem. 2006, 589(2), 212–218.
[108] Yamazaki, S., Siroma, Z., Fujiwara, N., & Ioroi, T. Electrochemical oxidation of glucose and gluconate by an electrode modified with a carbon-supported Rh phthalocyanine. J Mol Catal Chem. 2016, 425, 291–296.
[109] Yamazaki, S., Fujiwara, N., Takeda, S., & Yasuda, K. Electrochemical oxidation of sugars at moderate potentials catalyzed by Rh porphyrins. Chem Commun. 2010, 46(20), 3607–3609.
[110] Elouarzaki, K., Le Goff, A., Holzinger, M., Thery, J., & Cosnier, S. Electrocatalytic oxidation of glucose by rhodium porphyrin-functionalized MWCNT electrodes: application to a fully molecular catalyst-based glucose/O_2 fuel cell. J Am Chem Soc. 2012, 134(34), 14078–14085.
[111] Elouarzaki, K., Holzinger, M., Goff, A. L., Thery, J., S. Marks, R., & Cosnier, S. Glucose fuel cell based on carbon nanotube-supported pyrene–metalloporphyrin catalysts. J Mater Chem. A. 2016, 4(27), 10635–10640.
[112] Elouarzaki, K., Fisher, A. C., & Lee, J. M. Molecular Porphyrinic Freestanding Buckypaper Electrodes from Carbon Nanotubes for Glucose Fuel Cells. J Mater Chem A. 2017, 5(19), 8927–8932.
[113] Elouarzaki, K., Haddad, R., Holzinger, M., Le Goff, A., Thery, J., & Cosnier, S. MWCNT-supported phthalocyanine cobalt as air-breathing cathodic catalyst in glucose/O_2 fuel cells. J Power Sources. 2014, 255, 24–28.
[114] Haddad, R., Thery, J., Gauthier-Manuel, B., Elouarzaki, K., Holzinger, M., Le Goff, A., Gautier, G., El Mansouri, J., Martinent, A., & Cosnier, S. High performance miniature glucose/O_2 fuel cell based on porous silicon anion exchange membrane. Electrochem Commun. 2015, 54, 10–13.
[115] Elouarzaki, K., Goff, A. L., Holzinger, M., Agnès, C., Duclairoir, F., Putaux, J. L., & Cosnier, S. From gold porphyrins to gold nanoparticles: catalytic nanomaterials for glucose oxidation. Nanoscale. 2014, 6(15), 8556–8560.

Issei Otsuka, Redouane Borsali

3 Electrospun biomaterials

3.1 Electrospinning and its applications to soft matter

Electrospinning is a simple and versatile method that uses strong electrical fields to draw polymer solutions or melts for the production of continuous fibers with diameters ranging from several tens of nanometers to a few micrometers. Despite the fact that the process and apparatus for producing polymer filaments by taking advantage of the electrical field formed between electrodes ("electrostatic spinning") were patented in 1934 by Formhals [1], only modest academic attention had been paid to this technique for many years. In the 1990s, electrospinning was revived by several research groups and the fabrication of fibers for a wide range of polymers was studied both from fundamental and application points of view. Since then, nano-/microfibers produced by the electrospinning of soft matter have attracted ever increased attention for their diverse potential applications that include filtration membranes, functional textiles for biomedical applications such as tissue engineering, wound dressing, drug delivery, and so on. Indeed, the number of scientific publications concerning electrospinning has been exponentially increasing from the 2,000s to reach more than 2,500 publications per year in 2017 as shown in Fig. 3.1: The important advantages of the electrospun materials include their large surface area to volume ratio that arises from the thinness of the fibers (on the order of a few nanometers), their nano-/microscaled interstitial spaces, high porosity, and interconnectivity. These attributes result in superior mechanical properties. The simple electrospinning process enables large-scale productions of the final materials, making this technique very attractive for use in many applications.

3.2 Biopolymers and electrospinning

3.2.1 Introduction

Nano-/microfibers and their nonwoven materials produced by electrospinning of natural biopolymers such as proteins, nucleic acids, and polysaccharides have attracted ever-increasing attention for their diverse potential applications [2–4]. Some of the review articles have mentioned electrospun biomaterials for specific

Issei Otsuka, Redouane Borsali, CERMAV Univ. Grenoble Alpes, CNRS, Grenoble, France

https://doi.org/10.1515/9783110570526-003

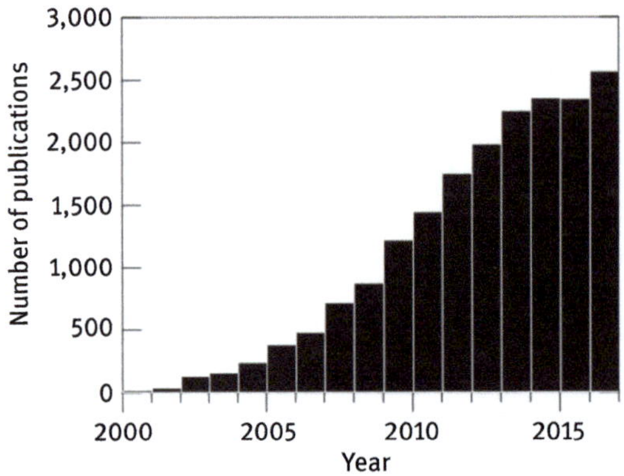

Fig. 3.1: Number of scientific papers and patents per year (2000–2017) with the keyword "electrospinning" (sources: SciFinder Scholar).

biomedical applications such as drug delivery [5–7], tissue engineering scaffolds [6, 8, 9], and wound dressing [10–12]. Among the electrospun biopolymers, polysaccharides, because of their abundance in nature, excellent biocompatibility, and biodegradability, have been widely studied to develop appropriate electrospinning techniques and understand their physical and biological properties. In this section, we introduce the electrospinning of polysaccharides and their derivatives via different methodologies and their potential applications.

3.2.2 Coelectrospinning of polysaccharides with other polymers

Preparation of defect-free homogeneous nano-/microfibers of polysaccharides via electrospinning is challenging in many cases due to their limited solubility in the solvents generally used for electrospinning. Blending polysaccharides with other biocompatible polymers often facilitates the processing by creating entanglements and physical bonds between the polymer chains. This is one of the major approaches to make the polysaccharides electrospinnable and to obtain nano-/microfibers and their nonwovens for biomedical applications. For instance, cellulose derivatives such as (hydroxypropyl)methyl cellulose, methyl cellulose, and carboxymethyl cellulose were mixed with poly(ethylene oxide) (PEO) and coelectrospun, followed by the extraction of PEO to form cellulose-based nanofibers [13]. Cellulose acetate was coelectrospun with poly(vinyl alcohol) (PVA) [14], hydroxyapatite (HA) [15], and polyvinylpyrolidone (PVP) [15] for potential use as biocompatible prosthetics. Chitin, which is poorly electrospinnable due to its low solubility in most organic solvents, was coelectrospun with poly(glycolic acid) (PGA) [16] and silk fibroin [17] to form composite nanofibers for tissue engineering scaffolds, wound healing, and skin regeneration purposes. Chitosan, a natural polycation that has strong repulsive forces between the chains preventing sufficient chain entanglement necessary for fiber formation, is difficult to electrospin directly. Hence, chitosan was often mixed and coelectrospun with biocompatible synthetic polymers and natural polymers such as PEO [18], PVA [19–21], poly(lactic acid) (PLA) [22], poly(caprolactone) (PCL) [23], silk fibroin [24], zein [25], and collagen [26] to produce chitosan-based composite nanofibers for various biomedical applications. For the same reason, alginate was coelectrospun with PEO [27–29] and PVA [29] to reduce the repulsive forces among the polyanionic chains. The presence of coelectrospum polymers, however, affects the properties of the resulting nanofibers such as their mechanical properties and biocompatibility by decreasing the polysaccharide content. The coaxial electrospinning technique [30, 31] provides an alternative and effective way of fabricating polysaccharide-based nanofibers. In this method, two different polymers are electrospun simultaneously through one spinneret composed of two coaxial capillaries to produce core–shell structured nanofibers. The first example of core–shell structured nanofibers consisting of polysaccharides was reported by Ojha et al. in 2008 for the coaxial electrospinning of chitosan used as the inner (core) layer and PEO as the outer (shell) layer [32]. The outer layer PEO segment was removed by washing the nanofibers with water to yield pure chitosan nanofibers. Later, core–shell nanofibers with reversed structure, that is, PEO (core)–chitosan (shell) nanofibers, was reported by Pakravan et al. [33]. Hollow nanofibers made of chitosan were obtained by the extraction of the PEO (core) segment from the core–shell nanofibers with water as shown in Fig. 3.2.

3.2.3 Electrospinning of polysaccharide derivatives

Modifying polysaccharides to increase their solubility in the solvents suitable for electrospinning is another major approach to improve electrospinnability. Cellulose, the most abundant natural polysaccharide, is poorly electrospinnable as it is because of its low solubility in most organic solvents due to the strong inter- and intra-molecular hydrogen bonds and highly crystalline nature. Hence, one has been focused on electrospinning of cellulose derivatives that are highly soluble in the solvents generally used for electrospinning such as acetone, dimethylacetamide (DMAc), dimethylformamide (DMF), and so on. Cellulose acetate is the most studied derivative due to its chemical resistance, stability, and good solubility in many organic solvents. In 1998, Jaeger et al. reported electrospinning of cellulose acetate for the first time using acetone as a solvent [34]. They obtained short fibers that were thinner than 1 µm having “beads on the string” morphology. This possibly was due to the low viscosity of the polymer solution or the low boiling point of acetone that causes fast evaporation and gelation of the polymer solution during the electrospinning process. Liu and Hsieh overcame this problem by using the solvent mixture

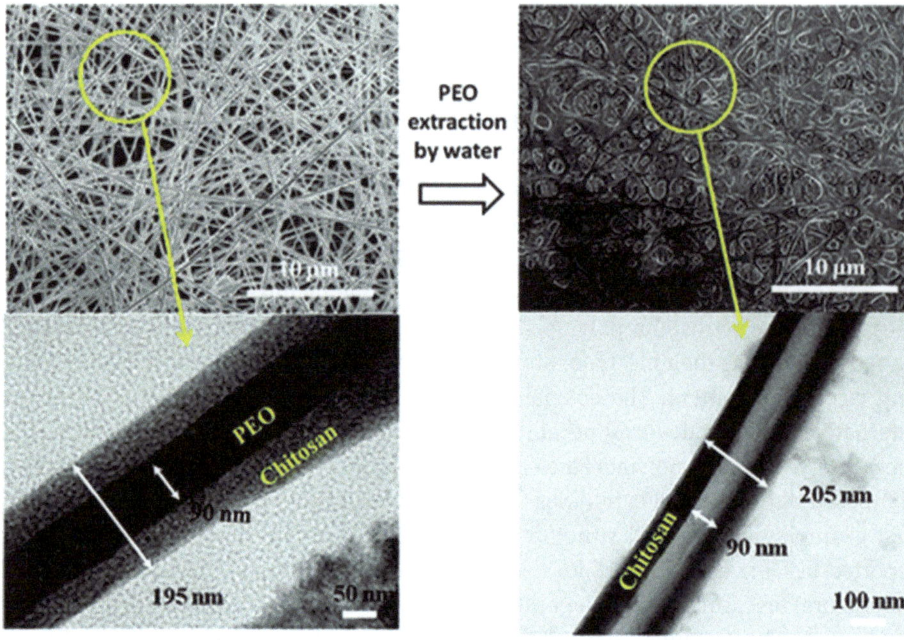

Fig. 3.2: SEM (top) and TEM (down) images of the PEO/chitosan core–shell nanofibers (left) and the chitosan hollow nanofibers (right). Reproduced from Pakravan et al. [33] with permission from American Chemical Society.

of acetone and DMAc for the electrospinning of cellulose acetate and successfully obtained defect-free homogeneous fibers with consistent fiber diameters ranging from 100 nm to 1 µm [35] A mixed solvent of acetone and water with basic pH conditions also resulted in successful electrospinning of cellulose acetate to form fine nanofibers [36]. The resulting cellulose acetate nanofibers were usually deacetylated by using alkaline solutions to regenerate cellulose nanofibers [36–40]. These regenerated cellulose nanofibers can be functionalized with biologically active substances via chemical modifications. One potential application of these nano-/microfibers and their nonwovens is membrane-based filtration. Electrospun nanofiber nonwovens offer a distinctly large surface area to volume ratio (typically ranging from 40 to 100 m^2/g, compared to 0.05 to 10 m^2/g for micron-sized spunbonded or melt blown nonwovens) [41], which can be beneficial in filter applications especially for affinity membrane-based filtration. For instance, Ma et al. functionalized regenerated cellulose nanofiber mats with Cibacron Blue F3GA, a general affinity dye ligand for separation of many biomolecules, and used them as affinity membranes to specifically capture bovine serum albumin or bilirubin via filtration [42]. The same group also reported affinity membrane filtration for immunoglobulin G (IgG) using regenerated cellulose nanofiber mats functionalized with protein A/G [43]. There have been only a few reports on the electrospinning of other polysaccharide derivatives. Ethyl-cyanoethyl cellulose was electrospun from its THF solution to form porous fibers [44]. Ethyl cellulose was dissolved in a solvent system of THF/DMAc and the effects of solvent ratios on the electrospun fiber size distribution and diameter were studied [45]. Hydroxypropyl cellulose could be electrospun using anhydrous ethanol and 2-propanol as solvents [46]. The obtained fiber mats were appropriate for use as templates for producing tin oxide nano-/macroporous fiber networks on microelectromechanical system devices. Hexanoyl chitosan was electrospun from its chloroform solution to form flat ribbon-like nano-/microfibers [47].

3.2.4 Direct electrospinning of native polysaccharides

Dextran, a branched glucan consists of α-1,6 and 1,3 glycosidic linkages between glucose units, shows good solubility in both water and some organic solvents. Hence, direct electrospinning of dextran using these solvents was investigated by Jian et al. [48]. It was found that uniform nanofibrous dextran membranes could be formed by electrospinning using water, DMSO/water, and DMSO/DMF mixtures as solvents by adjusting technical parameters such as concentration, voltage, and the distance between the electrodes. The wide spectrum of solvents for electrospinning of dextran allows direct incorporation of various bioactive agents such as DNA and proteins in the dextran nanofiber mats. On the other hand, most abundant polysaccharides such as cellulose, chitin, and chitosan are much less soluble in water and most organic solvents. Generally, their available solvents are highly dielectric and not very volatile, which is a disadvantage for the electrospinning process. Nevertheless, several research groups reported direct electrospinning of cellulose using known solvents such as *N*-methylmorpholine-*N*-oxide (NMMO) [49], DMAc with lithium chloride (LiCl) [50, 51], and NMMO/water [51]. These approaches, however, require high operation temperatures and removal of salt. Highly volatile organofluorine solvents were used for electrospinning of these poorly soluble polysaccharides. Park et al. reported successful electrospinning of chitin using hexafluoroisopropanol as a solvent [52]. Homogenous bead-free electrospun fibers of chitosan were first reported in 2004 by Ohkawa et al. using a mixed solution of trifluoroacetic acid (TFA)/dichloromethane [19]. Since then, electrospinning of chitosan using TFA has been widely studied [53–55]. It was proposed that TFA facilitates the electrospinning of chitosan because the amino groups of the chitosan form salts, preventing strong repulsive forces between the polycationic chains [56]. In addition, the high volatility of TFA can accelerate solidification of the chitosan during the electrospinning process. Recently, we have reported the simple fabrication of defect-free homogeneous cellulose nanofibers via electrospinning of both plant (soft and hardwood pulps)- and animal (tunicate)-derived cellulose as shown in Fig. 3.3 [57]. The plant- and animal-derived cellulose were dissolved in a mixture of TFA and dichloroethane (DCE) to give highly

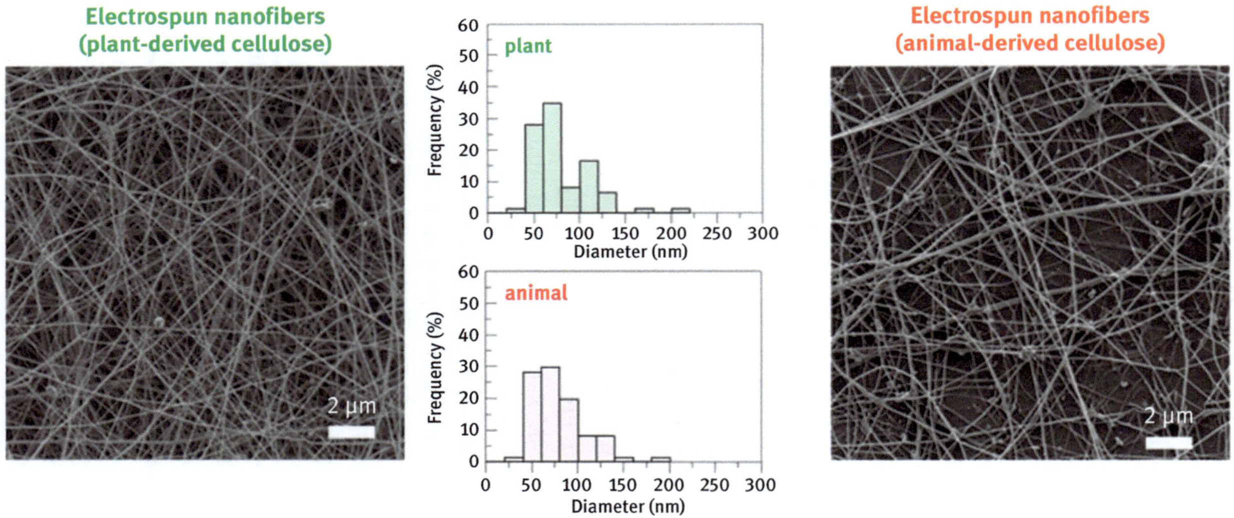

Fig. 3.3: SEM images and histograms of the diameter distribution of the nanofibers obtained via electrospinning of plant- and animal-derived cellulose solutions in TFA/DCE = 7/3 (v/v). Reproduced from Otsuka et al. [57] with permission from Springer Science Business Media B.V.

viscoelastic solutions. The unsubstituted cellulose nanofibers were obtained by the electrospinning of the cellulose dissolved in TFA/DCE by forming trifluoroacetyl ester groups, which were naturally hydrolyzed with moisture in the air during the electrospinning process. To the best of our knowledge, this was the first report on the direct electrospinning of dissolved plant- and animal-derived cellulose in mixtures of TFA and DCE. Notably, the nanofibers of tunicate-derived cellulose, which is an abundantly available biomass from the sea, were successfully obtained by electrospinning for the first time. These simply fabricated cellulose nanofibers will broaden the versatility of nanocellulose for various applications in interdisciplinary fields such as nanobiotechnology.

3.3 Block copolymers and electrospinning

3.3.1 Introduction

The interest in electrospinning of functional polymers has been growing for their potential applications in diverse fields including bio-, optoelectronic-, and nanotechnologies. Fabrication of complex architectures such as core–shell, hollow, and porous structures in the electrospun fibers could lead to much broader applications. In addition, electrospinning of block copolymers (BCPs) and their self-assembly within the confined fiber structure has received a lot of attention because the continuous fiber structure can direct well-known BCP self-assembly (e.g., sphere, cylinder, gyroid, and lamellar phases depending on the volume fraction, molecular weight, and the Flory–Huggins interaction parameter (χ) of the constituent blocks) along or perpendicular to the electrospun fiber axis. Such materials having long-range ordered periodic morphology within the fiber structure will be expected to be used in a wide range of applications such as optical waveguides, flexible electronic devices, biosensors, filters, and so on. In this section, we introduce the electrospinning of BCPs for different objectives and the phase-separation behavior of the BCP-based electrospun fibers.

3.3.2 Electrospinning of block copolymers

The studies on electrospinning of BCPs can be classified broadly into two groups in terms of the objective: one is for controlling functionality of the fibers such as hydrophilicity and biocompatibility mainly targeting biomedical applications, and the other is for controlling long-range ordered BCP phase-separated structures inner fibers. The first group includes electrospinning of the BCPs consisting of biocompatible polymers represented mainly by PLA and PCL. PLA and PCL are common biocompatible polymers that have been electrospun to fabricate fiber scaffolds by many research groups because of their biodegradability, good mechanical properties, and high solubility in common solvents for the electrospinning process. However, their hydrophobicity sometimes limits their biomedical applications. Therefore, the BCPs consisting of PLA or PCL and hydrophilic biocompatible polymers such as PEO and poly(*N*-vinylpyrrolidone) (PVP) were electrospun for the improvement of the hydrophilicity of the fibers. Examples include the electrospinning of PLA-*b*-PEO [58–60], PLA-*b*-PEO-*b*-PLA [61], PCL-*b*-PEO [62–64], PCL-*b*-PVP [65], and poly(lactide-*co*-glycolide)-*b*-PEO [66] that were designed for tissue engineering scaffolds, bio-sensing devices, affinity membranes, and drug delivery. The second group of studies focuses on phase separation of the electrospun BCP fibers. Despite the fact that the phase-separation behavior of BCP solutions, melts, and films has been extensively studied both theoretically and experimentally, there has been much fewer studies on the phase separation of electrospun BCP fibers. This type of BCP fiber consists of polystyrene-*b*-polybutadiene-*b*-polystyrene (PS-*b*-PBU-*b*-PS) [67], poly(styrene-*b*-dimethylsiloxane) (PS-*b*-PDMS) [68], poly(styrene-*b*-4-vinylpyridine) hydrogen-bonded pentadecylphenol (PS-*b*-P4VP(PDP)) [69, 70], poly(styrene-*b*-isoprene) (PS-*b*-PI) [71, 72], poly(styrene-*b*-isoprene-*b*-styrene) (PS-*b*-PI-*b*-PS) [73], poly(isoprene-*b*-dimethylaminoethyl methacrylate) (PI-*b*-PDMAEMA) [74], and poly (styrene-*b*-methyl methacrylate) (PS-*b*-PMMA) [75]. The phase-separation behavior of these BCP fibers are summarized in the following section.

3.3.3 Phase separation of electrospun block copolymer fibers

In recent years, numerous studies have been conducted for controlling self-assembly of BCPs to produce long-range ordered periodic phases such as sphere, cylinder, gyroid, and lamellar phases in the bulk and thin film states for applications in nanotechnology. Guiding or directing the self-assembly of BCPs using either chemically or topographically patterned substrates, so-called directed self-assembly (DSA), is the most established strategy to order phase-separated BCP domains periodically in long ranges [76]. The phase separation of BCPs within the elctrospun fiber structure along or perpendicular to the fiber axis is one of the promising methods in the DSA strategy. After the first discovery of weak and irregular phase separation on the surface of electrospun fibers of PS-*b*-PBU-*b*-PS by Fong et al. in 1990 [67], phase-separated cylindrical and spherical phases were observed in the electrospun fibers of PS-*b*-PDMS [68] and PS-*b*-P4VP(PDP) [69], respectively. However, the phase-separated BCP domains observed in these "as-spun" fibers are largely disordered because of the strong elongational deformation and rapid solvent evaporation during the electrospinning process. To improve the phase separation and self-organization of the BCP domains in long-range order, thermal or solvent vaper annealing of the fibers that provides mobility of the BCP chains to achieve the equilibrium morphology is necessary. At the same time, the fiber structure must be preserved from deformation during the annealing process. Core–shell structured fibers that can be prepared via coaxial electrospinning of two or more different polymers gave a solution for this challenge. In 2006, Kalra et al. [72] and Ma et al. [73] independently reported the first long-range ordered BCP nanostructures in thermally annealed electrospun core–shell nanofibers consisting of PS-*b*-PI and PS-*b*-PI-*b*-PS as the cores, which were enveloped by thermally stable SiO_2 or poly(methyl methacrylate-*co*-methacrylic acid) random copolymer (P(MMA-*ran*-MAA)) as the shells to protect the fibers from the loss of fiber morphology via the thermal annealing above the glass-transition temperature (T_g) of the BCP cores. The cross-sectional TEM images of the PS-*b*-PI-*b*-PS (core)–P(MMA-*ran*-MAA) (shell) nanofibers both along and perpendicular to the axis are shown in Fig. 3.4. The long-range ordering of the BCP nanostructures (lamellar

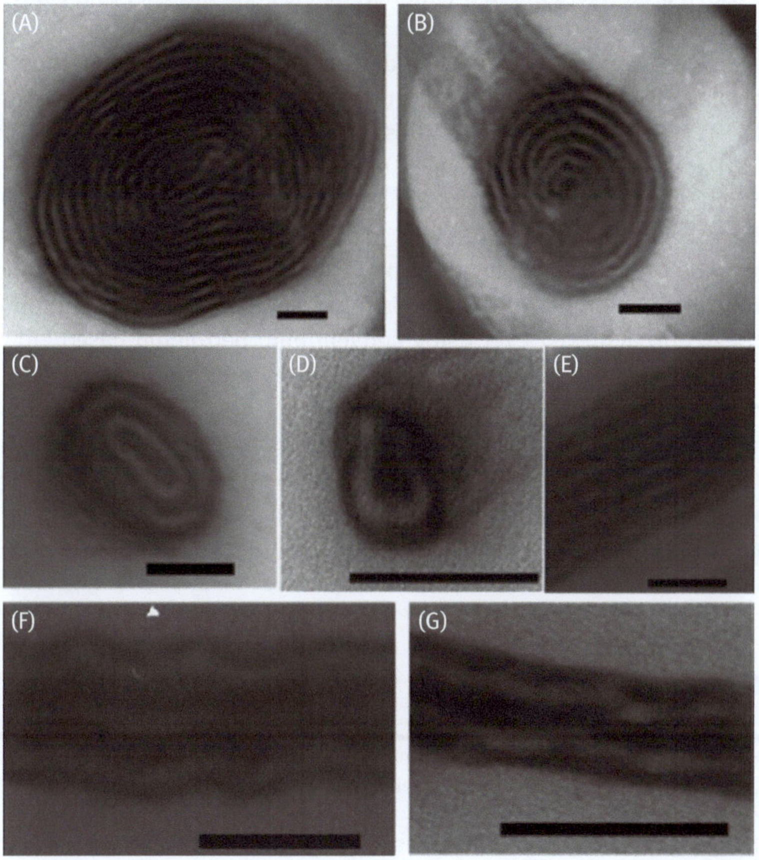

Fig.3.4: TEM images of coaxially spun the PS-*b*-PI-*b*-PS (core)–P(MMA-*ran*-MAA) (shell) fibers (only BCP cores are shown because of low contrast with the surrounding shell) of different diameters after thermal annealing in a vacuum oven at 140 °C for 10 days. (A)–(D) axial views; (E)–(G) longitudinal views. All scale bars are 100 nm. Reproduced from Ma et al. [73] with permission from American Chemical Society.

morphology parallel to the fiber axis) via thermal annealing was clearly observed. Kamperman et al. reported long-range ordered phase separation of PI-*b*-PDMAEMA in core–shell nanofibers consisting of PI-*b*-PDMAEMA/polymer-derived ceramic precursor nanocomposites as the cores and thermally stable polyacrylonitrile as the shells via thermal annealing [74]. Zhou et al. reported a strategy of utilizing solvent vapor annealing to achieve long-range ordered lamellar and cylindrical morphologies that are oriented perpendicular to the fiber axis in single-electrospun PS-*b*-PMMA fibers [75]. Although the PS-*b*-PMMA fibers were swelled during the annealing process in chloroform and toluene vapor, they kept the confined fiber structure and self-organized into lamellar and cylindrical phases. The authors mentioned that the swollen BCPs potentially flow toward the fiber edges due to the difference in height between the center and edge regions of the fibers, inducing the BCPs to form ordered nanostructures with a preferred domain orientation that is perpendicular to the fiber axis.

In recent years, our research group has developed a new class of natural-synthetic "hybrid" BCP systems, where one of the blocks consists of poly- or oligosaccharides [77–83]. These hybrid BCPs have very high incompatibility between saccharidic and synthetic polymer blocks from their hydrophilicity–hydrophobicity imbalance that is further enhanced sterically by the "rod"-like structure of the saccharidic blocks, leading to high χ-parameters of the BCP systems. Using these "high χ" BCP systems, we could achieve sub-10 nm scale phase separation where the BCP domain size is much smaller than the minimum domain size (ca. 20 nm) achieved ever using conventional petroleum-based "low χ" BCPs such as PS-*b*-PMMA and PS-*b*-PI. Such small features will lead to next-generation devices in various nanotechnology industries. Our next challenge is to produce long-range order of these BCP nanopatterns, which is essential for real applications. In addition to the well-known DSA approach [84], we are working on the electrospinning of hybrid BCPs and their long-range self-organization within the fiber structure [85]. Details will be discussed in our forthcoming publications.

References

[1] Formhals, A. Process and apparatus for preparing artificial threads. US1975504A 1934.

[2] Agarwal, S., Wendorff, J. H., & Greiner, A. Use of electrospinning technique for biomedical applications. Polymer. 2008, 49, 5603–5621.

[3] Schiffman, J. D., & Schauer, C. L. A review: electrospinning of biopolymer nanofibers and their applications. Polym Rev. 2008, 48, 317–352.

[4] Sridhar, R., Lakshminarayanan, R., Madhaiyan, K., Amutha Barathi, V., Lim, K. H. C., & Ramakrishna, S. Electrosprayed nanoparticles and electrospun nanofibers based on natural materials: applications in tissue regeneration, drug delivery and pharmaceuticals. Chem Soc Rev. 2015, 44, 790–814.

[5] Meinel, A. J., Germershaus, O., Luhmann, T., Merkle, H. P., & Meinel, L. Electrospun matrices for localized drug delivery: current technologies and selected biomedical applications. Eur J Pharm Biopharm. 2012, 81, 1–13.
[6] Sill, T. J., & von Recum, H. A. Electrospinning: applications in drug delivery and tissue engineering. Biomaterials. 2008, 29, 1989–2006.
[7] Balaji, A., Vellayappan, M. V., John, A. A., et al. An insight on electrospun-nanofibers-inspired modern drug delivery system in the treatment of deadly cancers. RSC Adv. 2015, 5, 57984–58004.
[8] Agarwal, S., Wendorff, J. H., & Greiner, A. Progress in the field of electrospinning for tissue engineering applications. Adv Mater. 2009, 21, 3343–3351.
[9] Martins, A., Reis, R. L., & Neves, N. M. Electrospinning: processing technique for tissue engineering scaffolding. Int Mater Rev. 2008, 53, 257–274.
[10] Zahedi, P., Rezaeian, I., Ranaei-Siadat, S. O., Jafari, S. H., & Supaphol, P. A review on wound dressings with an emphasis on electrospun nanofibrous polymeric bandages. Polym Adv Technol. 2010, 21, 77–95.
[11] Zhong, S. P., Zhang, Y. Z., & Lim, C. T. Tissue scaffolds for skin wound healing and dermal reconstruction. Wiley Interdisciplinary Reviews: Nanomed Nanobiotechnol. 2010, 2, 510–525.
[12] Rieger, K. A., Birch, N. P., & Schiffman, J. D. Designing electrospun nanofiber mats to promote wound healing: a review. J Mater Chem B. 2013, 1, 4531–4541.
[13] Frenot, A., Henriksson, M. W., & Walkenström, P. Electrospinning of cellulose-based nanofibers. J Appl Polym Sci. 2007, 103, 1473–1482.
[14] Ding, B., Kimura, E., Sato, T., Fujita, S., & Shiratori, S. Fabrication of blend biodegradable nanofibrous nonwoven mats via multi-jet electrospinning. Polymer. 2004, 45, 1895–1902.
[15] Bishop, A., Balázsi, C., Yang, J. H. C., & Gouma, P. I. Biopolymer-hydroxyapatite composite coatings prepared by electrospinning. Polym Technol Adv Technol. 2006, 17, 902–906.
[16] Park, K. E., Kang, H. K., Lee, S. J., Min, B.-M., & Park, W. H. Biomimetic nanofibrous scaffolds: preparation and characterization of PGA/Chitin blend nanofibers. Biomacromolecules. 2006, 7, 635–643.
[17] Park, K. E., Jung, S. Y., Lee, S. J., Min, B.-M., & Park, W. H. Biomimetic nanofibrous scaffolds: preparation and characterization of chitin/silk fibroin blend nanofibers. Int J Biol Macromol. 2006, 38, 165–173.
[18] Kriegel, C., Kit, K. M., McClements, D. J., & Weiss, J. Electrospinning of chitosan–poly(ethylene oxide) blend nanofibers in the presence of micellar surfactant solutions. Polymer. 2009, 50, 189–200.
[19] Ohkawa, K., Cha, D., Kim, H., Nishida, A., & Yamamoto, H. Electrospinning of chitosan. Macromol Rapid Commun. 2004, 25, 1600–1605.
[20] Li, L., & Hsieh, Y.-L. Chitosan bicomponent nanofibers and nanoporous fibers. Carbohydr Res. 2006, 341, 374–381.
[21] Zhang, Y., Huang, X., Duan, B., Wu, L., Li, S., & Yuan, X. Preparation of electrospun chitosan/poly(vinyl alcohol) membranes. Colloid Polym Sci. 2007, 285, 855–863.
[22] Xu, J., Zhang, J., Gao, W., Liang, H., Wang, H., & Li, J. Preparation of chitosan/PLA blend micro/nanofibers by electrospinning. Mater Lett. 2009, 63, 658–660.
[23] Shalumon, K. T., Anulekha, K. H., Girish, C. M., Prasanth, R., Nair, S. V., & Jayakumar, R. Single step electrospinning of chitosan/poly(caprolactone) nanofibers using formic acid/acetone solvent mixture. Carbohydr Polym. 2010, 80, 413–419.
[24] Park, W. H., Jeong, L., Yoo, D. I., & Hudson, S. Effect of chitosan on morphology and conformation of electrospun silk fibroin nanofibers. Polymer. 2004, 45, 7151–7157.
[25] Torres-Giner, S., Ocio, M. J., & Lagaron, J. M. Novel antimicrobial ultrathin structures of zein/chitosan blends obtained by electrospinning. Carbohydr Polym. 2009, 77, 261–266.

[26] Chen, Z., Mo, X., & Qing, F. Electrospinning of collagen–chitosan complex. Mater Lett. 2007, 61, 3490–3494.
[27] Lu, J.-W., Zhu, Y.-L., Guo, Z.-X., Hu, P., & Yu, J. Electrospinning of sodium alginate with poly (ethylene oxide). Polymer. 2006, 47, 8026–8031.
[28] Bhattarai, N., Li, Z., Edmondson, D., & Zhang, M. Alginate-based nanofibrous scaffolds: structural, mechanical, and biological properties. Adv Mater. 2006, 18, 1463–1467.
[29] Safi, S., Morshed, M., Ravandi, S. A. H., & Ghiaci, M. Study of electrospinning of sodium alginate, blended solutions of sodium alginate/poly(vinyl alcohol) and sodium alginate/poly (ethylene oxide). J Appl Polym Sci. 2007, 104, 3245–3255.
[30] Sun, Z., Zussman, E., Yarin, A. L., Wendorff, J. H., & Greiner, A. Compound core–shell polymer nanofibers by co-electrospinning. Adv Mater. 2003, 15, 1929–1932.
[31] Yu, J. H., Fridrikh, S. V., & Rutledge, G. C. Production of submicrometer diameter fibers by two-fluid electrospinning. Adv Mater 2004, 16, 1562–1566.
[32] Ojha, S. S., Stevens, D. R., Hoffman, T. J., et al. Fabrication and characterization of electrospun chitosan nanofibers formed via templating with polyethylene oxide. Biomacromolecules. 2008, 9, 2523–2529.
[33] Pakravan, M., & Heuzey M-C, A. A. Core–shell structured PEO-chitosan nanofibers by coaxial electrospinning. Biomacromolecules. 2012, 13, 412–421.
[34] Jaeger, R., Bergshoef, M. M., Batlle, C. M. I., Schönherr, H., & Vancso, G. J. Electrospinning of ultra-thin polymer fibers. Macromol Symp. 1998, 127, 141–150.
[35] Liu, H., & Hsieh, Y. L. Ultrafine fibrous cellulose membranes from electrospinning of cellulose acetate. J Polym Sci Part B: Polym Phys. 2002, 40, 2119–2129.
[36] Son, W. K., Youk, J. H., Lee, T. S., & Park, W. H. Electrospinning of ultrafine cellulose acetate fibers: studies of a new solvent system and deacetylation of ultrafine cellulose acetate fibers. J Polym Sci Part B: Polym Phys. 2004, 42, 5–11.
[37] Wang, M., Meng, G., Huang, Q., & Qian, Y. Electrospun 1,4-DHAQ-doped cellulose nanofiber films for reusable fluorescence detection of trace Cu2+ and further for Cr3+. Environ Sci Technol. 2012, 46, 367–373.
[38] Stephen, M., Catherine, N., Brenda, M., Andrew, K., Leslie, P., & Corrine, G. Oxolane-2,5-dione modified electrospun cellulose nanofibers for heavy metals adsorption. J Hazard Mater. 2011, 192, 922–927.
[39] Huang, X.-J., Chen, P.-C., Huang, F., Ou, Y., Chen, M.-R., & Xu, Z.-K. Immobilization of Candida rugosa lipase on electrospun cellulose nanofiber membrane. J Mol Catal B: Enzym. 2011, 70, 95–100.
[40] Dixit, V., Tewari, J., & Obendorf, S. K. Fungal growth inhibition of regenerated cellulose nanofibrous membranes containing Quillaja Saponin. Arch Environ Contam Toxicol. 2010, 59, 417–423.
[41] Desai, K., Kit, K., Li, J., Michael Davidson, P., Zivanovic, S., & Meyer, H. Nanofibrous chitosan non-wovens for filtration applications. Polymer. 2009, 50, 3661–3669.
[42] Ma, Z., Kotaki, M., & Ramakrishna, S. Electrospun cellulose nanofiber as affinity membrane. J Membr Sci. 2005, 265, 115–123.
[43] Ma, Z., & Ramakrishna, S. Electrospun regenerated cellulose nanofiber affinity membrane functionalized with protein A/G for IgG purification. J Membr Sci. 2008, 319, 23–28.
[44] Zhao, S., Wu, X., Wang, L., & Huang, Y. Electrospinning of ethyl–cyanoethyl cellulose/ tetrahydrofuran solutions. J Appl Polym Sci. 2004, 91, 242–246.
[45] Wu, X., Wang, L., Yu, H., & Huang, Y. Effect of solvent on morphology of electrospinning ethyl cellulose fibers. J Appl Polym Sci. 2005, 97, 1292–1297.
[46] Shukla, S., Brinley, E., Cho, H. J., & Seal, S. Electrospinning of hydroxypropyl cellulose fibers and their application in synthesis of nano and submicron tin oxide fibers. Polymer. 2005, 46, 12130–12145.

[47] Neamnark, A., Rujiravanit, R., & Supaphol, P. Electrospinning of hexanoyl chitosan. Carbohydr Polym 2006, 66, 298–305.
[48] Jiang, H., Fang, D., Hsiao, B. S., Chu, B., & Chen, W. Optimization and characterization of dextran membranes prepared by electrospinning. Biomacromolecules. 2004, 5, 326–333.
[49] Kulpinski, P. Cellulose nanofibers prepared by the N-methylmorpholine-N-oxide method. J Appl Polym Sci. 2005, 98, 1855–1859.
[50] Kim, C. W., Frey, M. W., Marquez, M., & Joo, Y. L. Preparation of submicron-scale, electrospun cellulose fibers via direct dissolution. J Polym Sci Part B: PolymPhys. 2005, 43, 1673–1683.
[51] Kim, C.-W., Kim, D.-S., Kang, S.-Y., Marquez, M., & Joo, Y. L. Structural studies of electrospun cellulose nanofibers. Polymer. 2006, 47, 5097–5107.
[52] Min, B.-M., Lee, S. W., Lim, J. N., et al. Chitin and chitosan nanofibers: electrospinning of chitin and deacetylation of chitin nanofibers. Polymer. 2004, 45, 7137–7142.
[53] Ohkawa, K., Minato, K.-I., Kumagai, G., Hayashi, S., & Yamamoto, H. Chitosan nanofiber. Biomacromolecules. 2006, 7, 3291–3294.
[54] Schiffman, J. D., & Schauer, C. L. Cross-linking chitosan nanofibers. Biomacromolecules. 2007, 8, 594–601.
[55] Sangsanoh, P., & Supaphol, P. Stability improvement of electrospun chitosan nanofibrous membranes in neutral or weak basic aqueous solutions. Biomacromolecules. 2006, 7, 2710–2714.
[56] Hasegawa, M., Isogai, A., Onabe, F., Usuda, M., & Atalla, R. H. Characterization of cellulose–chitosan blend films. J Appl Polym Sci. 1992, 45, 1873–1879.
[57] Otsuka, I., Njinang, C. N., & Borsali, R. Simple fabrication of cellulose nanofibers via electrospinning of dissolving pulp and tunicate. Cellulose. 2017, 24, 3281–3288.
[58] Luu, Y. K., Kim, K., Hsiao, B. S., Chu, B., & Hadjiargyrou, M. Development of a nanostructured DNA delivery scaffold via electrospinning of PLGA and PLA–PEG block copolymers. J Controlled Release. 2003, 89, 341–353.
[59] Buttaro, L. M., Drufva, E., & Frey, M. W. Phase separation to create hydrophilic yet non-water soluble PLA/PLA-b-PEG fibers via electrospinning. J Appl Polym Sci. 2014, 131, 41030 (online).
[60] Grafahrend, D., Calvet, J. L., Klinkhammer, K., et al. Control of protein adsorption on functionalized electrospun fibers. Biotechnol Bioeng. 2008, 101, 609–621.
[61] Kim, K., Yu, M., Zong, X., et al. Control of degradation rate and hydrophilicity in electrospun non-woven poly(d,l-lactide) nanofiber scaffolds for biomedical applications. Biomaterials. 2003, 24, 4977–4985.
[62] Choi, J. S., & Yoo, H. S. Electrospun nanofibers surface-modified with fluorescent proteins. J Bioact Compat Polym. 2007, 22, 508–524.
[63] Choi, J. S., Leong, K. W., & Yoo, H. S. In vivo wound healing of diabetic ulcers using electrospun nanofibers immobilized with human epidermal growth factor (EGF). Biomaterials. 2008, 29, 587–596.
[64] Dalton, P. D., Klinkhammer, K., Salber, J., Klee, D., & Möller, M. Direct in vitro electrospinning with polymer melts. Biomacromolecules. 2006, 7, 686–690.
[65] Cho, S. J., Jung, S. M., Kang, M., Shin, H. S., & Youk, J. H. Preparation of hydrophilic PCL nanofiber scaffolds via electrospinning of PCL/PVP-b-PCL block copolymers for enhanced cell biocompatibility. Polymer. 2015, 69, 95–102.
[66] Kim, T. G., & Park, T. G. Surface functionalized electrospun biodegradable nanofibers for immobilization of bioactive molecules. Biotechnol Prog. 2006, 22, 1108–1113.
[67] Fong, H., & Reneker, D. H. Elastomeric nanofibers of styrene–butadiene–styrene triblock copolymer. J Polym SciPart B: Polym Phys. 1999, 37, 3488–3493.

[68] Ma, M., Hill, R. M., Lowery, J. L., Fridrikh, S. V., & Rutledge, G. C. Electrospun poly(Styrene-block-dimethylsiloxane) block copolymer fibers exhibiting superhydrophobicity. Langmuir. 2005, 21, 5549–5554.
[69] Ruotsalainen, T., Turku, J., Heikkilä, P., et al. Towards internal structuring of electrospun fibers by hierarchical self-assembly of polymeric comb-shaped supramolecules. Adv Mater. 2005, 17, 1048–1052.
[70] Ruotsalainen, T., Turku, J., Hiekkataipale, P., et al. Tailoring of the hierarchical structure within electrospun fibers due to supramolecular comb-coil block copolymers: polystyrene-block-poly(4-vinyl pyridine) plasticized by hydrogen bonded pentadecylphenol. Soft Matter. 2007, 3, 978–985.
[71] Kalra, V., Kakad, P. A., Mendez, S., Ivannikov, T., Kamperman, M., & Joo, Y. L. Self-assembled structures in electrospun poly(styrene-block-isoprene) fibers. Macromolecules. 2006, 39, 5453–5457.
[72] Kalra, V., Mendez, S., Lee, J. H., Nguyen, H., Marquez, M., & Joo, Y. L. Confined assembly in coaxially electrospun block copolymer fibers. Adv Mater. 2006, 18, 3299–3303.
[73] Ma, M., Krikorian, V., Yu, J. H., Thomas, E. L., & Rutledge, G. C. Electrospun polymer nanofibers with internal periodic structure obtained by microphase separation of cylindrically confined block copolymers. Nano Lett. 2006, 6, 2969–2972.
[74] Kamperman, M., Korley, L. T. J., Yau, B., Johansen, K. M., Joo, Y. L., & Wiesner, U. Nanomanufacturing of continuous composite nanofibers with confinement-induced morphologies. Polym Chem. 2010, 1, 1001–1004.
[75] Zhou, Z., Cao, K., Chen, X., et al. Preferred domain orientation in block copolymer fibers after solvent annealing. Mol Syst Des Eng. 2018, 3, 357–363.
[76] Hu, H., Gopinadhan, M., & Osuji, C. O. Directed self-assembly of block copolymers: a tutorial review of strategies for enabling nanotechnology with soft matter. Soft Matter. 2014, 10, 3867–3889.
[77] Aissou, K., Otsuka, I., Rochas, C., Fort, S., Halila, S., & Borsali, R. Nano-organization of amylose-b-polystyrene block copolymer films doped with bipyridine. Langmuir. 2011, 27, 4098–4103.
[78] Cushen, J. D., Otsuka, I., Bates, C. M., et al. Oligosaccharide/silicon-containing block copolymers with 5 nm features for lithographic applications. ACS Nano. 2012,6,3424–3433.
[79] Otsuka, I., Isono, T., Rochas, C., et al. 10 nm Scale cylinder-cubic phase transition induced by caramelization in sugar-based block copolymers. ACS Macro Lett. 2012, 1, 1379–1382.
[80] Otsuka, I., Tallegas, S., Sakai, Y., et al. Control of 10 nm scale cylinder orientation in self-organized sugar-based block copolymer thin films. Nanoscale. 2013, 5, 2637–2641.
[81] Otsuka, I., Zhang, Y., Isono, T., et al. Sub-10 nm scale nanostructures in self-organized linear di- and triblock copolymers and miktoarm star copolymers consisting of maltoheptaose and polystyrene. Macromolecules. 2015, 48, 1509–1517.
[82] Sakai-Otsuka, Y., Zaioncz, S., Otsuka, I., Halila, S., Rannou, P., & Borsali, R. Self-assembly of carbohydrate-block-poly(3-hexylthiophene) diblock copolymers into sub-10 nm scale lamellar structures. Macromolecules (Washington, DC, U S). 2017, 50, 3365–3376.
[83] Liao, Y., Chen, W.-C., & Borsali, R. Carbohydrate-based block copolymer thin films: ultrafast nano-organization with 7 nm resolution using microwave energy. Adv Mater (Weinheim, Ger). 2017, 29, 1701645 (online)
[84] Otsuka, I., Nilsson, N., Suyatin, D. B., Maximov, I., & Borsali, R. Carbohydrate-based block copolymer systems: directed self-assembly for nanolithography applications. Soft Matter. 2017, 13, 7406–7411.
[85] Otsuka, I., Garg, G., & Borsali, R. Preparation of nanofibers having periodic internal structures made from carbohydrates via electrospinning of glycoconjugate polymers. Am Chem Soc. 2016, p. PMSE-604.

Seiya Tsujimura

4 Porous carbon materials for enzymatic fuel cells

4.1 Introduction: designing porous carbon electrodes

Enzymatic fuel cells (EFCs) are fuel cells that contain a purified enzyme as the electrocatalyst to convert the chemical energy of the oxidation reaction of reducing substances (such as sugars, organic acids, and alcohols) directly into electricity under ambient conditions, that is, 10–40 °C, atmospheric pressure, and a moderate pH range of 5–8. At the anode, electrons that are captured from an anode fuel by a specific enzyme are transferred to the electrode through the direct or mediated electron-transfer reaction. The electrons transferred through external resistance are passed to O_2 molecules with the aid of Cu-containing enzymes (multicopper oxidases) to form water at the cathode. EFCs minimally comprise anodic and cathodic enzymes and the carbon electrodes; in principle, they do not require a case to seal the cell components or a separator between the anode and the cathode. This simple structure facilitates straightforward applications of EFCs. Enzyme-catalyzed electrode reactions are classified into two categories: (1) direct electron transfer (DET), which involves DET between enzymes and electrodes; (2) mediated electron transfer (MET), in which redox molecules, called mediators, shuttle the electron between enzymes and electrodes.

EFCs are expected to become ubiquitous power devices in the field of communications and for medical (healthcare) applications; EFCs can power wearable or in vivo bioelectronics, such as biosensors (e.g., self-driven blood-glucose sensors and continuous glucose-monitoring devices), sensors (e.g., body thermometers, heart rate sensors, and accelerometers), and other implanted medical/healthcare devices (e.g., pacemakers and stimulators). EFCs are also expected to power microscale sensor/node systems for Internet of Things technologies. However, practical use of EFCs is hindered by their lower durability and lower output power density than conventional small power generators and batteries. Nevertheless, the highly favorable properties and the potential applications of EFCs have stimulated intense interest in the basic study and development of EFCs. In recent years, numerous review articles on EFCs, which serve as useful supplements to this review, have been published [1–8].

Seiya Tsujimura, Division of Materials Science, Faculty of Pure and Applied Sciences, University of Tsukuba, Tsukuba, Ibaraki, Japan

https://doi.org/10.1515/9783110570526-004

Two critical shortcomings – the short lifetime and low power density – prevent the application of EFCs in the real world; they are related to the instability of enzymes on the electrode surface, large barrier for electron transfer between the enzyme active site and electrode surface [2, 9–11], and limited amount of electroactive enzymes that can be loaded onto the electrode surface. Protein-engineered redox enzymes with a high activity and durability have been developed to improve the EFC performance, and numerous attempts have been made to immobilize enzymes using various three-dimensional (3D) nanostructured materials, including mesoporous materials, nanoparticles, and nanofibers [7, 12]. The large specific surface area is expected to enable a high enzyme loading. Moreover, enzyme stabilization in nanostructured materials can extend the lifetime of enzymes [13–15]. Enzymes can be stabilized by encapsulation in the pores of a support; this prevents enzyme removal from the support surface, aggregation, or degradation of the molecular structure. Enzyme–support interactions can be affected by pore characteristics, including the pore structure and morphology, and by surface chemical characteristics, such as hydrophobic/hydrophilic interactions, electrostatic interactions, and hydrogen bonding. Among these factors, the pore size can be an important parameter that affects enzyme immobilization. For EFC applications, designing the multidimensionality of the pore structure (hierarchical pore structure) is significant (Fig. 4.1). For the MET reaction, the mesopores should be optimized to immobilize the enzyme/mediator by considering the diffusion process and microinterface for fast electron transfer and stable immobilization of the redox mediator on the electrode. For the DET reaction, the mesopore size (structure and morphology) affects the enzyme loading and effectiveness of the electron transfer reaction by surrounding the enzyme by a conductive carbon wall. The modification of

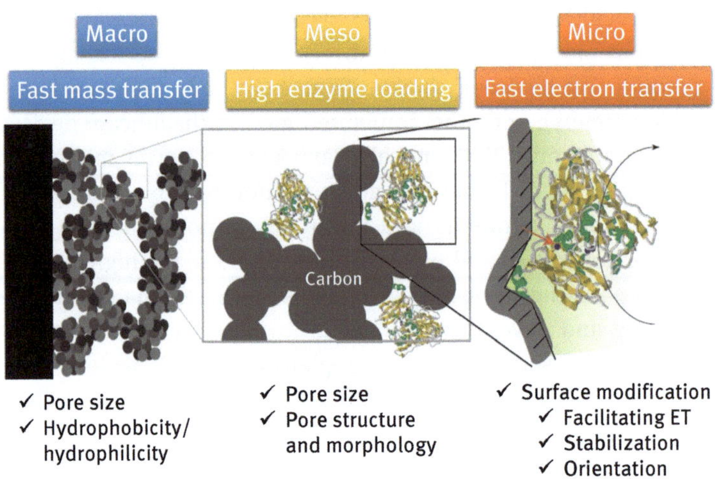

Fig. 4.1: Hierarchical structure of porous carbon for enzyme electrode.

functional chemicals on the electrode surface could facilitate the electron transfer rate by improving the orientation of the enzyme. In both cases, the macropores should be hydrophilic for liquid fuels, and the hydrophobicity should be modulated for gaseous fuels. This chapter reviews the basic concept for designing the electrode structure and recent developments in porous carbon electrodes featuring a controlled pore size for EFC applications.

4.2 Nanostructured carbon-based electrodes

4.2.1 Nanomaterials

Recently, numerous nanostructured carbon materials, including carbon nanotubes (CNTs), graphene, and carbon black, have become available for biofuel cells, and this has created remarkable progress in enzyme-electrode technologies (Fig. 4.2). These nanostructured materials are typically deposited on a current collector, such as carbon, Ti, Au, or stainless steel, and the material is selected by considering the potential window, corrosion of the substrates, and adhesiveness of the carbon materials on the substrate. Normally, a polymer binder such as polyvinylidene difluoride (PVDF; poly-1,1-difluoroethene), polytetrafluoroethylene (PTFE; poly (1,1,2,2-tetrafluoroethylene)), or styrene-butadiene rubber (SBR) is used to deposit carbon materials on the current collector. CNTs allow the formation of a binderless 3D

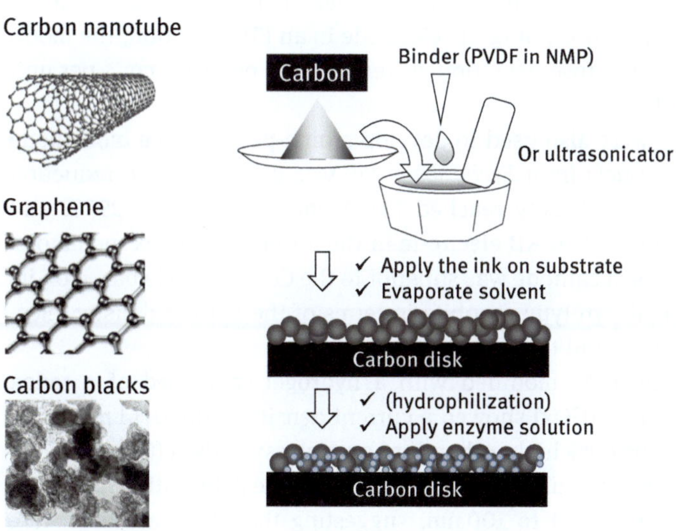

Fig. 4.2: Nanostructured carbons and electrode fabrication procedure.

structure by high-pressure pressing [8]. This chapter focuses on carbon blacks and mesoporous carbons as carbon materials describes their uses in EFCs.

4.2.2 Carbon blacks

Carbon blacks, such as Vulcan XC-72 (Cabot, USA) and Ketjen black (KB, Lion Specialty Chemicals Co. Ltd, Japan), have already been used in fuel cells, batteries, and capacitors to support a catalyst or to increase the electron conductivity. Especially KB provides high electron conductivity at smaller loading amounts. KB features a high specific Brunauer–Emmett–Teller (BET) surface area of 800–1,200 m^2 g^{-1} and a primary particle diameter of 40 nm, which provides a good platform for encaging enzymes by the carbon walls. The KB-modified electrode is fabricated by cast coating or printing a KB ink onto a substrate (current collector); the KB ink is typically prepared by mixing a KB powder with a binder using a mortar and pestle, planetary centrifugal mixer, or ultrasonicator. For example, a dispersion of KB mixed with 20 wt% PVDF in *N*-methylpyrrolidone (NMP) was drop-casted onto a carbon substrate; the porous KB-modified electrodes were obtained after evaporating the NMP [16].

KB-modified electrodes have been used for DET and MET reaction platforms to increase the current density per unit geometric surface area. The first reported use of a KB-modified electrode for DET-type bioelectrocatalysis was for fructose oxidation by D-fructose dehydrogenase (FDH) from acetic acid bacteria (Fig. 4.3A) [16]. The DET-type catalytic current density of FDH on a nonporous carbon electrode was as low as a few microamps per square centimeter [17]. When FDH was adsorbed on a KB-modified electrode prepared by dipping the electrode in an FDH solution, the modified electrode produced a fructose-oxidation current density of 10 mA cm^{-2} per unit geometric surface area [16, 18].

KB-based electrodes were also used to develop a high-performance biocathode based on DET, in which CueO from *Escherichia coli* was used as the O_2-reduction biocatalyst [19]. The current density reached 12 mA cm^{-2} at 1 atm O_2, 25 °C, and 10,000 rpm on the CueO-modified KB electrode in the absence of redox mediators. The CueO-modified KB-based cathode was superior to the CueO cathode based on its nonporous highly oriented pyrolytic graphite in terms of the current density, electrode kinetics, pH tolerance, and thermal stability.

Conversely, a KB electrode modified with a hydrogel composed of a redox polymer and glucose oxidase (GOx) showed a current density of up to 5.1 mA cm^{-2}. The current magnitude was 60% higher than that on a glassy carbon (GC) electrode [20]. KB features a large specific surface area, but the pore-size distribution of the KB-modified electrode ranges from 1 to 100 nm, suggesting that the parts that were smaller than the enzymes were inefficiently used for enzyme loading and electrochemical reactions (Fig. 4.3B). Moreover, the hydrogel that formed on the surface of

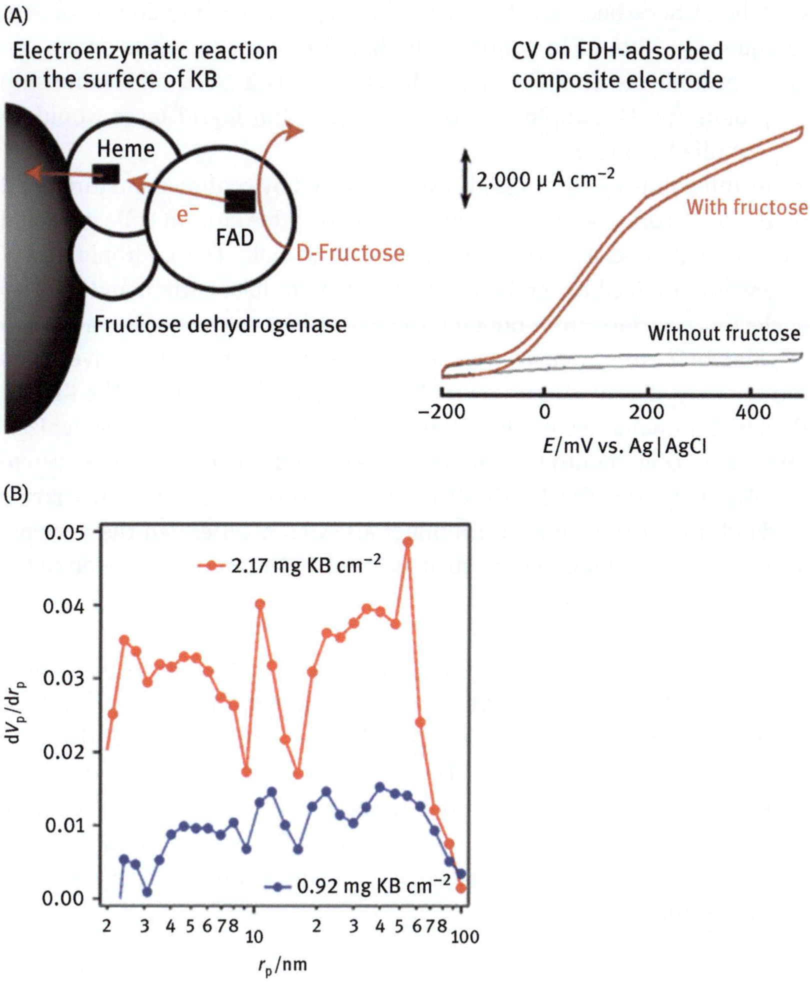

Fig. 4.3: (A) Fructose dehydrogenase modified on KB electrode and the pore-size distribution (from Kamitaka et al. [16]) and (B) pore-size distribution of KB modified on carbon support.

the KB layer could readily prevent mass transfer of glucose, GOx, and the redox polymer; sufficiently larger pores are required for hydrogel modification than for the DET reaction.

KB can also be applied to air-diffusion-type biocathodes, which are enzymatic cathodes that use O_2 from air; the cathode consists of a hydrophobic gas-diffusion layer and a hydrophilic catalytic layer including a proton conducting electrolyte layer, and forms a “pseudo”-tri-phase interface composed of a liquid electrolyte with the enzyme, solid electrode, and air phase. When the cathodes are fully dipped in the electrolyte solution, the cathode performance is usually limited by the O_2

supply toward the cathode because of the low solubility and low diffusion coefficient of O_2 in the aqueous solution. In contrast, the liquid electrolyte serves as an ion-transport medium from the anode to the cathode; there is a trade-off relationship between the proton and O_2 supplies to the enzyme; a thin liquid layer would be helpful for O_2 supplied from air.

The first air-diffusion-type biocathode was fabricated by coating a mixture of KB and PTFE binder onto Toray carbon paper (CP, gas-permeable current collector), and CueO from *E. coli* was used as the O_2-reduction catalyst [21]. The hydrophobicity/hydrophilicity was optimized by modulating the mixing ratio of carbon and PTFE to prevent the electrode surface from flooding the electrolyte solution. The measured steady-state catalytic current density was as high as 20 mA cm^{-2} in passive mode without a fan or pump. The modified electrode was prepared as follows. The KB and PTFE powders in 2-propanal were mixed thoroughly using a tip-type homogenizer, and the slurries were coated onto the surface of the CP and dried in a drying oven to remove the solvent. Lastly, the KB-CP electrodes were immersed in the enzyme solution to adsorb the enzyme on the KB materials, which generated the enzyme-adsorbed KB-CP electrode. A surfactant, such as Triton X-100, could be added to the enzyme solution to increase enzyme penetration through the hydrophobic carbon layer [22].

Based on these technologies, a flexible and high-power biofuel cell on a Japanese paper substrate using KB inks and a screen printer exhibited 0.12 mW cm^{-2} at 0.4 V [23]. The gas-diffusion-type O_2– reduction cathode was printed on a carbon ink containing a hydrophobic PTFE binder. The anode was fabricated with a carbon ink with an SBR binder, which was dispersed in water with carboxymethyl cellulose to create a hydrophilic carbon layer. The electrode modified with GOx and a redox mediator, tetrathiafulvalene, produced a glucose oxidation current of several milliamps per square centimeter.

4.3 Pore-size-controlled carbons

4.3.1 Mesoporous carbon

As noted in the previous section, nanostructured materials, including carbon blacks, have a high specific surface area and are promising materials for enzyme supports; however, controlling the hierarchical pore structures of the carbon electrode layer is challenging because they form a porous structure by aggregation, leading to a low usage efficiency of the porous structure. The strategies used to synthesize mesoporous carbons for enzyme immobilization can be categorized into several types (Fig. 4.4). The first strategy involves using the interstices of carbon particles, such as carbon gels [CGs; carbon aerogel (CAG) and carbon cryogel (CCG)]. Mesoporous carbons can be

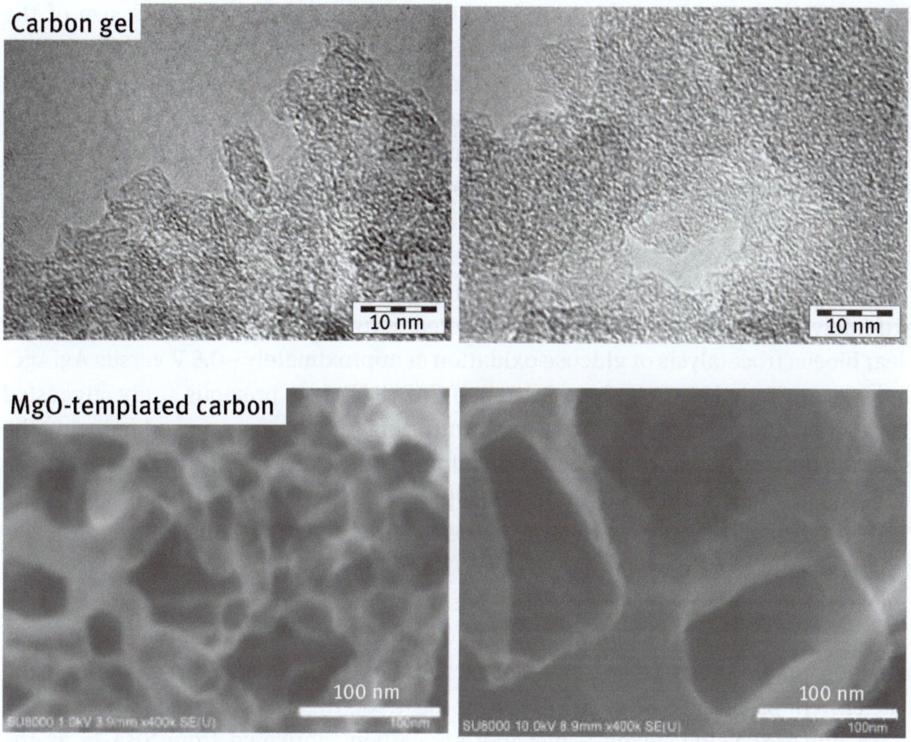

Fig. 4.4: Pore-size-controlled carbons, carbon gel, and MgO-templated carbon. From Tsujimura et al. [24] and Funabashi et al. [25].

prepared directly from carbon precursors through carbonization using hard templates (MgO, zeolites, mesoporous silica, etc.) and soft templates (metal-organic frameworks, block copolymer surfactants, etc.).

4.3.2 Mesoporous silica-templated carbon

The encapsulation of enzymes and other proteins in inorganic host materials, such as ordered mesoporous silica structures (e.g., MCM-48, MCM-41, and SBA-15), has attracted considerable attention for stabilizing the enzyme structure [14]. However, silica-based materials are electronic insulators from the viewpoint of electrochemical applications. The pores in the template silica are replicated in the carbons through either impregnation or chemical vapor infiltration of a carbon precursor, and this is followed by the carbonization and removal of the templates, mostly using HF [26, 27]. Highly ordered mesoporous carbons based on the mesoporous silica-templated carbon have been receiving substantial attention because of their extremely high surface area, defined pore size, and electronic conductivity. The pore size of the carbon material, as defined by the thickness of the silica wall of the mesoporous silica material, has a narrow distribution of approximately 5 nm; however, it is difficult to reliably obtain a pore size that is larger than the enzyme molecules.

A hydrophilic SBA-15-templated carbon prepared from glucose and poly(vinyl alcohol) as a carbon precursor has a mesopore diameter of about 4 nm [28]. However, the pore was not able to encapsulate GOx: the enzyme features a dimensional size of 5–10 nm. Direct electron exchange between an electrode and the enzyme's cofactor flavin adenine dinucleotide (FAD), which is buried in the protein shell of GOx, is extremely slow. GOx on a hydrophilic, ordered porous carbon surface began to show clear bioelectrocatalysis of glucose oxidation at approximately −0.5 V versus Ag|AgCl and was attributed to the redox reaction of FAD/$FADH_2$. A previous study suggested that CNTs could promote DET to GOx by protruding into the enzyme and reducing the electron transfer distance to the buried flrans, but direct electrocatalysis was not achieved at a reasonable overpotential [29].

The effect of the carbon structure on the electrochemistry of immobilized GOx was reported using two-dimensional (2D) CMK-3 and 3D CMK-1-like carbons, which were prepared using mesoporous silica FDU-5 as a template. Redox signals related to FAD were observed on both electrodes, and the electrochemical signal on the 3D electrode was more stable than that on the 2D electrode. However, the glucose-oxidation catalytic current based on DET from FAD to the electrode was not observed [30].

A glucose/O_2 EFC using a mesoporous carbon fabricated from SBA-15 as the template and sucrose as the carbon precursor showed 0.82 V of the open-circuit voltage and 38.7 μW cm^{-2} at 0.54 V, which is low considering its specific surface area [31, 32]. Meldola's blue, a redox catalyst to oxidize reduced nicotinamide adenine dinucleotide (NADH), was modified on the carbon surface, but oxidized NADH-dependent glucose dehydrogenase (GDH), which was considerably larger than the mesopores, might have capped the mesopores. The large surface was not efficiently used for bioelectrocatalysis.

Ordered mesoporous silica-templated carbon can create highly ordered mesopores, but the pore diameter is considerably lower than the diameter of the enzyme molecules. Thus, the material is not suitable for encapsulating enzymes within mesopores to increase the enzyme stability or enzyme loading; however, the nanostructured surface might enhance electron transfer by reducing the distance between the enzyme active site and the surface edge, or by increasing the loading of the redox mediator (redox catalysts).

4.3.3 Carbon gels

Previous studies have reported that the mean pore size of materials used for enzyme encapsulation must be at least 4–5 times larger than the mean molecular diameter of enzymes to guarantee unrestricted access [33, 34]. CGs, which are a type of pore-size-controllable porous carbons, can be a good carbon choice for enzyme loading because CGs possess several critical and useful properties: controllable pore sizes, high porosity, high surface area, and high electrical conductivity. CGs are synthesized using a resorcinol–formaldehyde sol–gel process, which is followed by either freeze–drying and carbonization (CCG) or CO_2-supercritical drying (CAG) [35–37]. The properties of CGs markedly depend on the synthesis and processing conditions [38]; notably, CGs harboring large pores (larger than the enzyme molecules) can be synthesized. The mean diameter of the pore formed in a CG can be increased by lowering the concentration of the CO_3^{2-} catalyst (C) solution at a given ratio of resorcinol to water (R/W) and by decreasing the R/W ratio at a given R/C ratio during gel synthesis. The average pore diameter (10–80 nm) was controlled by changing the molar R/C ratio [24]. However, CGs with a mean pore diameter >40 nm could not be readily prepared with a high degree of reproducibility because the extremely low catalyst concentration (Na_2CO_3) could be altered by certain contaminants. CCGs that harbored larger pores (14–108 nm) were synthesized using a sol–gel process from resorcinol and formaldehyde in carbonate buffer solutions by adjusting the total concentrations and pH of the catalyst components; the concentration of Na_2CO_3 in the sol–gel process was increased to stabilize the pH (buffer action), and the catalyst (CO_3^{2-}) concentration was controlled by adjusting the pH [39].

CGs have been investigated as electrode materials for constructing enzyme-based bioelectrochemical devices, including EFCs and biosensors, to improve the electron path and the enzyme immobilization process. FDH adsorbed on a CCG electrode exhibited large catalytic currents for fructose oxidation in the absence of redox mediators [24]. The catalytic current depends on the pore size, and the catalytic current density was very low when the CCG pore was smaller than the FDH molecule. Conversely, under the employed experimental conditions, the wider pores provided a higher catalytic current density. Thus, the enzymes caged within the carbon mesopores might be suitably positioned to directly communicate with the electrode (Fig. 4.5). The particle size of the carbon could also be a critical factor for determining the performance of an electrode; the measured catalytic current depends on the mass transfer of the substrate from the bulk solution to the surface of the carbon particle and inside the carbon particle [40]. The enzyme electrode harboring mesopores exhibited a high bioelectrocatalytic current when sufficient substrate was supplied to the FDH present inside the CG.

The CCG electrode allows a DET reaction of soluble pyrroloquinoline quinone-GDH from *Acinetobacter calcoaceticus* [41]. The response current depended on the glucose concentration, and the highest glucose-oxidation catalytic current density was 1 mA cm^{-2}, which was markedly higher than that on other materials.

CGs were also successfully used for an O_2– reducing biocathode, in which multi-copper oxidases were used as electrocatalysts [42–46]. Diffusion-controlled O_2 reduction into water was achieved without mediators using enzymes that were physically

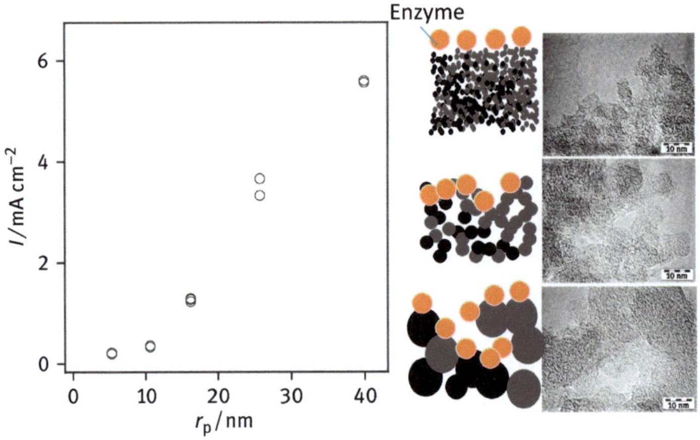

Fig. 4.5: Dependence of fructose oxidation catalytic current on the pore radius of CCG and schematic illustration of the relationship between pore size and enzyme distribution. From Tsujimura et al. [24].

adsorbed onto CAG particles; the following enzymes were used: laccase from *Trametes* sp., bilirubin oxidases (BODs) from *Myrothecium verrucaria* and *Bacillus pumilus*, and recombinant CueO from *E. coli*. The current density was predominantly controlled by the diffusion of dissolved O_2 in the rotating disk electrode experiments and reached 10 mA cm^{-2} at 1 atm O_2, 25 °C, and 8,000 rpm on the laccase-adsorbed electrode. The laccase-adsorbed CAG-based biocathode showed a stable O_2-reduction current in an air-saturated buffer for at least 10 days under a continuous-flow system; in contrast, the activity of laccase dissolved in a buffer solution decreased to half the initial activity during incubation for 5 days. Thermophilic BOD from *Bacillus pumilus* was also adsorbed onto a CCG electrode in the presence of a mediator, an Os polymer [46]. BOD adsorbed on a KB electrode also showed a DET-type electrochemical O_2-reduction activity, but the electrochemical response on the CCG electrode was more stable than that on the KB electrode, which could be attributed to the encapsulation of BOD by the mesoporous carbon wall.

4.3.4 MgO-templated carbon

CG can be a promising option for producing a desired porous carbon featuring a controlled pore size; however, there is a serious drawback – the production process is time-consuming and requires a large amount of organic solvents. Meanwhile, CG is not suitable for factory-scale mass production. A promising alternative porous carbon could be MgO-templated mesoporous carbon (MgOC, Figs. 4.4 and 4.6) [47]. One of the notable advantages of using MgOC is the simplicity of its production procedure

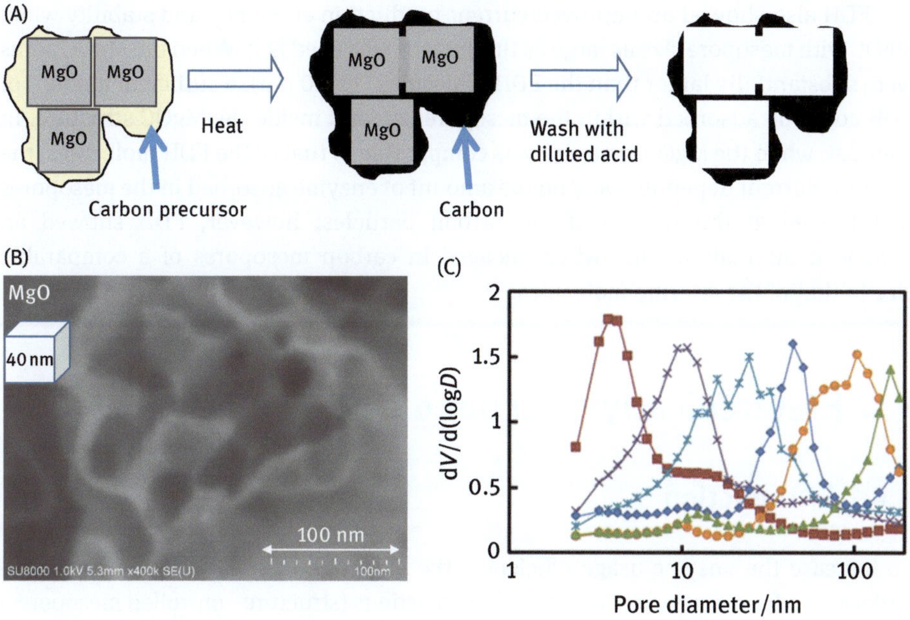

Fig. 4.6: (A) Typical production procedure of MgO-templated carbon, (B) FE-SEM image of MgO-templated carbon using MgO with 40 nm crystalline, and (C) pore-size distribution of MgOCs made from different MgO crystalline sizes.

over the procedures reported for mesoporous carbons, such as CCGs and CAGs. MgOC is synthesized as follows: a mixture of a carbon precursor, such as poly(vinyl alcohol), with the template MgO particles is heated in a furnace; then, the MgO template is removed by washing the carbon with a diluted acid, such as HCl or H_2SO_4, to obtain the porous carbon materials. MgO is a suitable template because it is both thermally and structurally stable during carbonization and is easy to remove from the resulting carbon by washing with a dilute acid. Another advantage of MgOC over other mesoporous carbons is its tunable pore-size distribution (2–150 nm), which can be modified by changing the crystalline structure of the MgO template. MgOC is the only carbon that is produced at the industrial level. Furthermore, the interconnected mesopores of MgOC provide a highly effective surface area per volume for enzyme immobilization.

MgOC featuring an average pore diameter of 38 nm and a BET-specific surface area of 580 $m^2\ g^{-1}$ was first used as the electrode material for a glucose-oxidizing anode [48]. The MgOC mixed with a binder was modified on a GC electrode using the cast-drop technique. Anodic catalysts composed of a redox polymer, GOx, and a crosslinker were applied to the MgOC-modified electrodes. The glucose-oxidation current density on MgOC was 11.3 mA cm^{-2} at 0.5 V, which was 5.3 times higher than that on GC.

FDH also showed an improved current-production efficiency and stability when MgOC with mesopore size as large as the enzyme was used [25]. When the MgOC pores were substantially larger than the FDH molecules (15–20 nm), a sufficient amount of FDH could be adsorbed within the mesopores on and inside the MgOC structure. In contrast, when the MgOC pore size was comparable to that of the FDH molecules, the catalytic current depended only on the amount of enzyme adsorbed in the mesopores that formed at the surface of the carbon particles; however, FDH showed an enhanced thermal stability when encaged in carbon mesopores of a comparable size to that of the enzyme molecules.

4.4 Hierarchically structured carbon electrodes

4.4.1 Introduction

To increase the enzyme-usage efficiency, the electrode material must have a large surface area for efficient bioelectrocatalytic reactions (structure-controlled mesopores) and provide rapid mass transport of reactants (macropores). Moreover, given that the enzymatic reaction near the electrode surface at high rate, the maximal catalytic current density would be suppressed by substrate consumption. To successfully satisfy the requirements of both a large specific surface area for enzyme loading and the rapid mass transport of fuel, a hierarchically structured carbon electrode would be desired: if the pores were considerably larger than the enzyme molecules, enzyme loading would be limited, but if the pores were small, enzyme penetration would be prevented. Distinct pore sizes can be obtained by mixing materials with pores of different sizes; three strategies could be considered to construct hierarchically structured carbon electrodes: (1) first, constructing the macroporous structure and then forming mesopores on the surface of the macroporous material (see Section 4.4.2), (2) building the macroporous structure upon mesoporous carbon materials (Section 4.4.3), and (3) mixing templates to form a hierarchically structured material (Section 4.4.4).

4.4.2 Mesoporous carbon formed on macrostructured carbon

Macroporous structures can be obtained using a carbon-fiber-based carbon texture, such as carbon cloth, carbon felt, CP, or carbon foam, all of which are templated carbons prepared using silica or polymer particles as the template. The mesopores can be formed on the macroporous structure by (1) depositing mesoporous carbon, (2) growing CNTs through chemical vapor deposition (CVD), and (3) forming pores on the substrate through activation treatment performed using CO_2 or H_2O with metal catalysts.

CNT modification on CP or carbon fiber by CVD led to a 100-fold increase in the surface area and a 10-fold increase in the current density (22 mA cm^{-2} of glucose current at 37.5 °C); CNT modification is a promising technique to fabricate a tailored hierarchical structure [49, 50]. Another study reported efficient O_2 reduction catalyzed by BOD using a redox hydrogel on a hierarchically structured carbon material obtained by growing carbon fibers on highly porous reticulated vitreous carbon [51]. Moreover, chitosan scaffolds doped with CNTs were developed for 3D hierarchically structured electrodes [52]. The porous and interconnected structures were formed by thermally induced phase separation followed by freeze–drying of an aqueous solution of chitosan acetic acid. Multiwalled CNTs provided electrical conductivity throughout the chitosan structure. A 3D macrocellular carbonaceous foam electrode is composed of Si-templated macropores, and the mesopores are introduced using a porous carbon precursor; 1,734 µg cm^{-2} of the GOx-hydrogel generated 18 mA cm^{-2} at 37 °C and 2,000 rpm [53].

4.4.3 Carbon particles deposited using electrophoretic deposition

Carbon particles can be deposited onto the current collector by electrophoretic deposition, in which carbon materials, such as CNT, CB, or MgOC, are dispersed in a solvent with a binder and are forced to migrate toward an electrode by applying an electric field (DC 10–200 V). The thickness of the carbon layer and the morphology (macropore size) can be controlled by tuning the applied voltage, carbon concentration, distance between the two electrodes, and electrophoresis time. Under moderate conditions, for example, at a low applied voltage (10 V), a smooth carbon layer is formed without any pores larger than 10 µm. However, above 50 V, the carbon particle was deposited on the substrate, forming a macropore. The electrode made from 40 nm MgOC was further coated with a biocatalytic hydrogel composed of a conductive redox polymer, deglycosylated FAD-dependent GDH (d-FAD-GDH), and a crosslinker [54]. The glucose-oxidation current density was 100 mA cm^{-2} at 25 °C and pH 7 with a hydrogel loading of 1.0 mg cm^{-2}. For the same hydrogel composition and loading, the current density on the MgOC-modified GC electrode was more than 30 times higher than that on a flat carbon electrode. For the drop-casted MgOC electrode, which lacked macropores larger than 10 µm, the maximal current was <40 mA cm^{-2}, even at an electrode rotation rate of 9,000 rpm. The maximal current was ascribed to the limitations of the biocatalyst and glucose transport in the carbon-layer structure on the GC support. Furthermore, the stability of the hydrogel electrode was enhanced using mesoporous carbon materials: >95% of the initial catalytic current remained after storage in phosphate buffer for 220 days at 4 °C, and 80% of the activity was observed after 7 days of continuous operation at 25 °C. The electrochemical response of FAD-GDH in the hydrogel on the GC electrode was diminished at 37 °C after 2 days of continuous operation in 0.1 M phosphate buffer at pH 7, but the response stability was improved by immobilizing the hydrogel on the porous carbon structure by encapsulating the hydrogel within the mesopores [55]. A glucose anode based on a GOx-hydrogel consisting of an Os-complex-tethered polymer produced a glucose oxidation current as high as 60 mA cm^{-2} at a loading of 1,000 µg cm^{-2} at 37 °C [56].

A macro-meso-hierarchical structure also enabled efficient BOD-catalyzed electrochemical O_2 reduction [57]. BOD was adsorbed on an MgOC electrode with a redox mediator – 2,2′-azinobis(3-ethylbenzothiazolin-6-sulfonate) (ABTS). The ABTS molecules adsorbed on the surface of MgOC exhibited an extremely stable redox behavior without disproportionation. The ABTS adsorbed on the electrode surface accelerated the DET and MET rates by improving the orientation of BOD on the carbon surface and by mediating the electron transfer from the electrode to the electroactive site of BOD. The maximal catalytic current was doubled in the presence of ABTS across a wide pH range.

4.4.4 Hierarchically structured porous carbon made from the dual-templated method

There are very few reports on meso-macroporous carbons that are worth mentioning. A 3D interconnected meso-macrobimodal porous silica sol–gel approach was developed using a solution of polystyrene sulfonate spheres as the macrotemplate and tetraethyl orthosilicate as the silica particle precursor for mesopores, followed by multimodal porous carbon fabricated by inverse replication of bimodal silica [58]. The carbon contains 6 nm mesopores and 200 nm macropores. MgO-templated carbon with a bimodal mesopore size distribution (with maxima at 2 and 10 nm) was prepared using two MgO precursors – Mg citrate and Mg gluconate – with poly(vinyl alcohol) as the carbon precursor [59]. The mesopores in these materials are too small with respect to the size of the enzyme molecules, and there are no strategic ways to control the mesopore size. There are no reports concerning bimodal carbon materials with mesopores larger than 20 nm and macropores. Based on an MgO-templated porous carbon framework, the 3D hierarchical electrode structure was fabricated from small to large, that is, 40–150 nm MgO templates [60] This strategy allows for easy control of the volume of meso-macropores, but it is difficult to control the morphology (arrangement) of the macro-mesopores. To overcome the drawback, the ratio of the mixing ratio of the two templates was varied to determine the best morphology. For a bilirubin-oxidase-catalyzed oxygen-reduction cathode, the optimal pore composition was a mixture of 33% macropores and 67% mesopores. The macropores improved the mass transfer inside the carbon material and the mesopores improved the electron transfer efficiency of the enzyme by surrounding the enzyme with carbon.

4.5 Conclusions

EFC construction is simplified by surface-confined bioelectrocatalytic electrodes because diffusional redox mediators and an ion-conducting separator are not needed. The removal of these inhibitory requirements enables the design of membrane-free EFCs, creating the possibility of further miniaturization for implantable/wearable/epidermal applications. The low stability and low current production efficiency of enzyme electrodes are insurmountable barriers to their application in EFCs. In combination with developing the engineering process for the enzyme, the use of a porous carbon material with a controlled pore structure and morphology allows simultaneous improvement in the stability and current density. In this chapter, the porous carbons to realize the efficient current production of an enzyme-based bioelectrocatalyst were reviewed, leading to a new strategy for designing porous carbon materials with a controlled macropore/mesopore morphology. Macropores increase the mass transfer of the biocatalyst as well as fuel, gas, and electrolyte. Mesopores can increase the current production efficiency by encaging the enzyme and reducing the electron transfer distance. To further improve the heterogeneous electron transfer rate, the process reported here must be combined with designing the nanostructures (nanointerface), and the chemical properties must be tuned by introducing specific molecules and heteroatom doping. The strategic approach reported here can be applied to other redox enzyme systems, including DET- and MET-type anodes and cathodes. Tailor-made porous carbon will unlock a new era in the fabrication and application of enzyme electrodes.

References

[1] Calabrese Barton, S., Gallaway, J., & Atanassov, P. Enzymatic biofuel cells for implantable and microscale devices. Chem Rev. 2004, 104, 4867–4886.

[2] Cracknell, J. A., Vincent, K. A., & Armstrong, F. A. Enzymes as working or inspirational electrocatalysts for fuel cells and electrolysis. Chem Rev. 2008, 108, 2439–2461.

[3] Palmore, G. T. R., & Whitesides, G. M. Microbial and enzymatic biofuel cells. In Enzymatic conversion of biomass for fuels production. ACS Symp Ser. 1998, 566, 271–290.

[4] Osman, M. H., Shah, A. A., & Walsh, F. C. Recent progress and continuing challenges in bio-fuel cells. *Part I: Enzymatic cells*. Biosens Bioelectron. 2011, 26, 3087–3102.

[5] Leech, D., Kavanagh, P., & Schuhmann, W. Enzymatic fuel cells: recent progress. Electrochim Acta. 2012, 84, 223–234.

[6] Minteer, S. D., Atanassov, P., Luckarift, H. R., & Johnson, G. R. New materials for biological fuel cells. Mater Today. 2012, 15, 166–173.

[7] De Poulpiquet, A., Ciaccafava, A., & Lojou, E. New trends in enzyme immobilization at nanostructured interfaces for efficient electrocatalysis in biofuel cells. Electrochim Acta. 2014, 126, 104–114.

[8] Holzinger, M., Le Goff, A., & Cosnier, S. Carbon nanotube/enzyme biofuel cells. Electrochim Acta. 2012, 82, 179–190.

[9] Marcus, R. A., & Sutin, N. Electron transfers in chemistry and biology. Biochimica et Biophysica Acta (BBA)-Rev Bioenergetics. 1985, 811, 265–322.
[10] Gray, H. B., & Winkler, J. R. Long-range electron transfer. Proc Natl Acad Sci U S A. 2005, 102, 3534–3539.
[11] Léger, C., & Bertrand, P. Direct electrochemistry of redox enzymes as a tool for mechanistic studies. Chem Rev. 2008, 108, 2379–2438.
[12] Kim, J., Jia, H., & Wang, P. Challenges in biocatalysis for enzyme-based biofuel cells. Biotechnol Adv. 2006, 24, 296–308.
[13] Lee, C. H., Lin, T. S., & Mou, C. Y. Mesoporous materials for encapsulating enzymes. Nano Today. 2009, 4, 165–179.
[14] Hartmann, M. Ordered mesoporous materials for bioadsorption and biocatalysis. Chem Mater. 2005, 17, 4577–4593.
[15] Takahashi, H., Li, B., Sasaki, T., Miyazaki, C., Kajino, T., & Inagaki, S. Catalytic activity in inorganic solvents and stability of immobilized enzymes depend on the pore size and surface characteristics of mesoporous silica. Chem Mater. 2000, 12, 3301–3305.
[16] Kamitaka, Y., Tsujimura, S., & Kano, K. High current density bioelectrolysis of D-fructose at fructose dehydrogenase-adsorbed and Ketjen black-modified electrodes without a mediator. Chem Lett. 2007, 36, 218–219.
[17] Ikeda, T., Matsushita, F., & Senda, M. Amperometric fructose sensor based on direct bioelectrocatalysis. Biosens Bioelectro. 1991, 6, 299–304.
[18] Tsujimura, S., Nishina, A., Kamitaka, Y., & Kano, K. Coulometric D-fructose biosensor based on direct electron transfer using D-fructose dehydrogenase. Anal Chem. 2009, 81, 9383–9387.
[19] Tsujimura, S., Miura, Y., & Kano, K. CueO-immobilized porous carbon electrode exhibiting improved performance of electrochemical reduction of dioxygen to water. Electrochim Acta. 2008, 53, 5716–5720.
[20] Suraniti, E., Vivès, S., Tsujimura, S., & Mano, N. Designing thin films of redox hydrogel for highly efficient enzymatic anodes. J Electrochem Soc. 2013, 160, G79–G82.
[21] Kontani, R., Tsujimura, S., & Kano, K. Air diffusion biocathode with CueO as electrocatalyst adsorbed on carbon particle-modified electrodes. Bioelectrochem. 2009, 76, 10–13.
[22] Asano, I., Hamano, Y., Tsujimura, S., Shirai, O., & Kano, K. Improved performance of gas-diffusion biocathode for oxygen reduction. Electrochem. 2012, 80, 324–326.
[23] Shitanda, I., Kato, S., Hoshi, Y., Itagaki, M., & Tsujimura, S. Flexible and high-performance paper-based biofuel cells using printed porous carbon electrodes. Chem Commun. 2013, 49, 11110–11112.
[24] Tsujimura, S., Nishina, A., Hamano, Y., Kano, K., & Shiraishi, S. Electrochemical reaction of fructose dehydrogenase on carbon cryogel electrodes with controlled pore sizes. Electrochem Commun. 2010, 12, 446–449.
[25] Funabashi, H., Murata, K., & Tsujimura, S. Effect of pore size of MgO-templated carbon on the direct electrochemistry of D-fructose dehydrogenase. Electrochem. 2015, 83, 372–375.
[26] Ryoo, R., Joo, S. H., Kruk, M., & Jaroniec, M. Ordered mesoporous carbons. Adv Mater. 2001, 13, 677–681.
[27] Lee, J., Han, S., & Hyeon, T. Synthesis of new nanoporous carbon materials using nanostructured silica materials as templates. J Mater Chem. 2004, 14, 478–486.
[28] Guo, C. X., Hu, F. P., Lou, X. W., & Li, C. M. High-performance biofuel cell made with hydrophilic mesoporous carbon as electrode material. J Power Sources. 2010, 195, 4090–4097.
[29] Ivnitski, D., Branch, B., Atanassov, P., & Apblett, C. Glucose oxidase anode for biofuel cell based on direct electron transfer. Electrochem Commun. 2006, 8, 1204–1210.

[30] You, C., Xu, X., Tian, B., Kong, J., Zhao, D., & Liu, B. Electrochemistry and biosensing of glucose oxidase based on mesoporous carbons with different spatially ordered dimensions. Talanta. 2009, 78, 705–710.
[31] Zhou, M., Deng, L., Wen, D., Shang, L., Jin, L., & Dong, S. Highly ordered mesoporous carbons-based glucose/O_2 biofuel cell. Biosens Bioelectron. 2009, 24, 2904–2908.
[32] Jun, S., Joo, S. H., Ryoo, R., Kruk, M., Jaroniec, M., Liu, Z., Ohsuna, T., & Terasaki, O. Synthesis of new, nanoporous carbon with hexagonally ordered mesostructured. J Am Chem Soc. 2000, 122, 10712–10713.
[33] Wang, Y. J., Wu, T. C., & Chiang, C. L. Effect of pore size distribution on enzyme immobilization in porous supports. AIChE J. 1998, 35, 1551–1554.
[34] Bosley, J. A., & Clayton, J. C. Blueprint for a lipase support: use of hydrophobic controlled-pore glasses as model systems. Biotechnol Bioeng. 1994, 43, 934–938.
[35] Pekala, R. W. Organic aerogels from the polycondensation of resorcinol with formaldehyde. J Mater Sci. 1989, 24, 3221–3227.
[36] Tamon, H., Ishizaka, H., Yamamoto, T., & Suzuki, T. Preparation of mesoporous carbon by freeze drying. Carbon. 1999, 37, 2049–2055.
[37] Yamamoto, T., Nishimura, T., Suzuki, T., & Tamon, H. Control of mesoporosity of carbon gels prepared by sol–gel polycondensation and freeze drying. J Non-Cryst Solids. 2001, 288, 46–55.
[38] Al-Muhtaseb, S. A., & Ritter, J. A. Preparation and properties of resorcinol–formaldehyde organic and carbon gels. Adv Mater. 2003, 15, 101–114.
[39] Hamano, Y., Tsujimura, S., Shirai, O., & Kano, K. Control of the pore size distribution of carbon cryogels by pH adjustments of catalyst solutions. Mater Lett. 2014, 128, 191–194.
[40] Hamano, Y., Tsujimura, S., Shirai, O., & Kano, K. Micro-cubic monolithic carbon cryogel electrode for direct electron transfer reaction of fructose dehydrogenase. Bioelectrochem. 2012, 88, 114–117.
[41] Flexer, V., Durand, F., Tsujimura, S., & Mano, N. Efficient direct electron transfer of PQQ-glucose dehydrogenase on carbon cryogel electrodes at neutral pH. Anal Chem. 2011, 83, 5721–5727.
[42] Tsujimura, S., Kamitaka, Y., & Kano, K. Diffusion-controlled oxygen reduction on multi-copper oxidase-adsorbed carbon aerogel electrodes without mediator. Fuel Cells. 2007, 7, 463–469.
[43] Miura, Y., Tsujimura, S., Kurose, S., Kamitaka, Y., Kataoka, K., Sakurai, T., & Kano, K. Direct electrochemistry of CueO and its mutants at residues to and near type I Cu for oxygen-reducing biocathode. Fuel Cells. 2009, 9, 70–78.
[44] Kamitaka, Y., Tsujimura, S., Setoyama, N., Kajino, T., & Kano, K. Fructose/dioxygen biofuel cell based on direct electron transfer-type bioelectrocatalysis. Phys Chem Chem Phys. 2007, 9, 1793–1801.
[45] Tsujimura, S., Asahi, M., Goda-Tsutsumi, M., Shirai, O., Kano, K., & Miyazaki, K. Direct electron transfer to a metagenome-derived laccase fused to affinity tags near the electroactive copper site. Phys Chem Chem Phys. 2013, 15, 20585–20589.
[46] Tsujimura, S., Suraniti, E., Durand, F., & Mano, N. Oxygen reduction reactions of the thermostable bilirubin oxidase from Bacillus pumilus on mesoporous carbon-cryogel electrodes. Electrochim Acta. 2014, 117, 263–267.
[47] Inagaki, M., Toyoda, M., Soneda, Y., Tsujimura, S., & Morishita, T. Templated mesoporous carbons: synthesis and applications. Carbon. 2016, 107, 448–473.
[48] Murata, K., Akatsuka, W., & Tsujimura, S. Bioelectrocatalytic oxidation of glucose on MgO-templated mesoporous carbon-modified electrode. Chem Lett. 2014, 43, 928–930.
[49] Calabrese Barton, S., Sun, Y., Chandra, B., White, S., & Hone, J. Mediated enzyme electrodes with combined micro- and nanoscale supports. Electrochem Solid-State Lett. 2007, 10, B96–B100.

[50] Wen, H., Nallathambi, V., Chakraborty, D., & Calabrese Barton, S. Carbon fiber microelectrodes modified with carbon nanotubes as a new support for immobilization of glucose oxidase. Microchim Acta. 2011, 175, 283–289.
[51] Little, S. J., Ralph, S. F., Mano, N., Chen, J., & Wallace, G. G. A novel enzymatic bioelectrode system combining a redox hydrogel with a carbon NanoWeb. Chem Commun. 2011, 47, 8886–8888.
[52] Lau, C., Cooney, M. J., & Atanassov, P. Conductive macroporous composite chitosan–carbon nanotube scaffolds. Langmuir. 2008, 24, 7004–7010.
[53] Flexer, V., Brun, M., Destribats, M., Backov, R., & Mano, N. A novel three-dimensional macrocellular carbonaceous biofuel cell. Phys Chem Chem Phys. 2013, 15, 6437–6445.
[54] Tsujimura, S., Murata, K., & Akatsuka, W. Exceptionally high glucose current of a hierarchically structured porous carbon electrode with “wired” flavin adenine dinucleotide-dependent glucose dehydrogenase. J Am Chem Soc. 2014, 136, 14432–14437.
[55] Suzuki, A., & Tsujimura, S. Long-term continuous operation of FAD-dependent glucose dehydrogenase hydrogel-modified electrode at 37 °C. Chem Lett. 2016, 45, 484–486.
[56] Suzuki, A., Murata, K., Mano, N., & Tsujimura, S. Redox hydrogel of glucose oxidase on MgO-templated carbon electrode. Bull Chem Soc Jpn. 2016, 89, 24–26.
[57] Tsujimura, S., & Murata, K. Electrochemical oxygen reduction catalyzed by bilirubin oxidase with the aid of 2,2′-azinobis(3-ethylbenzothiazolin-6-sulfonate) on a MgO-templated carbon electrode. Electrochim Acta. 2015, 180, 555–559.
[58] Kim, J.-H., et al. Facile synthesis of bimodal porous silica and multimodal porous carbon as an anode catalyst support in proton exchange membrane fuel cell, Electrochim Acta, 2010, 55, 7628–7633.
[59] Morishita, T., et al. Preparation of a carbon with a 2 nm pore size and of a carbon with a bi-modal pore size distribution. Carbon, 2007, 45, 209–211.
[60] Funabashi, H., Takeuchi, S., & Tsujimura, S. Hierarchical meso/macro-porous carbon fabricated from dual MgO templates for direct electron transfer enzymatic electrodes. Sci Rep. 2017, 7, 45147.

Huangxian Ju

5 DNA assembly for electrochemical biosensing

Deoxyribonucleic acid (DNA) is ubiquitous in biology because of its biological roles as carriers of genetic information, encoding amino acid sequences in protein synthesis. Its salient feature is the canonical Watson–Crick base-pairing interactions, which endow DNA to direct assembly with high selectivity and specificity, and the assembled DNA structures have been used as a kind of new biomaterials by scientists in chemistry and biology. Under well-designed sequences, the structural transformation of DNA assembly can serve as an electrochemical signal transducing method for constructing biosensors. Based on the programmable features, DNA assembly is also broadly used for signal amplification in electrochemical biosensing. Considering the advantages of rapidity, high signal-to-noise ratio, easy operation, low cost, and the specific interactions of DNA structures with a wide range of targets, electrochemical biosensors based on DNA assembly have attracted considerable attention for the detection of metal ions, small molecules, nucleic acids, and proteins. In this chapter, we introduce the general concept of DNA assembly applied in electrochemical biosensing techniques and summarize the recent applications of DNA assembly in electrochemical biosensors.

5.1 Introduction

Deoxyribonucleic acid (DNA) – a molecule that carries the genetic information of a living organism – is diverse in structure and function. Except from its biological significance, DNA structure is also used as a kind of biomaterial based on its biologically friendly, programmable, and relatively low-cost features. DNA can specifically recognize metal ions, small molecules, proteins, and even single cells. By integrating multiple organic or inorganic nanomaterials, DNA assembly shows more functions and applications. Furthermore, the development of proximity binding effect offers an easy and convenient way to integrate protein detection into DNA detection. Therefore, the application of DNA assembly has been tremendously broadened.

The programmability of Watson–Crick base pairing, combined with a decrease in the cost of synthesis, has made DNA a versatile engineering material for the assembly of molecular structures and dynamic molecular devices [1]. In the last 20 years, the

Huangxian Ju, State Key Laboratory of Analytical Chemistry for Life Science, Nanjing University, Nanjing 210023, China

https://doi.org/10.1515/9783110570526-005

DNA assembly technique has gone from an initial wave of making static structures of diffierent shapes and sizes, to the second wave of creating increasingly sophisticated and dynamic structures capable of carrying drugs, interacting with cell surface proteins and performing certain functions inside cells [2, 3]. Moreover, those sophisticated DNA superstructures can greatly facilitate the exponential signal amplification after DNA assembly in vitro biosensors, which can dissolve the bottleneck problem of life and biomedical science in detection of biomolecules with low abundance and acquisition of ultraweak biological signals [4]. Since DNA assembly guides the signal amplification, such as thermocycling amplification techniques and isothermal amplification techniques, it has witnessed the huge developments in biosensing fields. Due to the disadvantages of thermocycling amplification, such as easy contamination, high cost, time-consuming and requirement of a thermal cycler, many isothermal amplification strategies have been proposed and widely used in biosensors, such as rolling circle amplification (RCA), enzymatic cleavage amplification (ECA), strand displacement amplification (SDA), catalytic hairpin assembly (CHA), and hybridization chain reaction (HCR) [5]. Therefore, DNA assembly amplification techniques meet the practical demands in detection of trace amounts of various analytes and acquisition of ultraweak biological signals, which promotes the development of life science, biomedicine, and food and environmental science in analytical chemistry [4, 6].

Electrochemical biosensor is a powerful analytical tool due to their portability, miniaturization, and self-contained, which can produce digital electrochemical signal by transducing the biological-sensing element-target recognition event [5]. Typically, electrochemical biosensor consists of an analyte, receptor, signal, transducer, and data analysis system [7], and each of these elements has a different responsibility within the biosensor. Upon the recognition of analyte by the receptor, DNA assembly can be triggered to accomplish signal amplification to generate electrochemical signal for target detection, which offers electrochemical biosensors some advantages of high sensitivity and selectivity, high signal-to-noise ratio, simple equipment, low cost, and rapid response time compared to optical biosensors [5]. These remarkable characteristics and better performances make electrochemical biosensors attract considerable attention in bioelectrochemistry and be rapidly developed in recent years. In this chapter, we introduce the basic principles of DNA assembly for signal amplification and summarize their applications in electrochemical biosensors.

5.2 DNA structure transformation for electrochemical biosensing

DNA structures are mainly double helixes following the principle of Watson–Crick base-pairing rule, while in some special case G-quadruplexes, i-motif, or triplex are formed under certain sequences and conditions. One form of DNA structure can be transformed to another bonding form in response to some special stimuli or under different conditions. Through rational design, these transformations are able to change signal response in the electrochemical system. By attaching nanomaterial or small-molecule tag modification in the DNA sequence, the change of DNA structure can be easily transferred to electrochemical response by introducing or departing the electrochemical tag or changing the distance between electrochemical tag and the electrode surface. In this way, DNA assembly-triggered DNA structure transformation can be used as a practical way to construct electrochemical biosensors.

5.2.1 Stem-loop structure

Stem-loop structure [8] is a structure formed by a DNA strand, inside which two parts of DNA hybridize with each other as the rigid "stem" part and the bases between the stem part are circular as the flexible "loop" part. Quenching is keeping the single strand at a high temperature (mainly at 95 °C) for several minutes and slowly cooling it to the room temperature. By quenching, the stem-loop structure can be easily prepared and interstrand entanglement can be avoided. The application of stem-loop structure in biosensing is mainly designed to lock or unlock electrode-bound ones in response to stimuli to "turn on" or "turn off" the signal, which is a concept analogous to fluorescent "molecular beacons." A classic model of stem-loop structure in electrochemical biosensing is shown in Fig. 5.1. The existence of target tends to open the structure, making the electrochemical tag combined with it, and achieving electrochemical response by catalyzing the substrate [9]. With the signal amplification strategy, the stem-loop-based structure has been broadly used to construct various biosensors.

By integrating stem-loop structure with the ratiometric strategy, the sensitivity and the detection range of the electrochemical biosensor can be well improved. Based on a large amount of redox reporters [10], nonoverlapping redox potentials are available for ratiometric measurements. The transformation of stem-loop structure results in the release or approach of the redox reporters, which produces the "signal-off" or "signal-on" elements for ratiometric signal readout (Fig. 5.2). By integrating the proximity binding effect, the electrochemical ratiometric effect can be used for one step detection of proteins [11].

5.2.2 Tetrahedral DNA structure

Tetrahedral DNA structure is a rigid and well-confined 3D DNA nanostructure. Recently, tetrahedral DNA structures are extensively used in the electrochemical biosensing. With the help of this structure, defects on electrode surface, partial overcrowded modification, and unexpected ginterstrand entanglements can be perfectly avoided to achieve high reproducibility. The first tetrahedral DNA structure-

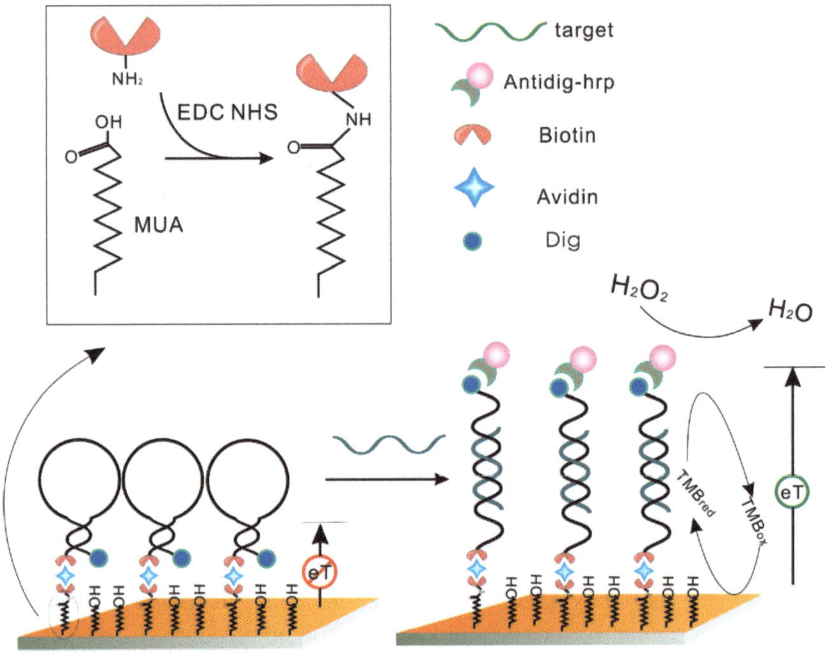

Fig. 5.1: Scheme of stem-loop structure for construction of biosensor. Reprinted with permission from Fan [8].

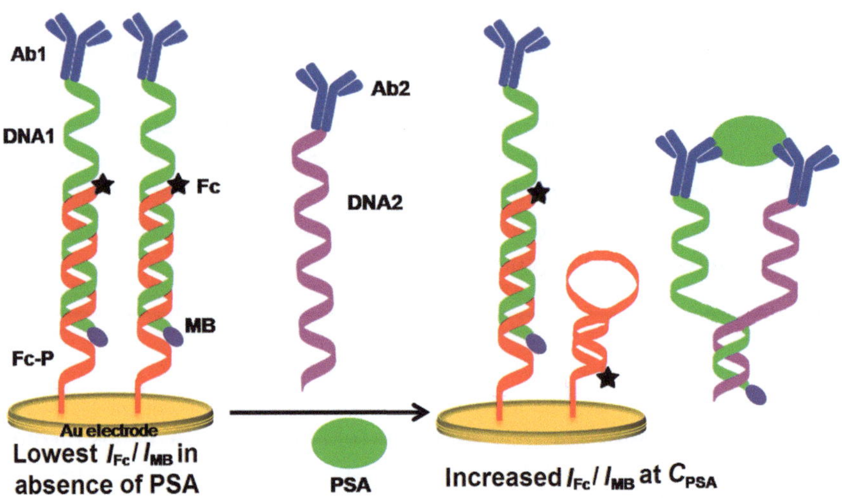

Fig. 5.2: A schematic illustration of ratiometric electrochemical proximity assay. Reprinted with permission from Ju and coworkers [11].

based biosensor was developed by Fan [12], in which three vertices were modified with thiol groups and one vertex was modified with the detecting probe (Fig. 5.3). The three thiol groups anchored the tetrahedral structure on the electrode surface via gold–thiol bond. The probe on the vertex sticking to the solution increased the accessibility of target molecules to probes and consequently increased the reaction efficiency. Because of these advantages, tetrahedral DNA structure-based biosensor has served as a platform to detect various targets, including microRNA (miRNA) [13] and exosome [14]. The structure change inside the tetrahedral DNA structure can also provide a practical way to transduce signal. Certain stimuli result in the switching or collapse of tetrahedral DNA structure, and change the distance between electrochemical tag and the electrode surface to impact the electrochemical response [15].

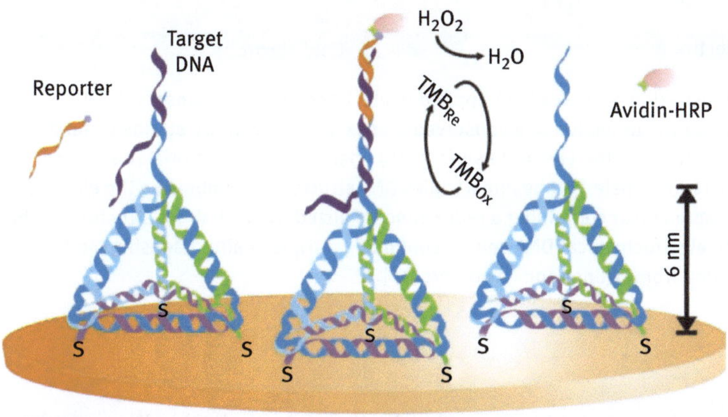

Fig. 5.3: A schematic illustration of tetrahedral DNA structure-based electrochemical sensor. Reprinted with permission from Fan and coworkers [12].

5.2.3 Triplex DNA

Triplex DNA assemblies are formed by a pair of double-helix DNA and an auxiliary single strand, termed triplex forming oligonucleotide through Hoogsteen interactions [16]. The pH value and the presence of additional binders or ions are able to affect the stability of triplex DNA and may transform the structure of triplex. Therefore, the triplex DNA can serve as a signal transducer for constructing biosensors (Fig. 5.4). The model of triplex DNA applied in the biosensors mainly includes: (1) changing the rigidity of the redox-labeled DNA strands by switching between the soft single-strand DNA and rigid double helix/triplex to change the electrical contact between the redox label and electrode

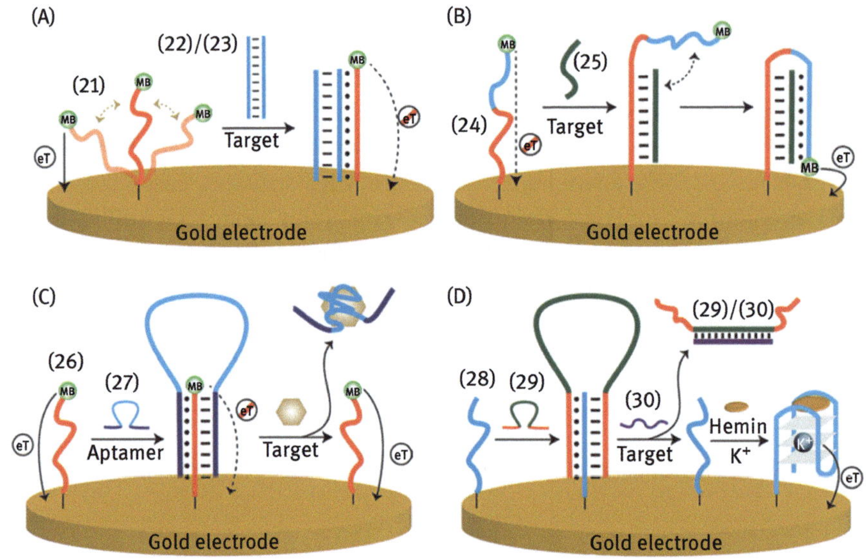

Fig. 5.4: A schematic illustration of triplex DNA applied in electrochemical biosensors. (A) An electrochemical sensor for the detection of a conserved duplex sequence in HIV strains through the assembly of a triplex structure on the electrode surface. (B) Analysis of a target DNA strand through the reconfiguration of a redox-labeled probe into a triplex DNA structure assembled on the electrode. (C) An electrochemical aptasensor containing a redox-functionalized triplex hairpin structure as the sensing module. (D) An electrochemical DNA sensor containing a triplex hairpin-sensing module. Reprinted with permission from Willner and coworkers [16].

surface, and (2) assembling a catalytic G-quadruplex/hemin complex by disassembling the triplex DNA.

5.2.4 G-quadruplex

G-quadruplexes are formed by guanine-rich DNA sequences by stacking two or more G-tetrads. The electrochemical discriminations between G-quadruplex and duplex make it possible to directly compare a G-quadruplex and its corresponding duplex through electrochemical methods [17]. Furthermore, by combining with hemin, G-quadruplex/hemin complex works as an efficient peroxidase to tremendously increase the catalysis ability of hemin alone. Under certain conditions, the G-quadruplexes are assembled into nanowires simultaneously as bienzymes and direct electron mediators to induce significant enhancement of the electrochemical signal. With the introduction of nicotinamide adenine dinucleotide (NADH) in electrolyte, hemin/G-quadruplex can oxidize NADH to NAD^+, accompanied with

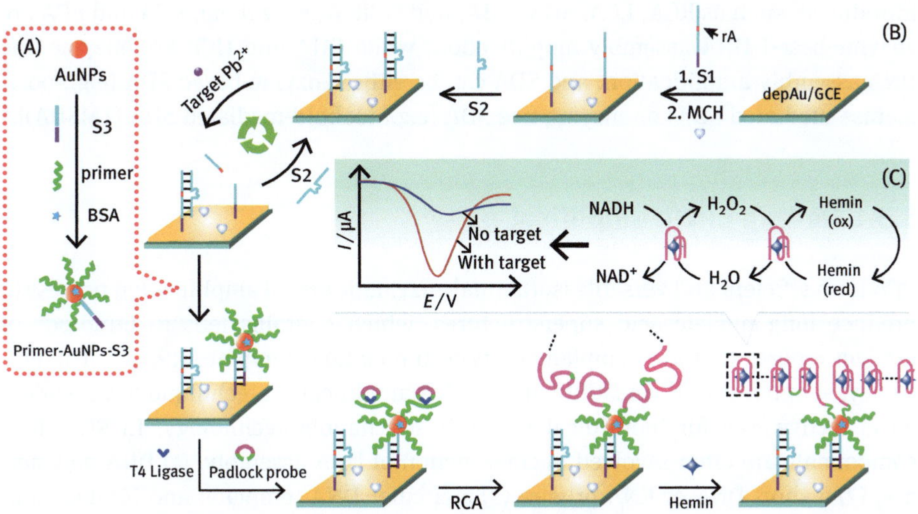

Fig. 5.5: A schematic illustration of the multiple amplification processes for Pb^{2+} detection. (A) The preparation procedure of primer–AuNPs–S3 complexes; (B) the process of hemin/G-quadruplex nanowire generation and Pb^{2+} detection; and (C) amplifying mechanism of pseudo-bienzyme cascade reaction of hemin/G-quadruplex. Reprinted with permission from Yuan [18].

generating lots of H_2O_2 for accelerating direct electron transfer of hemin to the electrode and significantly improving electrochemical response relating to the concentration of target Pb^{2+} (Fig. 5.5) [18].

5.3 Principles of DNA assembly amplification in electrochemical biosensing

Achieving high sensitivity is one of the major goals in developing novel electrochemical biosensing methods for the detection of biomarkers because a small variety of important biomarkers are sufficient to regulate the biological functions of cells and trigger disease process. Therefore, the ultrasensitive techniques based on DNA assembly signal amplification play essential roles not only in the elucidation of molecular mechanisms of life processes and many diseases, but also in promoting the early diagnosis of diseases and facilitating health care. Recently, tremendous advances have been achieved in exploration of signal amplification strategies for the development of ultrasensitive electrochemical biosensors. Here, some DNA assembly signal amplification strategies for constructing electrochemical biosensing are introduced, such as RCA, ECA, SDA, CHA, and HCR. Among them, RCA and ECA are enzyme-based DNA assembly amplification, while CHA and HCR are enzyme-free DNA assembly amplification, and SDA can be either enzyme-based SDA (e.g., polymerase-mediated SDA) or enzyme-free SDA [e.g., toehold-mediated SDA (TMSDA)].

5.3.1 Rolling circle amplification

RCA is an efficient and versatile isothermal enzymatic signal amplification process to produce long nucleic acid superstructures, which contains tens to hundreds of tandem repeats that are complementary to the circular template [19, 20]. The efficiency, simplicity, and versatility of this DNA amplification technique have made it an attractive tool for biomedical research and nanobiotechnology. In RCA, four components are often involved in construction of DNA assembly: (1) DNA polymerase; (2) a short DNA or RNA primer; (3) a circular DNA template; and (4) deoxynucleotide triphosphates (dNTPs). During the RCA process, a short DNA or RNA primer is amplified to form a long single-stranded DNA (ssDNA) or RNA concatemer using a circular DNA template and specific DNA or RNA polymerases (Fig. 5.6). Thus, the signal probe can hybridize or bind with the long nucleic acid concatemer to produce the strong signal with high sensitivity for target analysis.

Electrochemical impedance spectroscopy (EIS) is often used to characterize the fabrication and application of electrochemical biosensor through measuring the change of charge-transfer resistance (R_{et}) for redox molecule $[Fe(CN)_6]^{3-/4-}$. After the electrode was modified with DNA, the kinetics barrier between $[Fe(CN)_6]^{3-/4-}$ and negatively charged phosphate backbone of the DNA hampers charge transfer and the impedance signal accordingly increases significantly. Based on this mechanism and phenomenon, Guo group developed a novel versatile electrochemical platform for ultrasensitive adenosine detection [21]. In the presence of adenosine, a largely elongated RCA sequence was produced to hybridize with capture probes on electrode, causing difficult electron transfer to increase impedance signal significantly (Fig. 5.7). Except the electroactive species as the signal tag, AuNPs have attracted substantial research efforts on the basis of electrocatalytic activity. Gao and coworkers [22] have proposed that the target protein recognized its aptamer to trigger RCA to generate massive long DNA sequences containing many adenines, where a considerable amount of AuNPs are adsorbed on the RCA product through poly A–Au interactions. The excellent performance of the proposed electrochemical aptasensor is tested by recording the AuNP electrocatalytic reduction toward H_2O_2 to show a wide linear range and a low detection limit for thrombin.

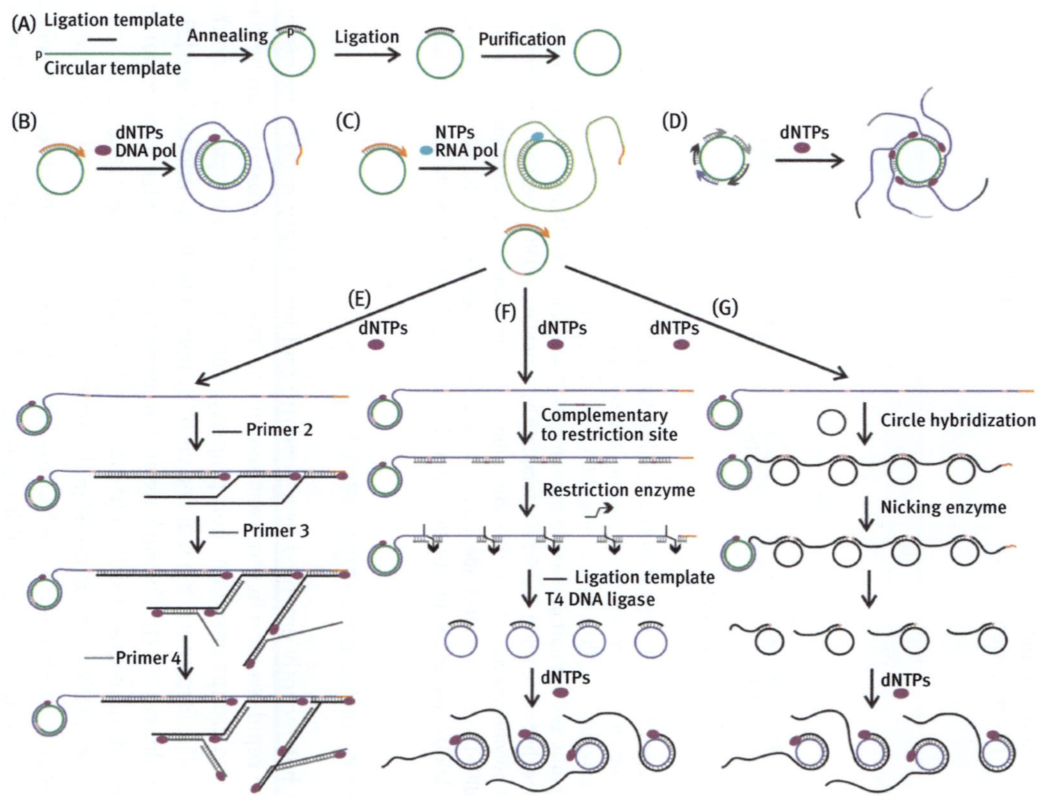

Fig. 5.6: A schematic illustration of the basic principles of RCA. (A) Making a DNA circle by template-mediated enzymatic ligation. (B) Linear RCA reaction to generate long ssDNA with a premade circle and target nucleic acids. (C) Linear RCA reaction to generate long RNA with a premade DNA circular template. (D) Multiprimed RCA to generate multiple copies of the RCA product from a single circle. (E) Exponential DNA amplification by hyperbranched RCA. (F) Exponential DNA amplification generating multiple circles from the RCA products. (G) Exponential DNA amplification by primer generation using nicking enzymes. Reprinted with permission from Zhao [19].

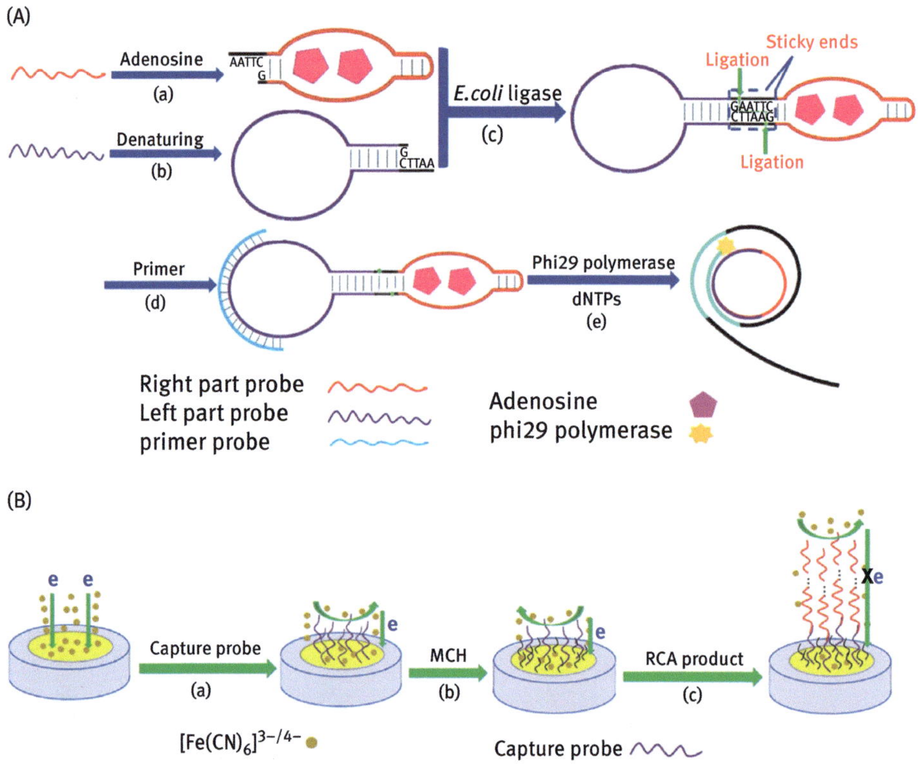

Fig. 5.7: A schematic illustration of electrochemical sensing system based on RCA. (A) Design of RCA process based on structure-switching aptamer and sticky end-based ligation: (a) Right part probe: target–aptamer binding; (b) left part probe: DNA denaturing to form a hairpin structure; (c) ligation of right part probe and left part probe with the same sticky ends, with assistance of *E. coli* DNA ligase; (d) the primer hybridized with the circular template by annealing; (e) RCA reaction in the presence of Phi29 DNA polymerase and dNTPs. (B) Fabrication of the electrochemical biosensor: (a) Capture probe modified to the bare electrode; (b) blocking the electrode with MCH; (c) capture probe–RCA product hybridization for EIS quantitative determination. Reprinted with permission from Guo [21].

5.3.2 Enzymatic cleavage amplification

The ECA has recently emerged as a powerful tool for its easy design, simple operation, and rapid reaction in ultrasensitive biosensing assays. Enzymes can be classified into two groups based on their property: natural protein enzyme and synthetic DNAzyme. The former is often realized with the help of nicking endonuclease [23–26] and exonuclease [27–31]. The nicking endonuclease can recognize a particular short sequence in double-stranded DNA (dsDNA) and cleave only one strand at the specific site to release ssDNA for further signal amplification [23–26, 32]. Ju and coworkers resorted to nicking endonuclease Nt.BbvCI for highly sensitive homogeneous electrochemical immunoassay by combining target-induced proximity hybridization with mesoporous silica nanoprobe (MSN) (Fig. 5.8). In the presence of target protein, the formed proximate complex hybridized with the ssDNA on MSN to form a rigid dsDNA structure and thus opened the biogate, which led to the release of electroactive methylene blue (MB) entrapped in the MSN. More importantly, the dsDNA structure contained a specific recognition site for Nt.BbvCI and was cleaved in one strand to in situ recycling of the proximate complex for the release of more MBs, thus amplifying the electrochemical signal to achieve a detection range of four orders of magnitude with a detection limit at pg mL^{-1} level for prostate-specific antigen.

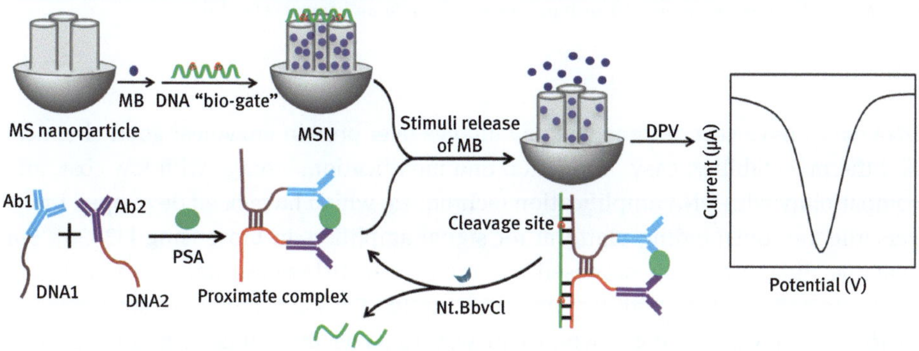

Fig. 5.8: A schematic illustration of homogeneous electrochemical immunoassay using proximity hybridization-responsive mesoporous silica nanoprobe. Reprinted with permission from Ju [23].

However, unlike nicking endonuclease, exonuclease does not require any specific recognition site and, thus, has received more and more attention for easy design without the limit of specific recognition sequence in DNA assembly amplification. For example, exonuclease III (Exo III) is a sequence-independent enzyme, which can only catalyze the stepwise removal of mononucleotides in dsDNA with blunt or recessed 3′-end in the direction from 3′- to 5′-terminus and is not active on ssDNA or dsDNA with blunt or recessed 3′-terminus [27–31]. Coupling with electrocatalysis of porphyrinic metal-organic framework (PCN-222) as a signal nanoprobe and Exo III signal amplification, Ju and coworkers [29] designed an electrochemical DNA sensor for DNA detection through the structure switch of triple-helix DNA (Fig. 5.9). Target DNA hybridized with the assistant DNA to trigger the Exo III cleavage process, followed by the target recycling and the release of the hairpin DNA. Afterward, the PCN-222@SA nanoprobe bound with hairpin DNA through biotin–streptavidin biorecognition to amplify the electrocatalytic current significantly toward oxygen reduction. Integrating with DNA recycling amplification of Exo III, the electrochemical biosensor showed a high sensitivity with a detection limit of 0.29 fM and was successfully used to detect DNA in a complex serum matrix.

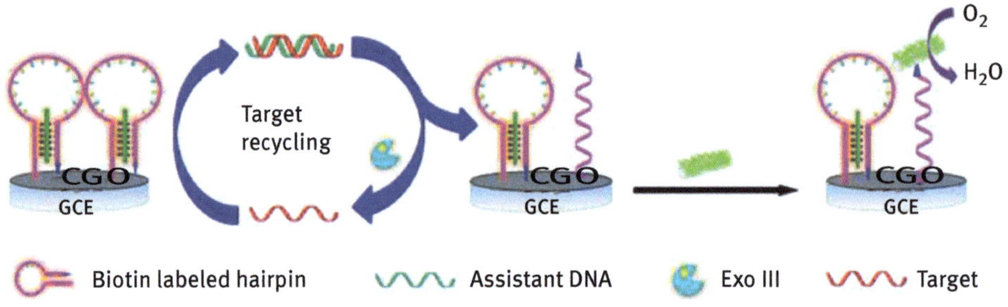

Fig. 5.9: A schematic illustration of electrochemical strategy coupling with target recycling amplification for DNA sensing. Reprinted with permission from Ju and coworkers [29].

DNAzymes have several practical advantages over protein enzymes: good chemical and thermal stability, easy production and modification, storage with low cost, and compatibility with DNA amplification techniques, which have been developed into a versatile and outstanding platform for signal amplifiers in biosensing [33–38]. For example, first, Ju group proposed a thermophilic DNAzyme that was capable of oxidizing substrates at high temperatures (up to 95 °C) and long reaction times (up to 18 h at 75 °C), offering a wonderful way to solve the limits of natural protein enzyme in sensitivity to thermal denaturation [35]. Considering these advantages of DNAzymes, Ju and coworkers [38] designed a simple electrochemical immunosensing for highly sensitive and selective detection of protein biomarker (Fig. 5.10). This method used a newly designed assembly of Mg^{2+}-dependent MNAzyme via target-driven triple-binder proximity hybridization to catalyze the cleavage of MB-labeled hairpin, which led to the departure of MB from the electrode surface and thus an amplified decrease of electrochemical signal for immunoassay of the target protein. The MNAzyme assembly was achieved by the simultaneous recognition of target protein with three DNA-labeled antibodies in the presence of Mg^{2+}, which greatly improved the detection sensitivity and selectivity. As a proof of concept, this strategy could detect carcinoembryonic antigen ranging from 0.002 to 500 ng mL^{-1} with a detection limit of 1.5 pg mL^{-1}, which showed a promising application in protein analysis. The excellent electrochemical immunosensor possessed good extensibility for large protein biomarkers by using corresponding antibodies.

5.3.3 Strand displacement amplification

SDA has been demonstrated as a powerful tool in the construction of isothermal autonomous signal amplification for ultrasensitive and selective detection of various analytes [39–46]. This strategy can be classified into enzyme-mediated SDA (e.g., polymerase-mediated SDA) [45] and enzyme-free SDA (e.g., TMSDA)

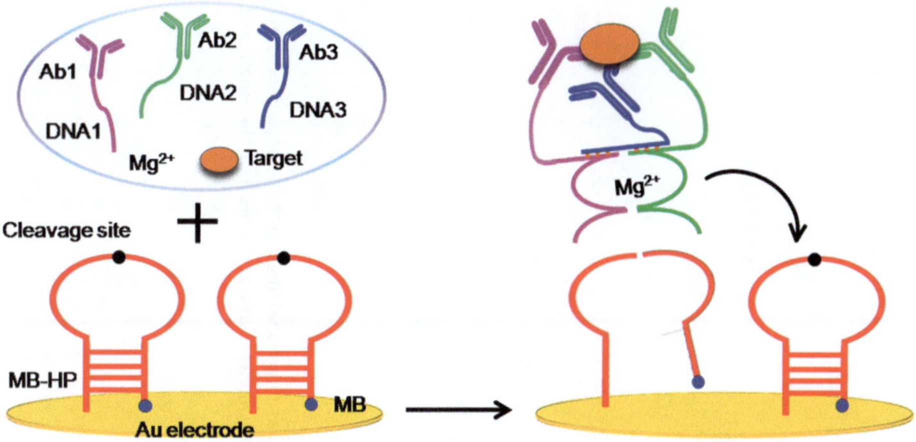

Fig. 5.10: A schematic illustration of triple-binder assembly of Mg^{2+}-dependent MNAzyme along with autocatalytic cleavage of MB-hairpin on electrochemical biosensor. Reprinted with permission from Ju and coworkers [38].

[46]. Different from enzyme-driven DNA strand displacement, TDSDA is initiated at complementary single-stranded domains termed "DNA toehold" in a dsDNA consisting of typically five to eight nucleotides and progresses through a branch migration process [41]. In a typical TDSDA, first, a long ssDNA hybridizes with the "toehold" region of a prehybridized DNA duplex, and the strand displacement is immediately initiated at toehold domains and progresses through a branch migration process, displacing one or more prehybridized strands and generating new ssDNA and dsDNA with more complementary base pairs in the process (Fig. 5.11) [39]. Thus, large numbers of repeated DNA units can be produced through the TDSDA process to generate and amplify detection signals. By varying the strength (length and sequence composition) of toehold domain, the rate of strand displacement reactions can be quantitatively controlled over a factor of 10^6, varying from 1 to $6 \times 10^6\,M^{-1}\,s^{-1}$, which allows engineering control over the kinetics of dynamic DNA nanotechnology [39].

The monitoring of miRNA expression levels is of great importance in cancer diagnosis. Xiang and coworkers [46] developed two cascaded TDSDA reactions for ultrasensitive electrochemical detection of miRNA-21 from human breast cancer cells (Fig. 5.12). The "Y" junction probes containing the locked G-quadruplexes are self-assembled on the electrode surface. The presence of the target miRNA-21 initiates the first TDSDA and results in the disassembly of the junction probes and the release of the active G-quadruplex sequences. Subsequently, the DNA fuel strand triggers the second TDSDA and leads to cyclic reuse of the target miRNA-21. The cascaded SDA thus generates many active G-quadruplexes to associate with hemin to produce

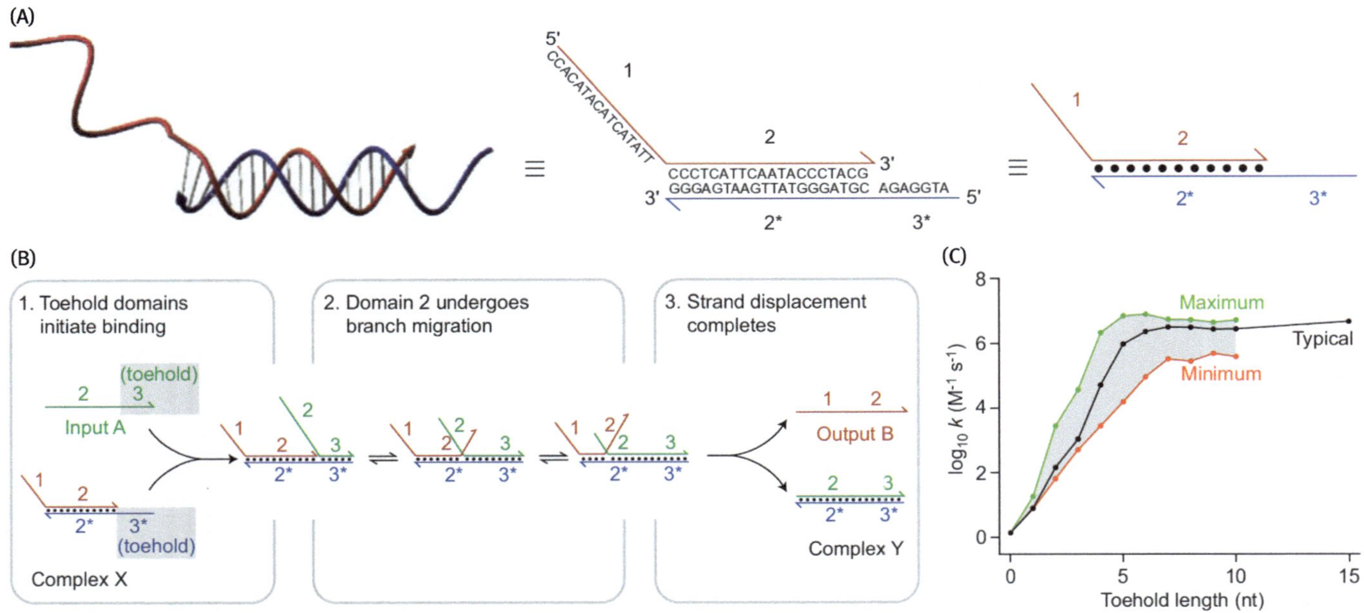

Fig. 5.11: A schematic representation of DNA strand displacement mechanism. (A) DNA represented as directional lines, with the hook denoting the 3′-end; (B) an example of this reaction; and (C) kinetics of strand displacement can be accurately modeled and predicted from the length and sequence of the toehold domain. Reprinted with permission from Zhang and Seelig [39].

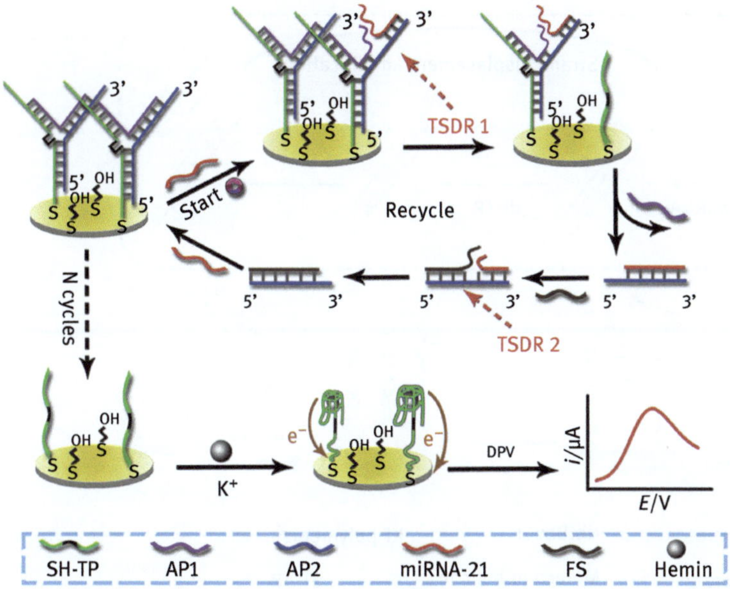

Fig. 5.12: A schematic representation of the junction probe-mediated nonenzyme TSDR target recycling amplification for label-free electrochemical detection of miRNA-21. Reprinted with permission from Xiang and coworkers [46].

significantly amplified electrochemical signal for miRNA-21 detection. For many TDSDA reactions, careful and rational sequence design is necessary in functional DNA domains that act as a unit in hybridization, branch migration, or dissociation, especially the toehold domain, which can greatly improve the efficiency of DNA assembly signal amplification.

In molecular biology, strand displacement frequently makes use of the strand displacement activity of the polymerase to displace a strand from dsDNA [34, 39]. Ding group established a simple electrochemical biosensor for highly sensitive and specific miRNA detection using polymerase-mediated SDA (Fig. 5.13) [45]. A target miRNA acted as the primer and hybridized with hairpin DNA to initiate the extension of the primer by the DNA polymerase. The extension formed recognition sequences of the nicking endonuclease, and thus enabling it to generate a nick within a strand of dsDNA. The primer then extended from the nick site and simultaneously displaced the newly nicking DNA triggers. The DNA triggers bound to the capture probes immobilized on the gold electrode to hybridize with the detection probes for producing an electrochemical signal. Thus, multiple copies of DNA triggers were produced through the repeating cycles of extension, cleavage, and SDA, resulting in the great amplification of detection signal.

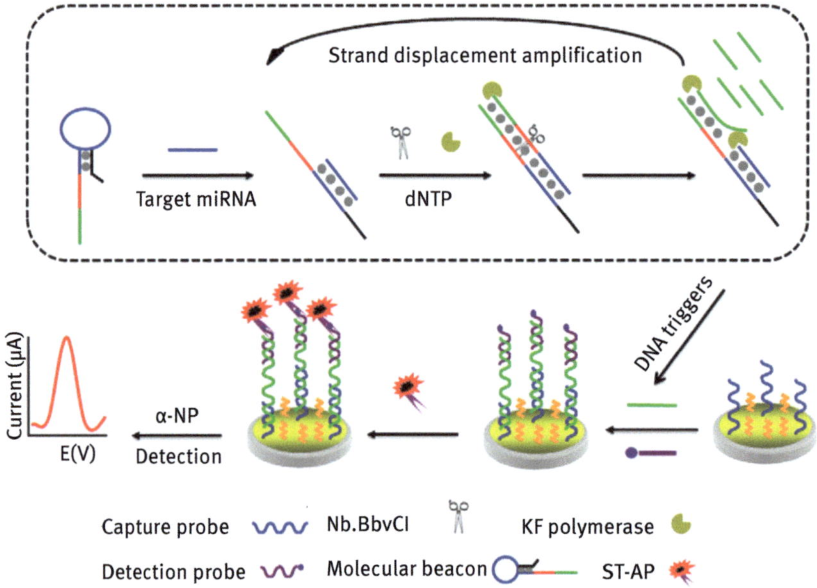

Fig. 5.13: A schematic representation of the electrochemical biosensor for target miRNA detection based on enzymatic and molecular beacon-mediated strand displacement amplification. Reprinted with permission from Ding and coworkers [45].

5.3.4 Catalytic hairpin assembly

CHA is a free-energy-driven isothermal autonomous process using hairpin DNA structures as catalytic energetic traps, which is activated by an initiator ssDNA to yield more stable duplex DNA assemblies for cyclic signal amplification [40, 47, 48]. First, an initiator ssDNA I as a catalytic role is introduced into the system containing two specific functional hairpins to hybridize with hairpin A to trigger the assembly reaction (Fig. 5.14). Then a disassembly reaction occurs when hairpin B initiates a branch migration to form more stable duplex (A + B), resulting in the displacement of I from A in each cycle of CHA to amplify the signal. Owing to its desirable programmability and hundreds-fold signal amplification ability, CHA has been successfully applied in engineering amplification systems for biosensors [49–53].

Ju group developed a simple electrochemical biosensor for highly sensitive and specific detection of target miRNA using mismatched CHA (Fig. 5.15) [50]. The target miRNA could trigger the CHA process, which led to the cyclic reuse of the target miRNA and the CHA products. Then the CHA products bound to capture probe to generate an amplified electrochemical signal. Compared with the traditional CHA, mismatched CHA decreased the nonspecific CHA products, which reduced the background signal significantly. Under the optimal experimental conditions and using

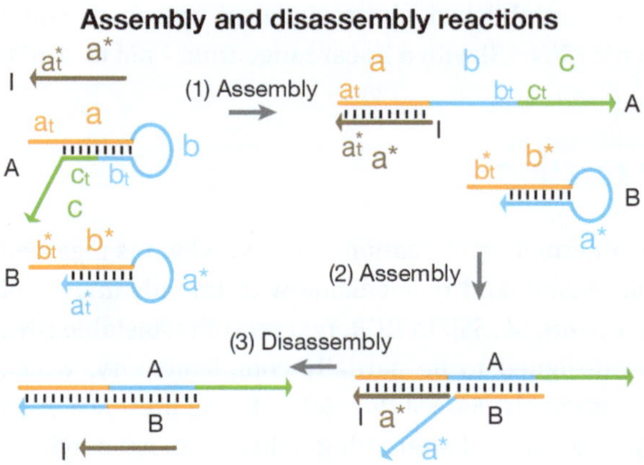

Fig. 5.14: A schematic representation of pathways of catalytic hairpin assembly. Reprinted with permission from Pierce and coworkers [47].

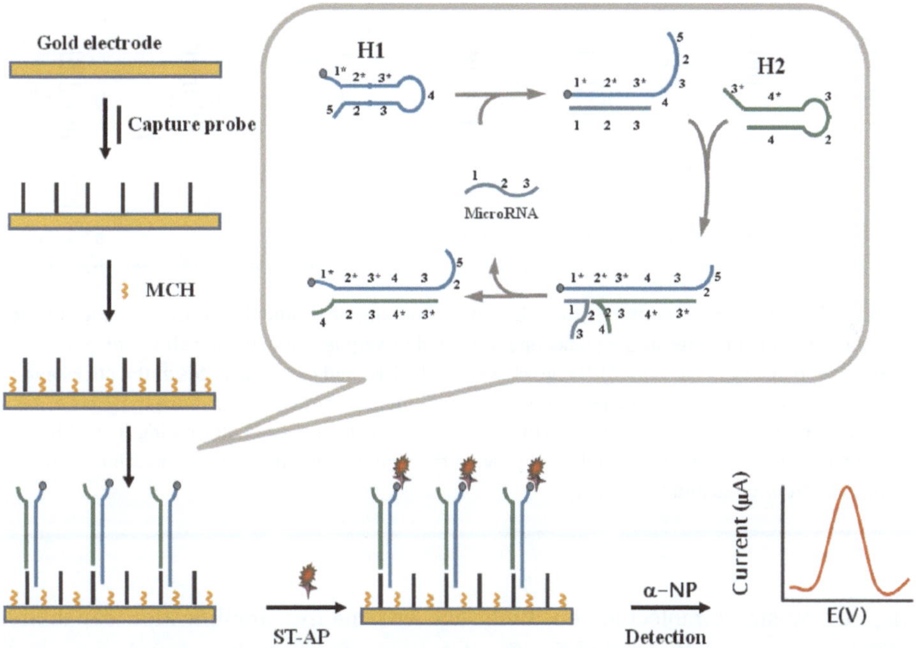

Fig. 5.15: A schematic representation of miRNA electrochemical detection based on mismatched catalytic hairpin assembly amplification. Reprinted with permission from Ju and coworkers [50].

differential pulse voltammetry, the established electrochemical biosensor could detect target miRNA down to 0.6 pM ($S/N = 3$) with a linear range from 1 pM to 25 nM.

5.3.5 Hybridization chain reaction

HCR is a powerful enzyme-free isothermal amplification strategy, which is regarded as an attractive technique for biosensing and biomedicine with the advantages of high simplicity and versatility [4, 40, 48, 54, 55]. In HCR, two sets of metastable DNA hairpin species (H1 and H2) are designed to be partially complementary, which provides building blocks for DNA assembly [54, 55] (Fig. 5.16). In the presence of H1 and H2, the initiator ssDNA triggers a chain of alternating hairpin molecule hybridization reaction, where H1 and H2 hairpins sequentially open to assemble into a long chain-like nicked dsDNA assemblies [4, 54, 55]. Since the HCR polymerization is mainly driven by the free energy of base pair formation, the key to this system is the storage of potential energy in short loops protected by long stems [38, 54].

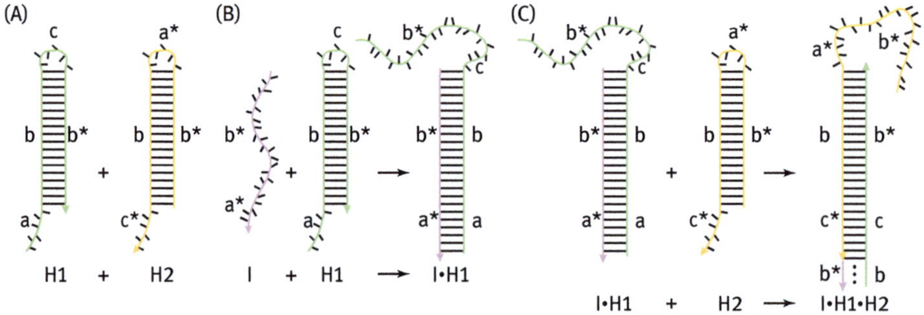

Fig. 5.16: The fundamental working principle of HCR. (A) Hairpins H1 and H2 are stable in the absence of initiator I. (B) I nucleates at the sticky end of H1 and undergoes an unbiased strand displacement interaction to open the hairpin. (C) The newly exposed sticky end of H1 nucleates at the sticky end of H2 and opens the hairpin to expose a sticky end on H2 that is identical in sequence to I. Hence, each copy of I can propagate a chain reaction of hybridization events between alternating H1 and H2 hairpins to form a nicked double helix, amplifying the signal of initiator binding. Reprinted with permission from Dirks and Pierce [54].

Inspired by single molecule sensitivity and enzyme-free amplification capability, HCR has been widely applied for the detection of various small molecules [56], nucleic acids [57, 58], proteins [59–61], and cells [62]. Ju group proposed a one-target-multitriggering HCR strategy for ultrasensitive electrochemical detection of DNA by combining SDA with the DNA self-assembly on the sensor surface (Fig. 5.17) [57]. The target DNA hybridized with the capture probe to trigger the SDA reaction

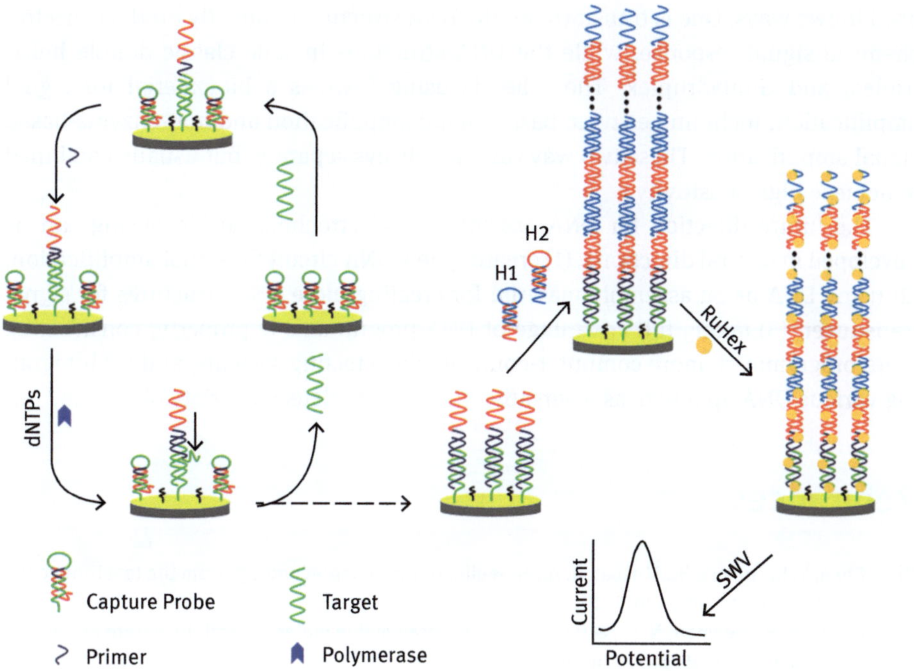

Fig. 5.17: A schematic illustration of the label-free electrochemical DNA sensing with a one-target-multitriggered hybridization chain reaction strategy. Reprinted with permission from Ju and coworkers [57].

and was subsequently displaced under the extension of the primer in the presence of dNTPs and polymerase. Afterward, the exposed stem of opened hairpin acted as an initiator to trigger the HCR assembly for producing numerous long-range nicked dsDNA. Using $[Ru(NH_3)_6]^{3+}$ as an electrochemically active indicator to interact with the product, the one-target-multitriggered HCR strategy demonstrated a perfect multiple signal amplification for highly sensitive DNA detection in a linear range from 0.1 fM to 10 pM.

5.4 Conclusions

The DNA assembly-based electrochemical biosensing is a well-developed, widely used, and highly promising sensing method. DNA assembly is programmable and predictable, and its applications in biosensing are low cost, highly sensitive, and biologically friendly. The DNA assembly for electrochemical biosensing is mainly used in two ways. One is transforming the DNA structure to directly produce electrochemical signal response, while the DNA structures include classic double helix, triplex, and G-quadruplex. The other is using DNA as a biomaterial for signal amplification, including enzyme-based signal amplification and non-enzyme-based signal amplification. These two ways are not always separate, but usually combined to achieve high sensitivity.

The future direction for DNA assembly in electrochemical biosensing can be developed in several directions: (1) creating new DNA circuit for signal amplification, (2) using DNA as an assembly material for creating more DNA structures for signal transducer, (3) taking full advantage of DNA-programmable property, constructing more programmed, more computer-controllable detecting systems, and (4) integrating various DNA aptamers as a versatile platforms to detect a variety of targets.

References

[1] Chen, Y. J., Groves, B., Muscat, R. A., & Seelig, G. DNA nanotechnology from the test tube to the cell. Nat Nanotechnol. 2015, 10, 748–760.
[2] Jones, M. R., Seeman, N. C., & Mirkin, C. A. Programmable materials and the nature of the DNA bond. Science. 2015, 347, 1260901.
[3] Ohta, S., Glancy, D., & Chan, W. C. DNA-controlled dynamic colloidal nanoparticle systems for mediating cellular interaction. Science. 2016, 351, 841–845.
[4] Ju, H. X. Signal amplification for highly sensitive immunosensing. J Anal Test. 2017, 1, 7.
[5] Qi, H. J., Yue, S. Z., Bi, S., Ding, C. F., & Song, W. L. Isothermal exponential amplification techniques: from basic principles to applications in electrochemical biosensors. Biosens Bioelectron. 2018, 110, 207–217.
[6] Ju, H. X. Functional nanomaterials and nanoprobes for amplified biosensing. Appl Mater Today. 2018, 10, 51–71.
[7] Ronkainen, N. J., Halsall, H. B., & Heineman, W. R. Electrochemical biosensors. Chem Soc Rev. 2010, 39, 1747–1763.
[8] Liu, G., Wan, Y., Gau, V., Zhang, J., Wang, L. H., Song, S. P., & Fan, C. H. An enzyme-based E-DNA sensor for sequence-specific detection of femtomolar DNA targets. J Am Chem Soc. 2008, 130, 6820–6825.
[9] Xiao, Y., Qu, X. G., Plaxco, K. W., & Heeger, A. J. Label-free electrochemical detection of DNA in blood serum via target-induced resolution of an electrode-bound DNA pseudoknot. J Am Chem Soc. 2007, 129, 11896–11897.
[10] Kang, D., Ricci, F., White, R. J., & Plaxco, K. W. Survey of redox-active moieties for application in multiplexed electrochemical biosensors. Anal Chem. 2016, 88, 10452–10458.
[11] Ren, K. W., Wu, J., Yan, F., & Ju, H. X. Ratiometric electrochemical proximity assay for sensitive one-step protein detection. Scientific Reports. 2014, 4, 4360.
[12] Pei, H., Lu, N., Wen, Y. L., Song, S. P., Liu, Y., Yan, H., & Fan, C. H. A DNA nanostructure-based biomolecular probe carrier platform for electrochemical biosensing. Adv Mater. 2010, 22, 4754–4758.
[13] Wen, Y. L., Pei, H., Shen, Y., Xi, J. J., Lin, M. H., Lu, N., Shen, X. Z., Li, J., & Fan, C. H. DNA nanostructure-based interfacial engineering for PCR-free ultrasensitive electrochemical analysis of microRNA. Scientific Reports. 2012, 2, 867.

[14] Wang, S., Zhang, L. Q., Wan, S., Cansiz, S., Cui, C., Liu, Y., Cai, R., Hong, C. Y., Teng, I. T., Shi, M. L., Wu, Y., Dong, Y. Y., & Tan, W. H. Aptasensor with expanded nucleotide using DNA nanotetrahedra for electrochemical detection of cancerous exosomes. ACS Nano. 2017, 11, 3943–3949.

[15] Abi, A., Lin, M. H., Pei, H., Fan, C. H., Ferapontova, E. E., & Zuo, X. L. Electrochemical switching with 3D DNA tetrahedral nanostructures self-assembled at gold electrodes. ACS Appl Mater Interfaces. 2014, 56, 8928–8931.

[16] Hu, Y. W., Cecconello, A., Idili, A., Ricci, F., & Willner, I. Triplex DNA nanostructures: from basic properties to applications. Angew Chem Int Ed. 2017, 56, 15210–15233.

[17] Rache, A. D., Doneux, T., & Herman, C. B. Electrochemical discrimination between G-quadruplex and duplex DNA. Anal Chem. 2014, 86, 8057–8065.

[18] Qing, M., Yuan, Y., Cai, W., Xie, S. B., Tang, Y., Yuan, R., & Zhang, J. An ultrasensitive electrochemical biosensor based on multifunctional hemin/G-quadruplex nanowires simultaneously served as bienzyme and direct electron mediator for detection of lead ion. Sens Actuators B: Chem. 2017, 263, 469–475.

[19] Ali, M. M., Li, F., Zhang, Z., Zhang, K., Kang, D. K., Ankrum, J. A., Le, X. C., & Zhao, W. Rolling circle amplification: a versatile tool for chemical biology, materials science and medicine. Chem Soc Rev. 2014, 43, 3324–3341.

[20] Feng, C., Mao, X. X., Yang, Y. C., Zhu, X. L., Yin, Y. M., & Li, G. X. Rolling circle amplification in electrochemical biosensor with biomedical applications. J Electroanal Chem. 2016, 781, 223–232.

[21] Yi, X. H., Li, L. D., Peng, Y., & Guo, L. A universal electrochemical sensing system for small biomolecules using target-mediated sticky ends-based ligation-rolling circle amplification. Biosens Bioelectron. 2014, 57, 103–109.

[22] Fan, T. T., Du, Y., Yao, Y., Wu, J., Meng, S., Luo, J. J., Zhang, X., Yang, D. Z., Wang, C. Y., Qian, Y., & Gao, F. L. Rolling circle amplification triggered poly adenine-gold nanoparticles production for label-free electrochemical detection of thrombin. Sens Actuators B: Chem. 2018, 266, 9–18.

[23] Ren, K. W., Wu, J., Zhang, Y., Yan, F., & Ju, H. X. Proximity hybridization regulated DNA biogate for sensitive electrochemical immunoassay. Anal Chem. 2014, 86, 7494–7499.

[24] Zong, C., Wu, J., Liu, M. M., Yan, F., & Ju, H. X. High-throughput imaging assay of multiple proteins via target-induced DNA assembly and cleavage. Chem Sci. 2015, 6, 2602–2607.

[25] Zong, C., Wu, J., Liu, M. M., Yang, L. L., Liu, L., Yan, F., & Ju, H. X. Proximity hybridization-triggered signal switch for homogeneous chemiluminescent bioanalysis. Anal Chem. 2014, 86, 5573–5578.

[26] Wang, Y. W., Guo, J. X., Guo, Y. H., Zhang, X. B., & Ju, H. X. Enzymatically driven formation of palindromic DNA-Au nanoparticles for snowball assembly and colorimetric biosensing. Sens Actuators B: Chem. 2018, 267, 328–335.

[27] Yang, K. L., Huo, M., Guo, Y. H., Yang, Y. Z., Wu, J., Ding, L., & Ju, H. X. Target-induced cyclic DNAzyme formation for colorimetric and chemiluminescence imaging assay of protein biomarkers. Analyst. 2017, 142, 3740–3746.

[28] Chen, Y. L., Ding, L., Song, W. Y., Yang, M., & Ju, H. X. Liberation of protein-specific glycosylation information for glycan analysis by exonuclease III-aided recycling hybridization. Anal Chem. 2016, 88, 2923–2928.

[29] Ling, P. H., Lei, J. P., & Ju, H. X. Porphyrinic metal-organic framework as electrochemical probe for DNA sensing via triple-helix molecular switch. Biosens Bioelectron. 2015, 71, 373–379.

[30] Luo, C. H., Tang, H., Cheng, W., Yan, L., Zhang, D. C., Ju, H. X., & Ding, S. J. A sensitive electrochemical DNA biosensor for specific detection of Enterobacteriaceae bacteria by Exonuclease III-assisted signal amplification. Biosens Bioelectron. 2013, 48, 132–137.

[31] Tao, C. Y., Yan, Y. R., Xiang, H., Zhu, D., Cheng, W., Ju, H. X., & Ding, S. J. A new mode for highly sensitive and specific detection of DNA based on exonuclease III-assisted target recycling amplification and mismatched catalytic hairpin assembly. Chem Commun. 2015, 51, 4220–4222.

[32] Yan, M. M., Bai, W. H., Zhu, C., Huang, Y. F., Yan, J., & Chen, A. L. Design of nuclease-based target recycling signal amplification in aptasensors. Biosens Bioelectron. 2016, 77, 613–623.

[33] Gong, L., Zhao, Z. L., Lv, Y. F., Huan, S. Y., Fu, T., Zhang, X. B., Shen, G. L., & Yu, R. Q. DNAzyme-based biosensors and nanodevices. Chem Commun. 2015, 51, 979–995.

[34] Peng, H. Y., Newbigging, A. M., Wang, Z. X., Tao, J., Deng, W. C., Le, X. C., & Zhang, H. Q. DNAzyme-mediated assays for amplified detection of nucleic acids and proteins. Anal Chem. 2017, 90, 190–207.

[35] Guo, Y. H., Chen, J. L., Cheng, M. P., Monchaud, D., Zhou, J., & Ju, H. X. A Thermophilic Tetramolecular GQuadruplex/Hemin DNAzyme. Angew Chem Int Ed. 2017, 56, 16636–16640.

[36] Zong, C., Wu, J., Liu, M. M., Yang, L. L., Yan, F., & Ju, H. X. Chemiluminescence imaging for a protein assay via proximity-dependent DNAzyme formation. Anal Chem. 2014, 86, 9939–9944.

[37] Zong, C., Wu, J., Xu, J., Ju, H. X., & Yan, F. Multilayer hemin/G-quadruplex wrapped gold nanoparticles as tag for ultrasensitive multiplex immunoassay by chemiluminescence imaging. Biosens Bioelectron. 2015, 43, 372–378.

[38] Ren, K. W., Wu, J., Ju, H. X., & Yan, F. Target-driven triple-binder assembly of MNAzyme for amplified electrochemical immunosensing of protein biomarker. Anal Chem. 2015, 87, 1694–1700.

[39] Zhang, D. Y., & Seelig, G. Dynamic DNA nanotechnology using strand-displacement reactions. Nat Chem. 2011, 3, 103.

[40] Wang, F., Lu, C. H., & Willner, I. From cascaded catalytic nucleic acids to enzyme–DNA nanostructures: controlling reactivity, sensing, logic operations, and assembly of complex structures. Chem Rev. 2014, 114, 2881–2941.

[41] Zhang, H. Q., Li, F., Dever, B., Li, X. F., & Le, X. C. DNA-mediated homogeneous binding assays for nucleic acids and proteins. Chem Rev. 2012, 113, 2812–2841.

[42] Wang, Q. B., Xu, N., Gui, Z., Lei, J. P., Ju, H. X., & Yan, F. Strand displacement activated peroxidase activity of hemin for fluorescent DNA sensing. Analyst. 2015, 140, 6532–6537.

[43] Guo, Y. H., Yang, K. L., Sun, J. C., Wu, J., & Ju, H. X. A pH-responsive colorimetric strategy for DNA detection by acetylcholinesterase catalyzed hydrolysis and cascade amplification. Biosens Bioelectron. 2017, 94, 651–656.

[44] Hu, Y. H., Xu, X. Q., Liu, Q. H., Wang, L., Lin, Z. Y., & Chen, G. N. Ultrasensitive electrochemical biosensor for detection of DNA from Bacillus subtilis by coupling target-induced strand displacement and nicking endonuclease signal amplification. Anal Chem. 2014, 86, 8785–8790.

[45] Wang, M. M., Shen, B., Yuan, R., Cheng, W., Xu, H. B., & Ding, S. J. An electrochemical biosensor for highly sensitive determination of microRNA based on enzymatic and molecular beacon mediated strand displacement amplification. J. Electroanal Chem. 2015, 756, 147–152.

[46] Shi, K., Dou, B. T., Yang, J. M., Yuan, R., & Xiang, Y. Cascaded strand displacement for non-enzymatic target recycling amplification and label-free electronic detection of microRNA from tumor cells. Anal Chim Acta. 2016, 916, 1–7.

[47] Yin, P., Choi, H. M., Calvert, C. R., & Pierce, N. A. Programming biomolecular self-assembly pathways. Nature. 2008, 451, 318.

[48] Chen, J. Y., Tang, L. J., Chu, X., & Jiang, J. H. Enzyme-free, signal-amplified nucleic acid circuits for biosensing and bioimaging analysis. Analyst. 2017, 142, 3048–3061.

[49] Zang, Y., Lei, J. P., Ling, P. H., & Ju, H. X. Catalytic hairpin assembly-programmed porphyrin–DNA complex as photoelectrochemical initiator for DNA biosensing. Anal Chem. 2015, 87, 5430–5436.

[50] Zhang, Y., Yan, Y. R., Chen, W. H., Cheng, W., Li, S. Q., Ding, X. J., Li, D. D., Wang, H., Ju, H. X., & Ding, S. J. A simple electrochemical biosensor for highly sensitive and specific detection of microRNA based on mismatched catalytic hairpin assembly. Biosens Bioelectron. 2015, 68, 343–349.
[51] Li, D. D., Cheng, W., Li, Y. J., Xu, Y. J., Li, X. M., Yin, Y. B., Ju, H. X., & Ding, S. Catalytic hairpin assembly actuated DNA nanotweezer for logic gate building and sensitive enzyme-free biosensing of microRNAs. Anal Chem. 2016, 88, 7500–7506.
[52] Zang, Y., Lei, J. P., Hao, Q., & Ju, H. X. CdS/MoS2 heterojunction-based photoelectrochemical DNA biosensor via enhanced chemiluminescence excitation. Biosens Bioelectron. 2016, 77, 557–564.
[53] Yan, L. W., Hui, J. J., Liu, Y. R., Guo, Y. H., Liu, L., Ding, L., & Ju, H. X. A cascade amplification approach for visualization of telomerase activity in living cells. Biosens Bioelectron. 2016, 86, 1017–1023.
[54] Dirks, R. M., & Pierce, N. A. Triggered amplification by hybridization chain reaction. Proc Natl Acad Sci U.S.A. 2004, 101, 15275–15278.
[55] Yang, D. W., Tang, Y. G., & Miao, P. Hybridization chain reaction directed DNA superstructures assembly for biosensing applications. TrAC Trends Anal Chem. 2017, 94, 1–13.
[56] Yang, B., Zhang, X. B., Kang, L. P., Shen, G. L., Yu, R. Q., & Tan, W. H. Target-triggered cyclic assembly of DNA–protein hybrid nanowires for dual-amplified fluorescence anisotropy assay of small molecules. Anal Chem. 2013, 85, 11518–11523.
[57] Zhu, Z., Lei, J. P., Liu, L., & Ju, H. X. Label-free electrochemical DNA sensing with a one-target-multitriggered hybridization chain reaction strategy. Analyst. 2013, 138, 5995–6000.
[58] Guo, Y. H., Wu, J., Li, J., & Ju, H. X. A plasmonic colorimetric strategy for biosensing through enzyme guided growth of silver nanoparticles on gold nanostars. Biosens Bioelectron. 2013, 78, 267–273.
[59] Tong, L., Wu, J., Li, J., Ju, H. X., & Yan, F. Hybridization chain reaction engineered DNA nanopolylinker for amplified electrochemical sensing of biomarkers. Analyst. 2013, 138, 4870–4876.
[60] Ge, Y. Q., Wu, J., Ju, H. X., & Wu, S. Ultrasensitive enzyme-free electrochemical immunosensor based on hybridization chain reaction triggered double strand DNA@ Au nanoparticle tag. Talanta. 2014, 120, 218–223.
[61] Xu, J., Wu, J., Zong, C., Ju, H. X., & Yan, F. Manganese porphyrin-dsDNA complex: a mimicking enzyme for highly efficient bioanalysis. Anal Chem. 2013, 85, 3374–3379.
[62] Chen, Y. L., Ding, L., Liu, T. T., & Ju, H. X. Arrayed profiling of multiple glycans on whole living cell surfaces. Anal Chem. 2013, 85,11153–11158.

Laurent Bouffier, Neso Sojic, Alexander Kuhn

6 Biochemical sensing based on bipolar electrochemistry

6.1 Introduction

6.1.1 Generalities

When an electrochemical reaction is performed at the interface between a solid electrode and a liquid electrolyte, a conventional electrochemical cell comprises typically a set of three electrodes. The electrode of main interest is the working electrode (WE), where a desired polarization with respect to the electrolyte is induced in order to promote a given electron transfer reaction (i.e., an oxidation or a reduction). The reference electrode (RE) has a known potential value and acts indeed as a reference to control accurately the potential of WE. Practically, the difference in potential is applied by the voltage source between the auxiliary electrode (AE) and the WE in order to compensate the charge added or removed at the WE and therefore to ensure the current flow. In this situation, when an oxidation does occur at the WE, a reduction takes place at the AE and conversely.

By comparing with this classic three-electrode system, bipolar electrochemistry (BPE) is a rather unconventional approach because the oxidation and reduction processes are in fact promoted on the same electrode. That is why the latter is called a bipolar electrode (BE). Also, the BE is not connected through a physical contact (i.e., cable) to the power supply and BPE can somehow be considered as a wireless electrochemical address. Such a situation can be practically observed in the case of corrosion science when a piece of non-noble metal is spontaneously oxidized at a given localization, whereas a reduction is driven further apart in order to compensate charges. Figure 6.1 describes the mechanism of corrosion of iron in the context of BPE. A piece of iron immersed in an aqueous solution behaves as a BE. On the left part, the localized anodic dissolution of iron releases ferrous cations, which precipitate in the presence of oxygen to form iron (III) oxide (i.e., Fe_2O_3). This oxidation is coupled with the reduction of oxygen on the right part, which behaves as a cathodic pole.

In fact, the concept of BPE has been explored experimentally for decades but remained confined to rather specific applications such as electrolysis, corrosion or batteries [1–3]. More recently, BPE has attracted a renewed interest in broader areas, especially related to analytical chemistry and materials science. The academic

Laurent Bouffier, Neso Sojic, Alexander Kuhn, ISM, Univ. Bordeaux, CNRS, Bordeaux INP, Talence, France

https://doi.org/10.1515/9783110570526-006

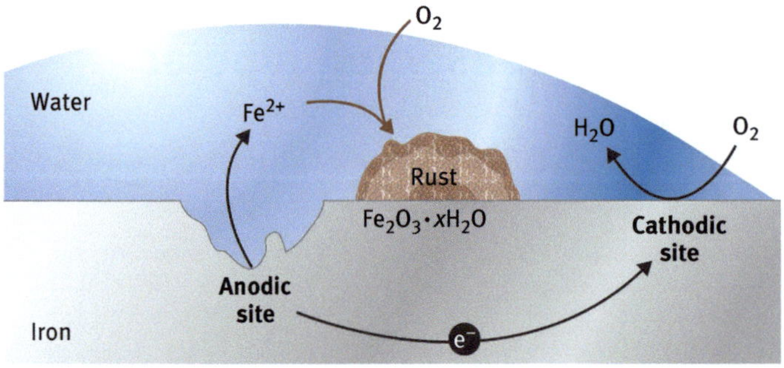

Fig. 6.1: Illustration of the formation of rust during the corrosion of iron in the frame of bipolar electrochemistry. The production of ferrous cations by oxidation of iron is coupled with the reduction of oxygen at two different locations of the same piece of conducting material. Adapted from https://chem.libretexts.org.

research in BPE has been performed primarily by several historical contributors such as the group of J.-C. Bradley at Drexel University (Philadelphia, USA) [4], J. Duval at Wageningen University (the Netherlands) [5], A. Manz at Imperial College (London, UK) [6], and R. M. Crooks at the University of Texas (Austin, USA) [7]. Following these ground-breaking contributions, there is since the last decade a real scientific emulation around BPE and this technique is now investigated by many other research teams [8–11]. Recently, several general reviews and book chapters dealing with BPE were published [12–18], as well as thematic reviews especially related to analytical BPE or surface gradients [19–23], and also a special issue of *ChemElectroChem* published in spring 2016 [24, 25]. Nowadays, BPE is truly a "hot topic" that is investigated worldwide by leading academic teams. This chapter does not intend to provide an exhaustive overview of the field but the aim is to highlight the main biorelevant applications of BPE in a tutorial fashion.

6.1.2 Principle of BPE

In conventional electrochemistry, the potential applied to WE is directly controlled by the potentiostat. On the contrary, in BPE, it is the potential of the solution which is controlled and the driving force is the electric field strength applied across the electrolyte solution. Let us consider the most common BPE setup which is the so-called open configuration. Two feeder electrodes are connected to the power supply and immersed in an ionically conducting solution while a piece of an electronic conductor is positioned inside the electrolyte as illustrated in Fig. 6.2A. When the electric field is applied across the ionic phase, a linear evolution of the potential (dE/dx) is established, which means that there is indeed a variation of the potential

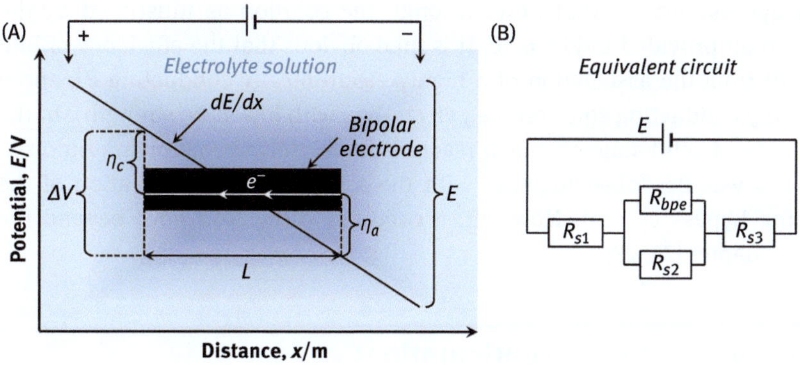

Fig. 6.2: Principle of bipolar electrochemistry. A conducting object is immersed inside an electrolyte solution and submitted to an external electric field (A). Corresponding electrical equivalent circuit (B). Adapted from Bouffier et al. [21].

value inside the solution. On the other hand, as the electronic conductor is by definition at an equipotential value, an interfacial polarization potential is established alongside the object, leading to cathodic and anodic overpotentials experienced on both sides. The cathodic pole is in fact facing the feeder anode, whereas the anodic one is in front of the feeder cathode with a maximum amplitude at both extremities of the object. When the driving force becomes sufficient, faradaic reactions are coupled across the object with a reduction reaction occurring on one side and an oxidation on the other side, justifying thus the denomination of BE. In other words, BPE can be considered as an original way for a controlled break of symmetry by an electrochemical means as the BE is indeed split gradually into two sections that undergo a different electrochemical process.

The main parameters governing BPE have been extensively discussed in the literature and the reader can be referred essentially to the series of above-cited review articles. In a first approximation, the polarization potential of the BE (ΔV) is directly proportional to the electric field strength (E/D, where E is the applied potential and D the distance between both driving electrodes) and also to the length (L) of the BE according to the following equation:

$$\Delta V = E \times \frac{L}{D} \tag{6.1}$$

It is noteworthy that ΔV is very often referred to as the fraction of E that drops over the BE. Also, a direct consequence of eq. (6.1) is that the shorter the BE, the larger is the required electric field. This is indeed an intrinsic physical limitation of BPE. Another key factor that should be considered in BPE is the convenience to use a rather resistive supporting electrolyte in order to favor BPE processes. This is related to the competition between the faradaic current flowing across the BE and the ionic current (or bypass current) that flows through the solution as illustrated by the equivalent circuit provided in Fig. 6.2B. It is then obvious that the efficiency of BPE would benefit from the association of a highly electronically conducting BE and a poorly ionically conducting solution (i.e., electrolyte with low ionic strength). In the past, the design of solid-state chemical reactors or electrolyzers involving fluidized bed electrodes was described together with the appropriate determination of the corresponding bypass current. However, a complete discussion goes beyond the scope of this chapter [26, 27].

6.1.3 Open versus closed configurations

Historically, most of the reports about BPE employed experimentally the "open-configuration" described earlier, meaning that a single compartment filled with the electrolyte is used where both driving electrodes and also the BE are immersed. It is however noteworthy that more recently another possible experimental setup was described in the framework of BPE and referred to as a "closed configuration" [28–31]. In that case, both poles of the BE are located in separate compartments, which behave like half-cells and the BE is in fact the only possible current path between them (Fig. 6.3A). Experimentally, each of the driving electrodes is immersed inside one of the two compartments bridged together by the BE. As a consequence, the corresponding equivalent circuit can be simplified as there is no more competition between the faradaic current (i.e., bipolar current) and the ionic one, avoiding thus any bypass pathway (Fig. 6.3B). This configuration equally offers the key advantage of coupling electrochemical reactions across a BE involving chemicals dissolved in two separated compartments. Consequently, this configuration allows to use

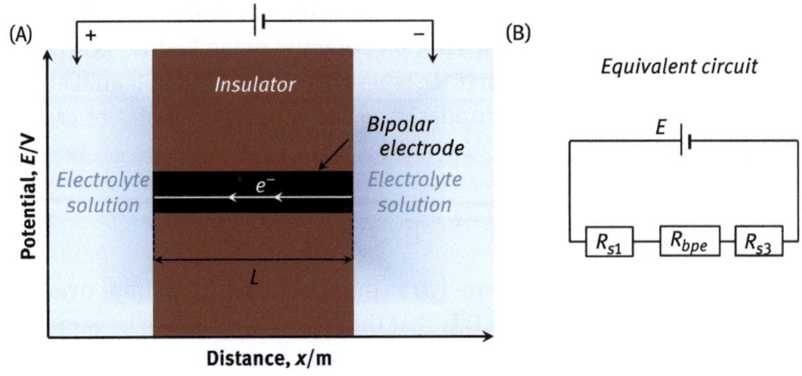

Fig. 6.3: Illustration of a closed-cell configuration (A) and corresponding equivalent circuit (B). Adapted from Bouffier et al. [21].

oxidants and reductants that are eventually unstable when mixed together to use two different (immiscible) solvents or to adjust independently the pH value in anodic versus cathodic compartments. In that context, the group of R. A. W. Dryfe (University of Manchester, UK) described the voltammetric response of a bipolar cell during electroless deposition [28]. B. Zhang et al. at the University of Washington (Seattle, USA) demonstrated that the coupling of electrochemical reactions in this "closed configuration" can readily explain the behavior of micro- and nanoscale electrodes that have been used for decades for analytical sensing.

6.1.4 Analytical strategies

Over the last years, BPE has found many new applications, especially for the fabrication of asymmetric or *Janus* particles [32, 33], for the wireless modification of surfaces [10, 34–37], for the development of microswimmers [22], or for the screening of catalytic materials [38, 39]. But among all the possible fields of application, the use of BEs is mostly investigated for analytical purposes. This may first appear surprising because the direct recording of the current which is very often measured in electroanalytical sensing is not straightforward with a BE. On the other hand, BPE offers several key advantages by comparison to classic electrochemistry. First of all, there is no need for any physical connexion between BEs and the power supply, making BPE a true wireless approach. As a direct consequence, several individual BEs can be electrochemically addressed simultaneously with a single potentiostat avoiding the necessity to use expensive multichannel apparatus. Therefore, BPE enables the simple powering of multiple BEs such as patterned conducting arrays deposited onto insulating surfaces [40], or even ensembles of conducting particles simply dispersed inside the electrolyte [41]. The second advantage of BPE arises also from its intrinsic principle as it is based on the coupling of an electrochemical reduction occurring at one pole of the BE (cathode) and an oxidation taking place at the opposite anodic side. Due to charge neutrality, the number of electrons necessary for the reduction process is strictly equal to the amount of electrons involved in the oxidation, leading to the bipolar current flowing across the BE. This is why a typical strategy is the use of one pole of the BE for electrochemical sensing, whereas the other one is employed as a reporting pole, facilitating thus a simple visual detection. If, for example, the reporting reaction is an electrochemical oxidation, then the electrons required at the cathodic pole where a given molecule dissolved in solution can be readily reduced travel across the BE from the anodic side. The strength of the reporting reaction does correlate with the amount of electroactive target present in solution. On the other hand, when the same electric field is applied without the target analyte, the electrochemical reactions cannot be coupled anymore. In that case, the bipolar current remains nil and the reporting reaction does not take place. In the early and late 2000s, the two principal reporting strategies were proposed by Manz et al. and Crooks et al., respectively [6, 7, 42]. Both strategies are based on a reporting oxidation process that offers a very straightforward readout in order to monitor a given reduction reaction. The first one which is by far the most popular is electrogenerated chemiluminescence or more simply electrochemiluminescence (ECL). This is a special mode of luminescence which is achieved when the excited state of a luminophore is populated according to a multistep mechanism that starts with an initial electron transfer step occurring at the electrode surface [43, 44]. ECL is known for several decades and is nowadays widely used in electroanalytical chemistry [45–48]. Here, the readout is the light emission at a given wavelength depending on the choice of the luminophore which proceeds at the anodic pole of the BE. By comparison, the other reporting reaction is based on anodic dissolution of a metal layer acting as BE [42]. Typically, the anode part of the BE is oxidized and generates the corresponding metal ions that diffuse toward the bulk of the solution. This leads to a time-dependent shortening of the length of the BE which can be simply measured with a ruler in order to provide a very convenient readout of the complementary reduction reaction occurring on the other side.

6.2 Open configuration

6.2.1 Detection by ECL

ECL generation at the anodic pole of a BE is a very simple and useful way to detect and quantify a reduction reaction taking place at the opposite side (cathode) of a bipolar cathode. This strategy was initially presented as a novel detector applied for electrokinetic chromatographic separation on a microfabricated glass device [6]. The setup involved a U-shaped floating platinum electrode where ECL is generated by the electric field available in the separation channel during electrophoretic separation. This was applied to the separation and detection of ECL-active luminophores and also for the detection of several amino acids, namely alanine, aspartic acid and valine.

The following year, a more general report on electrochemical sensing in microfluidic systems using ECL as a photonic reporter was published [7]. The authors rationalized the indirect detection strategy where the electrochemical sensing reaction does not participate in the ECL emission but involves the same amount of electrons. They also investigated the key parameters enabling the ECL reporting such as the influence of the driving voltage on ECL intensity and the role of the size of the anodic and cathodic poles, respectively. This versatile analytical platform was employed to design a wireless electrochemical DNA microarray sensor [49]. The gold BE is locally modified with a DNA probe (i.e., single strand). The hybridization occurs with a DNA target conjugated with a Pt nanoparticle. Thus, the coupling between anodic ECL (using $Ru(bpy)_3^{2+}$ and tri-*n*-propylamine (TPrA)) and oxygen reduction is directly conditioned by DNA recognition (Fig. 6.4A). One of the key features of BPE is to control the sensor solely with the two feeder electrodes regardless of the number of individual sensing electrodes. This allows designing (large)-scale bipolar microarrays that gather an ensemble of parallel BEs (Fig. 6.4B). The fabrication of a high density array with up to 2,000 sensing elements per cm^2 that can be addressed simultaneously was achieved [40]. It is also noteworthy that electrochemical logic gates and integrated circuits with an ECL output can be operated by means of BPE [50].

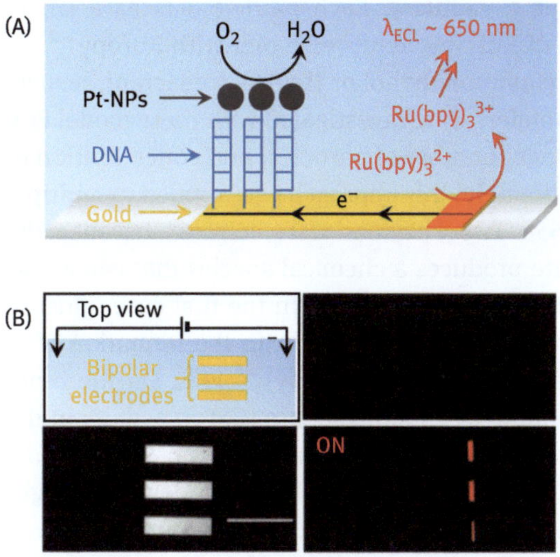

Fig. 6.4: Typical sensing strategy where a single-stranded DNA probe is immobilized for the specific recognition of a complementary DNA target labeled with a nanoparticle. Oxygen reduction at the cathodic pole (sensing) is coupled with ECL at the anodic pole (reporting) enabling a simple visual readout. Adapted from Chow et al. [49].

The use of a U-shaped bipolar microelectrode embedded inside a fluidic channel is a very practical experimental setup that is very often chosen. In such a configuration, the anodic pole of a split BE is modified for instance with an antisense DNA used as recognition element and labeled with a Ru-based ECL-active particle. In presence of the target DNA, the hybridization state controls the distance between the electrode surface and the particle, modulating therefore the ECL intensity [51]. This approach was successfully applied for the detection of intracellular nucleic acid targets in tumor cells demonstrating thus the practical application to point-of-care diagnostic. A comparable biosensing platform was proposed for the detection of a specific protein expressed at the surface of cell lines. In this case, the authors use a recognition aptamer labeled with an ECL quencher that is immobilized on the BE surface through hybridization. The presence of MCF cancer cells displaces the quencher away from the surface and restore the ECL signal [52].

Again, one of the key features of BPE remains the simultaneous addressing of several BEs regardless of their number. This advantage was first used with the design of thin metal layer BE arrays prepared by microchip technology combined with microfluidics [53]. Another possibility is to employ an ensemble of conducting particles dispersed inside the electrolyte as discrete BEs. Such a bulk approach allows a wireless addressing in the whole electrolytic volume (i.e., 3D) instead of a 2D interface [41]. The proof of principle was established by using carbon microbeads or carbon nanotubes dispersed inside a capillary. Each particle acts as a single emitter and the collective generation of ECL was achieved either with $Ru(bpy)_3^{2+}$ or luminol as luminophore and 2-(dibutylamino)ethanol or H_2O_2 as coreactant, respectively [54]. The key experimental parameters were investigated with these model ECL systems, particularly the applied voltage, number of particles, and concentration of both luminophore and coreactant. Ultimately, this approach was applied to multiple enzymatic detection by bulk ECL [55]. Two enzymes were selected because the reaction with their respective substrate produces a chemical species that can act as ECL coreactant for the emission of $Ru(bpy)_3^{2+}$ or luminol. In the first case, glucose consumption by glucose dehydrogenase (GDH) is coupled with the formation of β-nicotinamide adenine dinucleotide, promoting thus $Ru(bpy)_3^{2+}$ ECL. On the other hand, the enzymatic production of betaine by choline oxidase (ChOx) activity in the presence of oxygen generates H_2O_2, enabling thus luminol ECL (Fig. 6.5A, B). A typical result is given in Fig. 6.5C, where the image collected under white light prior to applying the electric field is compared to the ECL recorded in the dark in the case of ChOx enzyme. Also, both ECL reactions could be analyzed simultaneously with a spatially separated quantitative determination of glucose and choline over a wide concentration range.

In most of the reported BPE-based assays, the BE is fixed and/or cannot be moved inside the experimental setup. However, the motion of a freestanding conducting particle can also be driven by the electric field [22, 56]. A spherical carbon bead is positioned inside a capillary of about the same diameter. The mechanism involves the production of gas bubbles at the bottom of the BE where they accumulate and push the bead upward. Typically, H_2 evolution was selected to take place at the cathodic pole, whereas either water or a sacrificial species is oxidized at the anode. On the other hand, ECL could equally be promoted at the anode as demonstrated with $Ru(bpy)_3^{2+}$/TPrA model system for the design of a light-emitting vertical swimmer [57]. However, such an approach could be used for dynamic sensing because the ECL reaction could be driven by an enzymatic reaction allowing to quantify a given substrate with a concentration-dependent light emission. This was exemplified in a capillary format filled with glucose, GDH and also the luminophore. It was also possible to eventually observe a vertical glucose concentration gradient because

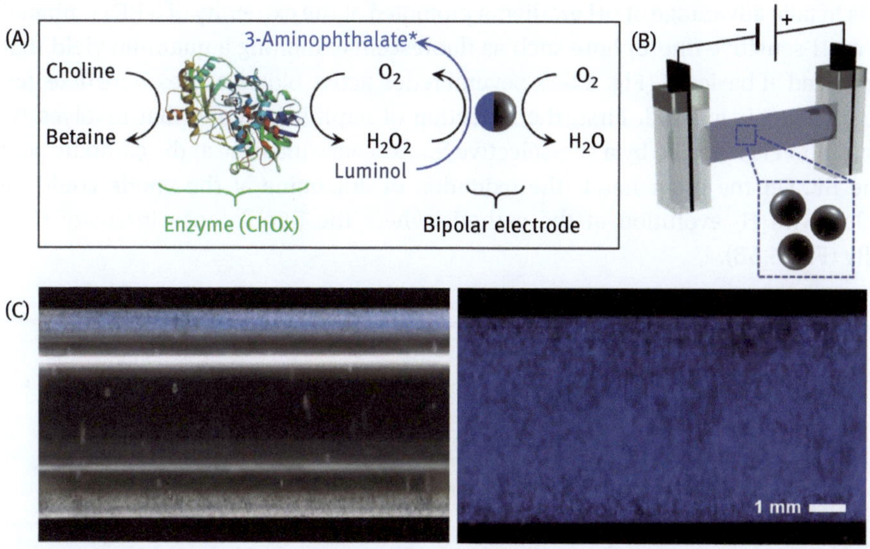

Fig. 6.5: Cascade reaction combining the enzymatic reaction where the substrate is converted by choline oxidase with concomitant production of H_2O_2 (A). The latter acts as a coreactant for luminol ECL. Scheme of the experimental setup (B). The capillary is filled with a suspension of carbon beads (Ø ~ 25 μm) as well as luminol and the enzymatic system. The microbeads are polarized by the electric field generated between the feeder electrodes, enabling anodic and cathodic reactions at the opposite poles. Images collected under white light before applying the electric field and ECL recorded in the dark (C). Adapted from de Poulpiquet et al. [55].

the moving BE could explore the entire capillary and probe the substrate concentration with spatial resolution [58].

6.2.2 Other reporting strategies

Even if ECL is by far the most employed reporting reaction, it is noteworthy that other approaches were also investigated from an academic point of view. For example, the electrodissolution of the cathodic side of a BE made out of silver was proposed as a visual readout [42]. Typically, the cathodic side of the BE was modified with a DNA probe that is hybridized with a complementary strand labeled with biotin. A bottom-up assay enables the recognition with avidin and subsequent immobilization of horseradish peroxidase. In such a configuration the electroenzymatic reduction of H_2O_2 is therefore coupled across the BE with Ag dissolution. A simple measurement of the length of the BE prior and after applying the driving voltage allows the transduction of DNA hybridization, thanks to the shortening of the BE. Another reporting strategy based on local fluorescence modulation was equally proposed [59]. Here, the idea is to take advantage of pH gradients promoted at the extremity of a BE combined with a pH-sensitive fluorophore such as fluorescein exhibiting a quantum yield that is enhanced at basic pH (Fig. 6.6A). Several redox active biomolecules were detected by applying this method. First, the reduction of naphthoquinone that involves H^+ consumption is revealed by a site-selective fluorescence increase at the cathodic pole of the BE. On the other hand, the oxidation of dopamine at the anode could be coupled with H_2 evolution at the cathode where the fluorescence intensity rises locally (Fig. 6.6B).

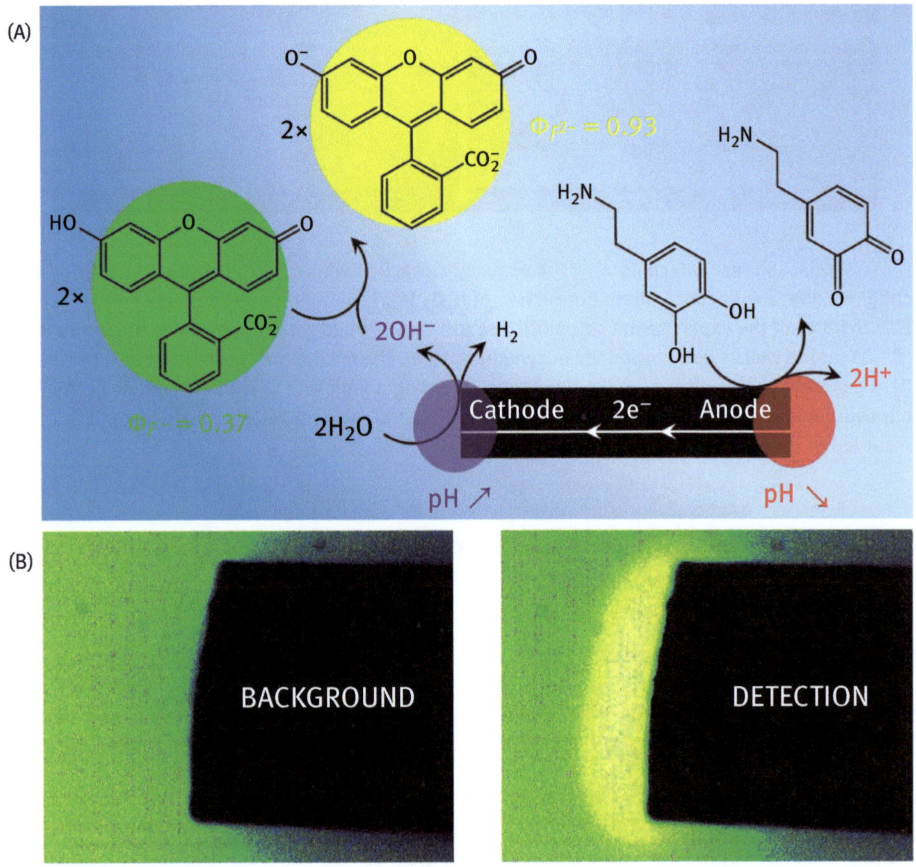

Fig. 6.6: Mechanism enabling the detection of dopamine by localized pH-triggered fluorescence enhancement (A). The electrochemical oxidation of dopamine is coupled across the bipolar electrode with water reduction, whereas fluorescein is used as a pH-sensitive reporter. Fluorescence microscopy revealing the detection of 1 mM dopamine in the presence of 10 μM fluorescein at the vicinity of the cathodic pole of the bipolar electrode. Adapted from Bouffier et al. [59].

A completely different additional readout mechanism has been proposed recently based on the electromechanical bending of a conducting polymer triggered by BPE [60, 61]. A freestanding strip of a polypyrrole membrane in its oxidized form is addressed by the electric field. On its anodic extremity, an analyte molecule such as glucose is oxidized, and this reaction is coupled with the reduction of the conducting polymer at the opposite side. The latter one is accompanied by the uptake of cations inside the polymer matrix, resulting in a partial swelling with the final consequence that the strip bends. The degree of bipolar bending could be quantitatively correlated with the concentration of analytes.

6.2.3 Interplay with electrophoretic properties

BPE is also a great analytical tool for the concentration and separation of analytes. This phenomenon is studied in detail and is referred to as BPE focusing. It takes place when employing narrow capillaries exhibiting intrinsic electrophoretic properties. BPE focusing is based on the direct influence of the BE on the spatial distribution of the electric field [62]. The driving force of this focusing is due to the competition between the electrophoretic flow and electroosmotic flow, which are indeed antagonist forces. BPE focusing was demonstrated with a model fluorescent dye whose concentration appears to be enhanced locally at the edge of the BE embedded inside a microchannel. The concentration gradient as well as the position of the concentration zone are time dependent as revealed by fluorescence microscopy. If several organic dyes exhibiting a different electrophoretic mobility are mixed in solution, each one is focused at a different location inside the microchannel, leading thus to spatiotemporal concentration with high enrichment factor (up to 500,000 fold) [63]. The theoretical framework was clearly established with the identification of key parameters that control the concentration enrichment. Also the design and engineering of a BE array instead of a single microband BE enables the stepwise relocation of the concentration gradient from one position to another [64]. This was later applied to ion depletion based on enhanced water electrolysis due to electric field focusing that occur at both edges of the BE. From a technological point of view, it can be considered as a membrane-less filtration procedure for the separation of charged molecules from a solution containing also neutral species that remain unaffected [65]. The initial experiments on BPE focusing were performed with anionic species before being extended to the concentration enrichment and separation of cations [66]. The simultaneous addressing of both anions and cations was finally achieved by controlling the interplay between enrichment and depletion zones inside a dual-channel configuration [67]. Even if most reports used model molecules and dyes, the principle is fully transposable to small biorelevant molecules as well as charged biological macromolecules.

More recently, the capture of circulating tumor cells by dielectrophoresis was achieved by applying a high-frequency alternating current (AC) field across an electrochemical cell comprising an array of BEs. Several designs were proposed and modeled (Fig. 6.7A) with the ability to capture cell lines depending on their dielectric properties. In particular, breast cancer cells (MDA-MB-231) can be discriminated from white blood cells (Jurkat T cells) when applying a 40 kHz AC electric field (Fig. 6.7B, C) [68]. Such devices enable the capture of rare cells with the possibility of single-cell analysis.

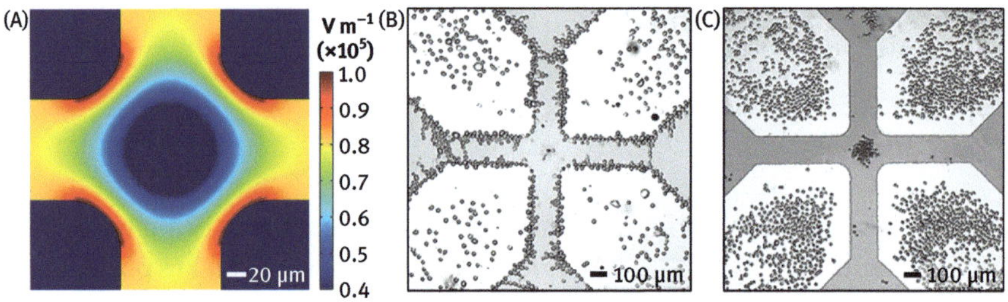

Fig. 6.7: Simulation of the electric field strength at the center of a quadrupole electrode (A). The maximum field is located at the electrode edges, whereas a local minimum is found at the center region. Cancer cells are captured at the bipolar electrode tips (B) whereas white blood cells retained in fluid flow (C). Adapted from Li et al. [68].

6.3 Closed configuration

The closed-cell configuration is very different from the open configuration. As already mentioned, the reason is simply because the piece of conducting material used as BE offers the only pathway between the two separated compartments where the feeder anode and cathode are placed, respectively. Historically, most of the reports in bipolar electroanalytical chemistry used a single electrochemical cell configuration in which the BE is simply immersed in the electrolyte solution or located inside an open microchannel (i.e., open configuration). However, it was reported in early 2012 that carbon fiber microelectrodes backfilled with an electrolyte just behave as closed BEs even if it was not previously described as such [30]. The voltammetric response was studied in detail within the framework of BPE. The coupling between an oxidation reaction triggered in the analytical solution and the reduction of soluble oxygen occurring in the fiber inner compartment was proposed to be responsible for the conductivity. This is indeed a major understanding that allows to achieve quantitative measurement when considering the intrinsic behavior of both poles of the BE. The theory was rationalized and confronted with the corresponding experimental results revealing that the voltammetric response is likely distorted and often slower than that recorded in a conventional two-electrode setup. Indeed, the shape of the response strongly depends on the ratio between the limiting current at each pole [31]. This has a major consequence when using ECL reporting in a closed configuration which is very often employed because the surface areas of the sensing versus reporting poles strongly influence the collected ECL intensity [69].

The closed configuration was implemented to the previously investigated microchannel approach simply by positioning the screen-printed microband electrode across two independent fluidic channels [70]. The analytical possibilities offered by such an approach are significantly extended because the sensing and reporting channels are separated in space (Fig. 6.8A). This is indeed a key advantage when dealing with instability issues between redox active species that may not be thermodynamically stable when mixed together. For example, when the sensing microchannel is filled with a solution containing a given target molecule ($Fe(CN)_6^{3-}$), the ECL intensity recorded in the reporting channel containing a typical ($Ru(bpy)_3^{2+}$/TPrA) mix correlates linearly with the analyte concentration (Fig. 6.8B). Sensing application of a biorelevant target was established for the detection of HbA_{1c}, which is an important blood glucose level marker. For that, a gold electrode positioned inside the sensing

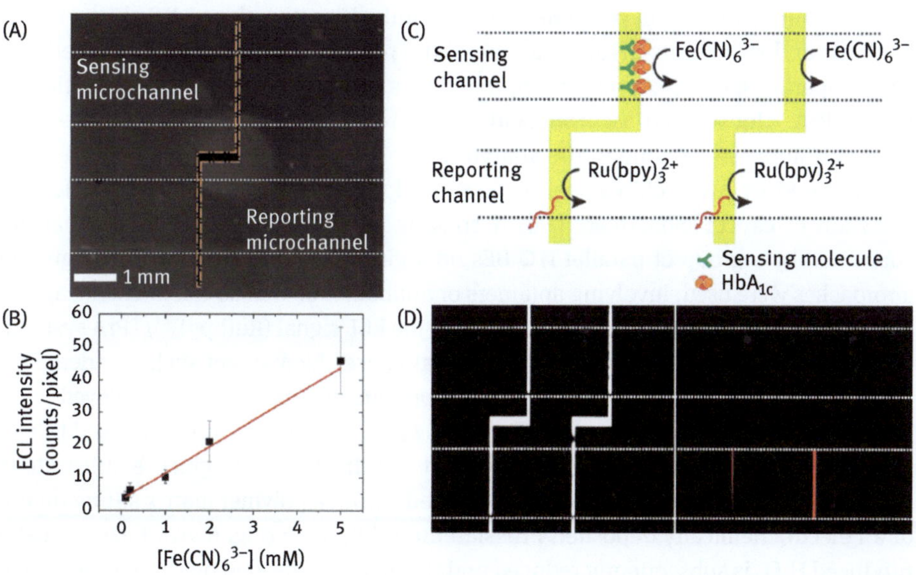

Fig. 6.8: An optical image of a closed bipolar electrode connecting two parallel sensing and reporting channels (A). A calibration curve of the electrochemiluminescence intensity recorded in the reporting channel as a function of the analyte concentration inside the sensing channel (B). A scheme illustrating the detection of a marker for blood glucose level (HbA_{1c}) based on an extinction of light emission due to an impedance increase (C) and corresponding micrographs (D). Adapted from Chang et al. [70].

channel is modified by a specific capture probe for HbA_{1c} recognition. Therefore, the detection of HbA_{1c} is monitored as a decrease in ECL emission intensity on the reporting pole because of the increase in impedance and thus lowering of electron transfer capability (Fig. 6.8C, D). A recent study in the same configuration also suggested that besides the electrochemical reactions promoted at the poles of the BE, the reaction occurring at the feeder electrodes may also be taken into account [71]. Therefore, the solution composition filling each compartment and the size of both bipolar poles and feeder electrodes may affect the overall coupling.

A comparable strategy was later adopted with an ITO-based BE interconnecting two independent reservoirs. Again, ECL was chosen as a readout signal and it was found that the location of the feeder electrode far away from the BE is crucial in order to eliminate completely the ECL background generated also at the vicinity of the driving electrodes [72]. By using $Ru(bpy)_3^{2+}$ and TPrA in the reporting channel, several analytical strategies were proposed. The detection of the luminophore or the ECL coreactant is obvious but the presence of dopamine could equally be quantified because its oxidation product quenches ECL. H_2O_2 and $K_3Fe(CN)_6$ were also detected as model targets because the presence of redox-active species inside the sensing channel does affect the coupling between both poles of the BE. An implementation of the two-channel design was proposed with a multichannel configuration exhibiting three independent channels linked by two closed BEs [73]. All the oxidants and reductants involved at one of the poles of the BEs can be detected in a single operating device. Such an approach was illustrated by the determination of H_2O_2, TPrA, glucose and ascorbic acid by recording the evolution of ECL intensity upon the stepwise addition of the analytes.

A closed BE array chip was also developed as an ECL imaging platform for the detection of cancer biomarkers [74]. It consists of two separated channels that are connected by a group of parallel ITO BEs on a glass substrate. Several biorecognition approaches were used, involving aptamers or antibodies as well as electrochemical tags as signal amplification indicators that mediate the ECL signal ($Ru(bpy)_3^{2+}$/TPrA system). As a result, the multiplex detection of several cancer biomarkers such as adenosine triphosphate, prostate-specific antigen, α-fetoprotein and thrombin was achieved.

Besides the classic ($Ru(bpy)_3^{2+}$/TPrA) ECL system, the use of luminol-based ECL in a closed-cell configuration was also reported [75]. In this work, glucose substrate is oxidized by glucose oxidase that is immobilized within a polymer matrix sitting on top of an electrochemically deposited Prussian blue (PB) layer (Fig. 6.9A). Enzymatically produced H_2O_2 is subsequently reduced and the electrons needed on this cathodic side are shuttled from the anodic pole where the oxidation of luminol/H_2O_2 results in a proportional ECL emission. Such a combination where the same reactant, namely, H_2O_2 is involved on both sides of the BE is not usable in the open configuration. Therefore, employing the closed configuration is crucial. Experimentally, a carbon rod is used as BE to link the two compartments and it is noteworthy that in addition a salt bridge was also positioned between both sides (Fig. 6.9B). The H_2O_2 calibration was

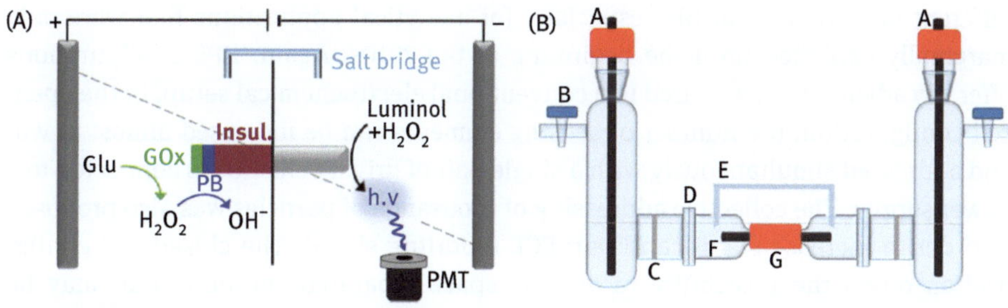

Fig. 6.9: Illustration of the coupling between glucose detection and luminol-based electrochemiluminescence across a closed bipolar electrode (A). Scheme of the experimental setup: A, feeder electrode; B, valve; C, frit; D, flange; E, salt bridge; F, bipolar electrode; G, double screw-cap with O-rings. Adapted from Eßmann et al. [75].

adjusted in order to optimize the detection of glucose that was achieved in the millimolar range up to 200 mM.

More recently, the coupling between two oxygenic enzymes across a closed bipolar cell was also proposed [76]. The authors selected photosystems 1 and 2 (namely, PS1 and PS2) and immobilized both of them in one of the compartments of a photoelectrochemical half-cell, respectively. The driving force for the electron transfer reactions is applied by BPE, whereas the actual potential of each pole of the BE being self-regulated by the redox processes. The principle parameters of the coupling are discussed in terms of limiting poles. Also, a complementary scanning electrochemical probe was employed to monitor the time-dependent changes in enzymatic performances due to processes of inactivation of both protein complexes.

Apart from ECL reporting, it is noteworthy that an electrochromic sensing platform using PB as the colored indicator was also proposed [77]. Such a bipolar electrochromic strategy expands the scope of the closed configuration. The principle is based on the modulation of heterogeneous electron transfer using $Fe(CN)_6^{3-/4-}$ redox probes that are affected by steric hindrance effects. It was applied to the signaling of carcinoembryonic antigen (CEA) which is a 180 kDa glycoprotein that is an important tumor marker with a typical concentration of 5 ng mL^{-1} in normal human serum. Thus, a higher or lower CEA concentration can be directly observed by the naked eye based on the color change of the BPE device.

6.4 Conclusion and perspectives

Despite the fact that BPE has been known for several decades, it remains nowadays a very active area of research. The reason for its recent revival is obviously due to the different potentialities of BPE especially for analytical applications that were only marginally exploited until the beginning of the 2010s. Again, BPE configurations offer key advantages compared to a conventional electrochemical setup. In the open-cell configuration, the number of sensing elements can be increased almost at will and addressed simultaneously with a single pair of driving electrodes connected to a power supply. The collective addressing of thousands of particles was also proposed in order to increase significantly an ECL reporting signal. The closed-cell configuration offers the possibility to employ space-separated solutions that may be chemically incompatible or to avoid interferences between chemicals. Also the combination with microfluidic technology in channels and capillaries could be proposed to enable in-flow sensing based on BPE.

We are still at a time where fundamental studies remain necessary to fully address the advantages and drawbacks, but we anticipate that in a near future, BPE will be increasingly chosen as a versatile and adaptable platform for biosensing. A considerable amount of electroanalytical strategies that have been used for years in conventional electrochemistry are potentially transposable to BPE. Among all the possible approaches, the coupling of BPE with ECL appears to be one of the most attractive and promising strategies. The detection of many different biomolecules such as DNA, aptamers, proteins, and antibodies has already been reported. Now, the evaluation of real blood samples which are more complex than model analytical solutions should be carried out prior to the development of point-of-care devices for practical clinical diagnostic.

In future, enzymatic coupling across a bipolar cell that has just been reported or eventually the study of entire living systems such as bacteria or cells may also be possible by using BPE.

References

[1] Backhurst, J. R., Coulson, J. M., Goodridge, F., Plimley, R. E., Fleischmann, M. A preliminary investigation of fluidized bed electrodes. J Electrochem Soc. 1969, 116(11), 1600–1607.

[2] Eardley, D. C., Handley, D., Andrew, S. P. S. Bipolar electrolysis with intra phase conduction in two phase media. Electrochim Acta. 1973, 18(11), 839–848.

[3] Fleischmann, M., Ghoroghchian, J., Rolison, D., Pons, S. Electrochemical behavior of dispersions of spherical ultramicroelectrodes. J Phys Chem. 1986, 90(23), 6392.

[4] Bradley, J.-C., Chen, H.-M., Crawford, J., Eckert, J., Ernazarova, K., Kurzeja, T., Lin, M. McGee, M., Nadler, W. Stephens, S. G. Creating electrical contacts between metal particles using directed electrochemical growth. Nature. 1997, 389(6648), 268–271.

[5] Duval, J. Kleijn, J. M. van Leeuwen, H. P. Bipolar electrode behaviour of the aluminium surface in a lateral electric field. J Electroanal Chem. 2001, 505(1), 1–11.

[6] Arora, A., Eijkel, J. C. T., Morf, W. E., Manz, A. A wireless electrochemiluminescence detector applied to direct and indirect detection for electrophoresis on a microfabricated glass device. Anal Chem. 2001, 73(14), 3282–3288.

[7] Zhan, W., Alvarez, J., Crooks, R. M. Electrochemical sensing in microfluidic systems using electrogenerated chemiluminescence as a photonic reporter of redox reactions. J Am Chem Soc. 2002, 124(44), 13265–13270.
[8] Ulrich, C., Andersson, O., Nyholm, L., Björefors, F. Formation of molecular gradients on bipolar electrodes. Angew Chem Int Ed. 2008, 47(16), 3034–3036.
[9] Warakulwit, C., Nguyen, T., Majimel, J., Delville, M.-H., Lapeyre, V., Garrigue, P., Ravaine, V., Limtrakul, J., Kuhn, A. Dissymmetric carbon nanotubes by bipolar electrochemistry. Nano Lett. 2008, 8(2), 500.
[10] Inagi, S., Ishiguro, Y., Atobe, M., Fuchigami, T., Bipolar patterning of conducting polymers by electrochemical doping and reaction. Angew Chem Int Ed. 2010, 49(52), 10136–10139.
[11] Ramakrishnan, S., Shannon, C. Display of solid-state materials using bipolar electrochemistry. Langmuir. 2010, 26(7), 4602–4606.
[12] Loget, G., Kuhn, A. Shaping and exploring the micro- and nanoworld using bipolar electrochemistry. Anal Bioanal Chem. 2011, 400(6), 1691–1704.
[13] Loget, G., Kuhn, A. Bipolar electrochemistry in the nanosciences. In Electrochemistry: Nanosystems Electrochemistry. Royal Soc Chem. 2012, Vol(11), pp 71–103.
[14] Fosdick, S. E., Knust, K. N., Scida, K., Crooks, R. M. Bipolar Electrochemistry. Angew Chem Int Ed. 2013, 52(40), 10438–10456.
[15] Loget, G., Zigah, D., Bouffier, L., Sojic, N., Kuhn, A. Bipolar electrochemistry: from materials science to motion and beyond. Acc Chem Res. 2013, 46(11), 2513–2523.
[16] Sequeira, C. A. C., Cardoso, D. S. P., Gameiro, M. L. F. Bipolar electrochemistry, a focal point of future research. Chem Eng Commun. 2016, 203(8), 1001–1008.
[17] Koefoed, L., Pedersen, S. U., Daasbjerg, K. Bipolar electrochemistry – A wireless approach for electrode reactions. Curr Opin Electrochem. 2017, 2(1), 13–17.
[18] Bouffier, L., Zigah, D., Sojic, N., Kuhn, A. Recent advances in bipolar electrochemistry. Taylor & Francis. 2017, Vol. 27, pp 27–118.
[19] Mavré, F., Anand, R. K., Laws, D. R., Chow, K.-F., Chang, B.-Y., Crooks, J. A., Crooks, R. M. Bipolar electrodes: a useful tool for concentration, separation, and detection of analytes in microelectrochemical systems. Anal Chem. 2010, 82(21), 8766–8774.
[20] Inagi, S. Fabrication of gradient polymer surfaces using bipolar electrochemistry. J Polym. 2016, 48(1), 39–44.
[21] Bouffier, L., Arbault, S., Kuhn, A., Sojic, N. Generation of electrochemiluminescence at bipolar electrodes: concepts and applications. Anal Bioanal Chem. 2016, 408, 7003–7011.
[22] Bouffier, L., Ravaine, V., Sojic, N., Kuhn, A. Electric fields for generating unconventional motion of small objects. Curr Opin Colloid Interface Sci. 2016, 21, 57–64.
[23] Zhang, X., Zhai, Q., Xing, H., Li, J., Wang, E. Bipolar electrodes with 100% current efficiency for sensors. ACS Sensors. 2017, 2, 320–326.
[24] Kuhn, A., Crooks, R. M., Inagi, S. A compelling case for bipolar electrochemistry. ChemElectroChem. 2016, 3(3), 351–352.
[25] Crooks, R. M. Principles of Bipolar Electrochemistry. ChemElectroChem. 2016, 3(3), 357–359.
[26] Kusakabe, K., Morooka, S., Kato, Y. Current paths and electrolysis efficiency in bipolar packed-bed electrodes. J Chem Eng Jpn. 1982, 15(1), 45–50.
[27] Wodiunig, S., Bokeloh, F., Comninellis, C. Electrochemical promotion of bipolar electrodes: an estimation of the current bypass. Electrochimica Acta. 2000, 46(2), 357–363.
[28] Plana, D., Shul, G., Stephenson, M. J., Dryfe, R. A. W. The voltammetric response of bipolar cells: Mechanistic investigations of electroless deposition. Electrochem Commun. 2009, 11(1), 61–64.
[29] Plana, D., Jones, F. G. E., Dryfe, R. A. W. The voltammetric response of bipolar cells: Reversible electron transfer. J Electroanal Chem. 2010, 646(1–2), 107–113.

[30] Guerrette, J. P., Oja, S. M., Zhang, B. Coupled electrochemical reactions at bipolar microelectrodes and nanoelectrodes. Anal Chem. 2012, 84(3), 1609–1616.

[31] Cox, J. T., Guerrette, J. P., Zhang, B. Steady-state voltammetry of a microelectrode in a closed bipolar cell. Anal Chem. 2012, 84, 8797–8804.

[32] Loget, G., Roche, J., Kuhn, A. True bulk synthesis of janus objects by bipolar electrochemistry. Adv Mater. 2012, 24(37), 5111–5116.

[33] Loget, G., Roche, J., Gianessi, E., Bouffier, L., Kuhn, A. Indirect bipolar electrodeposition. J Am Chem Soc. 2012, 134(49), 20033–20036.

[34] Dorri, N., Shahbazi, P., Kiani, A. Self-movement of water droplet at the gradient nanostructure of Cu fabricated using bipolar electrochemistry. Langmuir. 2014, 30(5), 1376–1382.

[35] Tisserant, G., Gillion, J., Lannelongue, J., Fattah, Z., Garrigue, P., Roche, J., Zigah, D., Kuhn, A., Bouffier, L. Single-step screening of the potential dependence of metal layer morphologies along bipolar electrodes. ChemElectroChem. 2016, 3(3), 387–391.

[36] Kayran Yasin, U., Eßmann, V., Grützke, S., Schuhmann, W. Selection of highly SERS-active nanostructures from a size gradient of Au nanovoids on a single bipolar electrode. ChemElectroChem. 2016, 3(3), 399–403.

[37] Bouffier, L., Reculusa, S., Ravaine, V., Kuhn, A. Modulation of wetting gradients by tuning the interplay between surface structuration and anisotropic molecular layers with bipolar electrochemistry. ChemPhysChem. 2017, 18(19), 2637–2642.

[38] Fosdick, S. E., Crooks, R. M. Bipolar electrodes for rapid screening of electrocatalysts. J Am Chem Soc. 2012, 134(2), 863–866.

[39] Tan, S. M., Pumera, M. Composition-graded MoWSx hybrids with tailored catalytic activity by bipolar electrochemistry. ACS Appl Mater Interfaces. 2017, 9(48), 41955–41964.

[40] Chow, K.-F., Mavré, F., Crooks, J. A., Chang, B.-Y., Crooks, R. M. A Large-Scale, Wireless electrochemical bipolar electrode microarray. J Am Chem Soc. 2009, 131(24), 8364–8365.

[41] Sentic, M., Arbault, S., Bouffier, L., Manojlovic, D., Kuhn, A., Sojic, N. 3D electrogenerated chemiluminescence: from surface-confined reactions to bulk emission. Chem Sci. 2015, 6, 4433–4437.

[42] Chow, K.-F., Chang, B.-Y., Zaccheo, B. A., Mavré, F., Crooks, R. M. A sensing platform based on electrodissolution of a Ag bipolar electrode. J Am Chem Soc. 2010, 132(27), 9228–9229.

[43] Visco, R. E., Chandross, E. A. Electroluminescence in solutions of aromatic hydrocarbons. J Am Chem Soc. 1964, 86(23), 5350–5351.

[44] Santhanam, K. S. V., Bard, A. J. Chemiluminescence of electrogenerated 9,10-diphenylanthracene anion radical. J Am Chem Soc. 1965, 87(1), 139–140.

[45] Richter, M. M. Electrochemiluminescence (ECL). Chem Rev. 2004, 104(6), 3003–3036.

[46] Miao, W. Electrogenerated chemiluminescence and its biorelated applications. Chem Rev. 2008, 108(7), 2506–2553.

[47] Hu, L., Xu, G. Applications and trends in electrochemiluminescence. Chem Soc Rev. 2010, 39(8), 3275–3304.

[48] Liu, Z., Qi, W., Xu, G. Recent advances in electrochemiluminescence. Chem Soc Rev. 2015, 44(10), 3117–3142.

[49] Chow, K.-F., Mavré, F., Crooks, R. M. Wireless electrochemical DNA microarray sensor. J Am Chem Soc. 2008, 130(24), 7544–7545.

[50] Chang, B.-Y., Crooks, J. A., Chow, K.-F., Mavré, F., Crooks, R. M. Design and operation of microelectrochemical gates and integrated circuits. J Am Chem Soc. 2010, 132(43), 15404–15409.

[51] Wu, M.-S., Qian, G.-s., Xu, J.-J., Chen, H.-Y. Sensitive electrochemiluminescence detection of c-Myc mRNA in breast cancer cells on a wireless bipolar electrode. Anal Chem. 2012, 84(12), 5407–5414.

[52] Wu, M.-S., Yuan, D.-J., Xu, J.-J., Chen, H.-Y. Sensitive electrochemiluminescence biosensor based on Au-ITO hybrid bipolar electrode amplification system for cell surface protein detection. Anal Chem. 2013, 85(24), 11960–11965.
[53] Bouffier, L., Sojic, N., Kuhn, A. Capillary-assisted bipolar electrochemistry: a focused mini review. Electrophoresis. 2017, 38(21), 2687–2694.
[54] dePoulpiquet, A., Diez-Buitrago, B., Milutinovic, M., Goudeau, B., Bouffier, L., Arbault, S., Kuhn, A., Sojic, N. Dual-color electrogenerated chemiluminescence from dispersions of conductive microbeads addressed by bipolar electrochemistry. ChemElectroChem. 2016, 3(3), 404–409.
[55] de Poulpiquet, A., Diez-Buitrago, B., Dumont Milutinovic, M., Sentic, M., Arbault, S., Bouffier, L., Kuhn, A., Sojic, N. Dual enzymatic detection by bulk electrogenerated chemiluminescence. Anal Chem. 2016, 88(12), 6585–6592.
[56] Loget, G., Kuhn, A. Electric field-induced chemical locomotion of conducting objects. Nat Commun. 2011, 2, 535.
[57] Sentic, M., Loget, G., Manojlovic, D., Kuhn, A., Sojic, N. Light-emitting electrochemical "Swimmers". Angew Chem Int Ed. 2012, 51(45), 11284–11288.
[58] Sentic, M., Arbault, S., Goudeau, B., Manojlovic, D., Kuhn, A., Bouffier, L., Sojic, N. Electrochemiluminescent swimmers for dynamic enzymatic sensing. Chem Commun. 2014, 50(71), 10202–10205.
[59] Bouffier, L., Doneux, T., Goudeau, B., Kuhn, A. Imaging redox activity at bipolar electrodes by indirect fluorescence modulation. Anal Chem. 2014, 86(8), 3708–3711.
[60] Gupta, B., Goudeau, B., Kuhn, A. Wireless electrochemical actuation of conducting polymers. Angew Chem Int Ed. 2017, 56(45), 14183–14186.
[61] Gupta, B., Goudeau, B., Garrigue, P., Kuhn, A. Bipolar conducting polymer crawlers based on triple symmetry breaking. Adv Funct Mater. 2018, 28(25), 1705825.
[62] Dhopeshwarkar, R., Hlushkou, D., Nguyen, M., Tallarek, U., Crooks, R. M. Electrokinetics in microfluidic channels containing a floating electrode. J Am Chem Soc. 2008, 130(32), 10480–10481.
[63] Anand, R. K., Sheridan, E., Knust, K. N., Crooks, R. M. Bipolar electrode focusing: faradaic ion concentration polarization. Anal Chem. 2011, 83(6), 2351–2358.
[64] Perdue, R. K., Laws, D. R., Hlushkou, D., Tallarek, U., Crooks, R. M. Bipolar electrode focusing: the effect of current and electric field on concentration enrichment. Anal Chem. 2009, 81(24), 10149–10155.
[65] Sheridan, E., Knust, K. N., Crooks, R. M. Bipolar electrode depletion: membraneless filtration of charged species using an electrogenerated electric field gradient. Analyst. 2011, 136(20), 4134–4137.
[66] Sheridan, E., Hlushkou, D., Knust, K. N., Tallarek, U. Crooks, R. M. Enrichment of cations via bipolar electrode focusing. Anal Chem. 2012, 84(17), 7393–7399.
[67] Knust, K. N., Sheridan, E., Anand, R. K., Crooks, R. M. Dual-channel bipolar electrode focusing: simultaneous separation and enrichment of both anions and cations. Lab on a Chip. 2012, 12(20), 4107–4114.
[68] Li, M., Anand, R. K. High-throughput selective capture of single circulating tumor cells by dielectrophoresis at a wireless electrode array. J Am Chem Soc. 2017, 139(26), 8950–8959.
[69] Oja, S. M., Zhang, B. Electrogenerated chemiluminescence reporting on closed bipolar microelectrodes and the influence of electrode size. ChemElectroChem. 2015, 3(3), 457–464.
[70] Chang, B.-Y., Chow, K.-F., Crooks, J. A., Mavré, F., Crooks, R. M. Two-channel microelectrochemical bipolar electrode sensor array. Analyst. 2012, 137(12), 2827–2833.
[71] Zhang, J.-D., Zhao, W.-W., Xu, J.-J., Chen, H.-Y. Electrochemical behaviors in closed bipolar system with three-electrode driving mode. J Electroanal Chem. 2016, 781, 56–61.

[72] Zhang, X., Chen, C., Li, J., Zhang, L., Wang, E. New insight into a microfluidic-based bipolar system for an electrochemiluminescence sensing platform. Anal Chem. 2013, 85(11), 5335–5339.
[73] Zhang, X., Li, J., Jia, X., Li, D., Wang, E. Full-featured electrochemiluminescence sensing platform based on the multichannel closed bipolar system. Anal Chem. 2014, 86(11), 5595–5599.
[74] Wu, M.-S., Liu, Z., Shi, H.-W., Chen, H.-Y., Xu, J.-J. Visual electrochemiluminescence detection of cancer biomarkers on a closed bipolar electrode array chip. Anal Chem. 2015, 87(1), 530–537.
[75] Eßmann, V., Jambrec, D., Kuhn, A., Schuhmann, W. Linking glucose oxidation to luminol-based electrochemiluminescence using bipolar electrochemistry. Electrochem Commun. 2015, 50, 77–80.
[76] Eßmann, V., Zhao, F., Hartmann, V., Nowaczyk, M. M., Schuhmann, W., Conzuelo, F. In operando investigation of electrical coupling of photosystem 1 and photosystem 2 by means of bipolar electrochemistry. Anal Chem. 2017, 89(13), 7160–7165.
[77] Zhai, Q., Zhang, X., Xia, Y., Li, J., Wang, E. Electrochromic sensing platform based on steric hindrance effects for CEA detection. Analyst. 2016, 141(13), 3985–3988.

Dan Shan, Wen-Rong Cai

7 Biomaterials for electrochemiluminescence

7.1 Introduction

Electrochemiluminescence (ECL), also called electrogenerated chemiluminescence, is a process whereby species generated at electrodes undergo high-energy electron-transfer reactions to form excited states that emit light. Although the phenomenon of light emission during electrolysis was observed in the late 1920s, the first detailed ECL studies were reported by Hercules [1] and Santhanam and Bard [2]. in the mid-1960s. This technique represents the marriage of electrochemistry with spectroscopy, which was applied for bioassay purposes in 1989. After many years development, ECL has evolved as a powerful analytical technique featured with rapidity, high sensitivity, and selectivity. In this chapter, we pay more attention on a detailed discussion of ECL mechanisms, luminophor of ECL, ECL enhancement strategies, and ECL for assay in biological matrices. Conjugation strategies for biomolecules are also included.

7.2 Principle of ECL

ECL is a means of converting electrical energy into radiative energy. It involves the production of reactive intermediates from stable precursors at the surface of an electrode. These intermediates then react under a variety of conditions to form excited states that emit light. In most of the ECL systems, there are generally four steps as shown in Fig. 7.1 [3], namely, (1) redox reactions at electrode, (2) homogeneous chemical reactions, (3) excited state species formation, and (4) light emission. In general, any analyte that can inhibit or promote the processing of the aforementioned steps will result in the change of ECL emission. ECL can be produced by two dominant pathways, that is, annihilation and coreactant pathways, although most of the ECL applications are almost based on the latter one.

7.2.1 Annihilation pathway

Ion annihilation involves the formation of electrochemically generated intermediate species at the electrode, which then interacts and undergoes the formation of both ground and electronically excited states, which consequently emit light by

Dan Shan, Wen-Rong Cai, School of Environmental and Biological Engineering, Nanjing University of Science and Technology, Nanjing, China

https://doi.org/10.1515/9783110570526-007

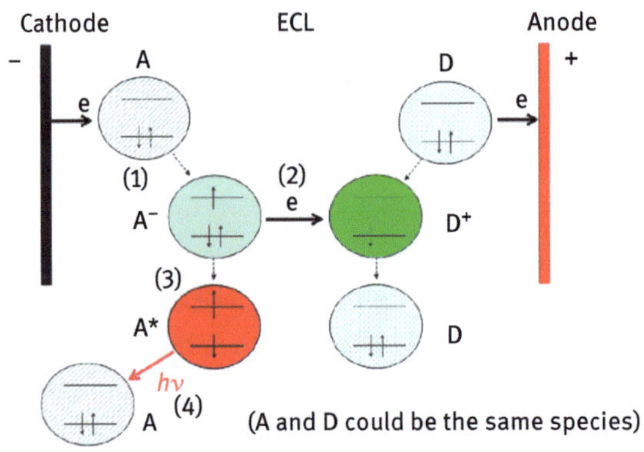

Fig. 7.1: Schematic representation of ECL process (from Miao [3]).

relaxation. A typical example of ion annihilation process is the ECL of 9,10-diphenylanthracene (DPA), which is based on the following processes [4]:

$$\mathrm{DPA} - \mathrm{e}^- \rightarrow \mathrm{DPA}\bullet^+ \text{(oxidation at the electrode)} \tag{7.1}$$

$$\mathrm{DPA} + \mathrm{e}^- \rightarrow \mathrm{DPA}\bullet^- \text{(reduction at the electrode).} \tag{7.2}$$

$$\mathrm{DPA}\bullet^- + \mathrm{DPA}\bullet^+ \rightarrow \mathrm{DPA} + \mathrm{DPA}^* \text{(annihilation process)} \tag{7.3}$$

$$\mathrm{DPA}^* \rightarrow \mathrm{DPA} + \mathrm{h}\nu(\lambda = 425\ \mathrm{nm}) \text{(emission of light)} \tag{7.4}$$

Annihilation processes may also occur in systems in which the radical cations and radical anions are from different molecules. The Gibbs free energy related to the annihilation process is calculated from the redox potentials of eqs. (7.1) and (7.2) using the following equation:

$$\Delta G = -nF(E^{\mathrm{o}}{}_{\mathrm{reduction}} - E^{\mathrm{o}}{}_{\mathrm{oxidation}}) \tag{7.5}$$

where $\triangle G$ is the Gibbs free energy and F is the Faraday constant, while $E^{\mathrm{o}}{}_{\mathrm{reduction}}$ – $E^{\mathrm{o}}{}_{\mathrm{oxidation}}$ are the formal potentials for the reduction and oxidation, respectively. From the Gibbs free energy it is possible to calculate the enthalpy, which is directly related to the Gibbs free energy based on the following equation:

$$\Delta G = \Delta H - T\Delta S \tag{7.6}$$

If the enthalpy exceeds the energy required to produce the lowest excited states from the ground state, then the reaction is defined as "energy sufficient" or is defined as to follow the singlet route "S-route," because $_1R^*$ will be directly generated. DPA is one of the systems that follows the S-route. In contrast, if the enthalpy is lower than the energy required to produce the lowest excited state but still exceeding the triple state energy, $_3R^*$, then $_1R^*$ will be generated by subsequent annihilation of $_3R^*$ (triplet–triplet annihilation, TTA). A typical example of TTA annihilation is the ECL of ruthenium tris-bipyridyl-type derivatives. In addition, ion annihilation can also lead to the formation of excimers (excited dimers) and exciplexes (excited complexes), and in this case the system is said to follow the "E-route." The major advantage of the annihilation process is that it requires only the ECL species, solvent, and supporting electrolyte to generate light.

For example, the potential of the working electrode (WE) is quickly changed between two different values to generate the oxidized, $DPA^{\bullet+}$, and reduced, $DPA^{\bullet-}$, species (eqs. (7.1) and (7.2), respectively) that will react near the electrode surface to form the emissive state, DPA (eq. (7.3)).

A classic example involves rubrene [5–8]. ECL is generated when a double-potential step is applied to an electrode (such as platinum, gold, or glassy carbon), producing the radical cation ($RUB^{\bullet+}$) upon anodic oxidation and the radical anion ($RUB^{\bullet-}$) upon cathodic reduction. The resulting electrogenerated products can then react and undergo annihilation to produce an excited state (RUB•) that is then able to emit light.

7.2.2 Coreactant pathway

The second dominant pathway to produce ECL is defined as "coreactant ECL," involving the reaction between luminophore and an additionally added reagent. The major advantage of using the coreactant approach is that the more intense generation of ECL emission than that obtained in an annihilation reaction due to the larger potential window of a solvent. The coreactant can be reduced or oxidized under the potential supply to form an intermediate and produce the excited states. When a potential (positive or negative) is applied at the electrode, both the luminophore species and the coreactant undergo oxidation or reduction with concomitant formation of radicals and intermediate states. Then, the intermediate states will decompose with the formation of highly reactive oxidizing or reducing species, which will interact with the oxidized or reduced luminophore to produce the excited state and subsequent ECL emission. Till date, the following chemical compounds are used as coreactants, such as peroxydisulfate ($S_2O_8^{2-}$), tripropylamine (TPA), H_2O_2, 2-(dibutylamino)ethanol (DBAE), $C_2O_4^{2-}$, SO_3^{2-}, and so on.

Generally, the coreactant ECL mechanism can be classified as "oxidation–reduction" ECL and "reduction–oxidation" ECL according to the oxidation and reduction processes on the surface of the electrode [9]. The general mechanisms of coreactant ECL systems are summarized in Table 7.1. A typical example is the ECL of TPA as the coreactant. TPA is present in the solution and is either oxidized or reduced in the same potential step as the luminophore species. Through electrons transfer or chemical reactions, the coreactant generates a product that reacts with the ECL luminophore to generate an excited state. In the case of $Ru(bpy)_3^{2+}$-based ECL, tertiary amines such as TPA are the dominant coreactants. However, others

Table 7.1: General mechanisms of coreactant ECL systems (from Miao [3]).

Reaction process	Oxidative-reduction ECL	Reductive-oxidation ECL
Redox reaction at electrode	$R{-}e{\rightarrow}R^{\bullet +}$ $C{-}e{\rightarrow}C^{\bullet +}$	$R{+}e{\rightarrow}R^{\bullet -}$ $C{+}e{\rightarrow}C^{\bullet -}$
Homogeneous chemical reactions	$R^{\bullet +}+C{\rightarrow}R{+}C^{\bullet +}$ $C^{\bullet +}{+}{\rightarrow}C{\neq}Red$ $C{\neq}Red{+}R{\rightarrow}R^{\bullet -}{+}P$	$R^{\bullet -}{+}C{\rightarrow}R{+}C^{\bullet -}$ $C^{\bullet -}{+}{\rightarrow}C{\neq}Ox$ $C{\neq}Ox{+}R{\rightarrow}R^{\bullet +}{+}P$
Exited state species formation	$R^{\bullet +}{+}R^{\bullet -}{\rightarrow}R{+}R^*$ or $R^{\bullet +}{+}R{\neq}Red{\rightarrow}R^*{+}P$	$R^{\bullet +}{+}R^{\bullet -}{\rightarrow}R{+}R^*$ or $R^{\bullet -}{+}C{\neq}Ox{\rightarrow}R^*{+}P$
Light emission	$R^*{\rightarrow}R{+}hu$	$R^*{\rightarrow}R{+}hu$

Here, R, luminophore; C, coreactant; $C^{\neq}$, coreactant intermediate with subscript "Red" for reducing agent and "Ox" for oxidizing agent; and P, product associated with $C^{\neq}$ reaction.

species include persulfate, hydrazine, and hydrogen peroxide ions, which operate in the reductive-oxidation mode, while oxalate and pyruvate ions operate in the oxidative-reduction mode. Coreactants are often used when $R^{+\bullet}$ (or $R^{-\bullet}$) are unstable, when $R^{+\bullet}$ (or $R^{-\bullet}$) cannot be formed because of the narrow potential window of the solvent or when the annihilation process is not particularly efficient. The coreactant-based approach is particularly useful when it is necessary to avoid the quenching of oxygen (often encountered in the ion annihilation process), allowing the analysis to be performed without the need to deoxygenate the samples. It is important to note that only the luminophore species can be regenerated at the electrode, while the coreactant is consumed during the chemical (or electrochemical) reactions. Suitable coreactants can be easily oxidized or reduced and then undergo a rapid chemical reaction to form an intermediate species with sufficient oxidizing or reducing power to create the excited state of the luminophore.

7.3 Luminophore for ECL

7.3.1 Luminol

5-Amino-2,3-dihydrophthalazine-1,4-dione (luminol) is a classical organic compound used for ECL. The ECL of luminol is often produced in an alkaline solution in the presence of hydrogen peroxide. Since luminol is a pH-dependent substance with pK_a of 6.00, the deprotonated ion undergoes subsequent cascade of luminescence reactions. Many biological activities involve the transformation or metabolism of hydrogen peroxide, which can be detected by luminol ECL with high sensitivity.

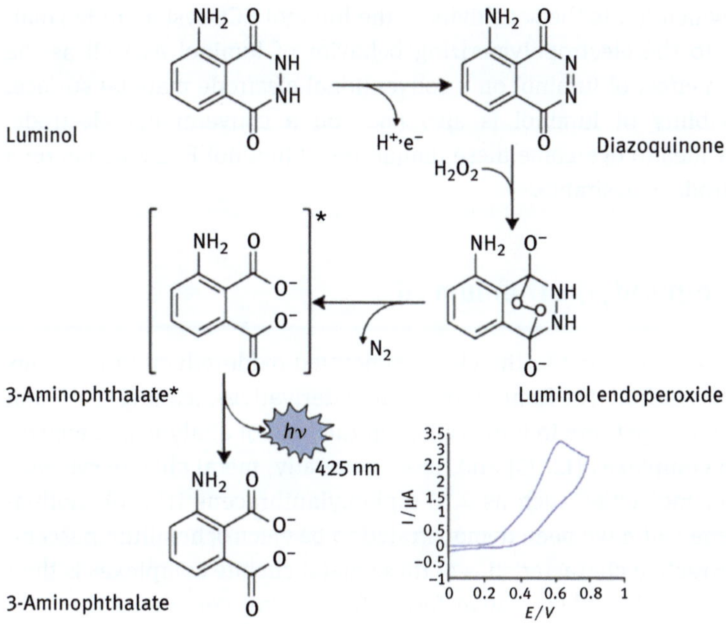

Fig. 7.2: Schematic representation of the electrocatalyzed chemiluminescent reaction; inset: a typical cyclic voltammogram of luminol in aqueous solution (from Marquette and Blum [11]).

For this reason, luminol-based ECL has applications in various biosensors. Figure 7.2 shows overall reactions in aqueous medium. The luminol oxidation leads to the formation of an aminophthalate ion in an excited state, which emits light when returning to the ground state. The quantum yield of the reaction is low (~0.01) and the emission spectrum shows a maximum at 425 nm [10].

The electrochemical oxidation of luminol is usually considered as the second most efficient way of triggering the reaction, after the horseradish peroxidase-biocatalyzed one. In a mechanistic study of this ECL reaction, Sakura proposed that luminol was first oxidized at the electrode surface and then reacted, mole to mole, with hydrogen peroxide (Fig. 7.2). The theoretical ratio (photon produced)/(H_2O_2 consumed) is then 1, while it is only 0.5 for the peroxidase-catalyzed reaction.

Luminol has good water solubility; more importantly, compared to other ECL systems, the dissolved oxygen can be used as the coreactant of the luminol ECL reaction on conventional electrode materials. It is an environmental-friendly coreactant and does not present any other side effects such as the higher background noise and the side interaction of the coreactant with DNA probes or proteins as in the case while using H_2O_2 as an ECL coreactant for bioassays.

However, the luminol ECL system still has several drawbacks. First, compared to the hydrophobic organic phase, the ECL quantum yield of luminol in the water phase is obviously low, which limits the sensitivity of the luminol ECL system in bioanalysis. Second, due to the electropolymerizing behavior of luminol as well as the stronger adsorption effect of luminol on a conventional electrode material surface, the ECL reproducibility of luminol is also poor on a conventional electrode. Therefore, the new idea to overcome these limitations of luminol ECL reaction on a conventional electrode is desirable.

7.3.2 Tris(2,2′-bipyridyl)ruthenium (II)

Electroluminescence here refers to the electrogenerated oxidoreduction reactions of compounds different from the luminol or luminol derivatives, leading to photon emission. The main competitors to luminol ECL in the field of analytical chemistry are the ruthenium complexes [12, 13] and, more generally, metal chelate systems. Nevertheless, other molecules such as 9,10-diphenylanthracene [14], phenothiazine [15], and pyrene [16] have been demonstrated to be electrochemiluminescent. One of the most attractive characteristics of these metal chelate complexes is their ability to be regenerated in their native form after having completed the light emission reaction sequence. A single molecule could then theoretically generate more photons than the luminol in destructive ECL. The most widely used and studied of these metal chelate complexes is tris (2,2′-bipyridyl)ruthenium (II), also named $Ru(bpy)_3^{2+}$ (Fig. 7.3a). Indeed, since it was first reported as an electrochemiluminescent compound in 1972 [17], $Ru(bpy)_3^{2+}$ has become the most thoroughly studied ECL-active molecule [12, 18]. This domination of the field is mainly due to its strong luminescence, its solubility in both aqueous and nonaqueous media at room temperature, and, of course, its ability to undergo, as mentioned earlier, a reversible one-electron transfer reaction. Figure 7.4 shows the most widely used system for triggering ECL of $Ru(bpy)_3^{2+}$ in aqueous solution. First, the ruthenium complex is electro-oxidized in a one-electron reaction at the electrode surface (polarized at a potential between +1 and +1.5 V). Concomitantly, the classical coreagent TPA [19] is also oxidized and deprotonated to generate a radical species that will reduce the oxidized metal complex ($Ru(III)(bpy)_3^{2+}$), leading to an excited state of the reduced ruthenium complex that emits a photon (617 nm) while returning to the ground state.

7.3.3 Porphyrin

Porphyrins are 18π aromatic azaannulenes consisting of regularly arranged four methine carbons. As a visible-light photosensitizer, porphyrins received much attention because of their widespread occurrence in nature, strong optical absorption and emission, and electrochemical properties applicable for electron transfer reactions

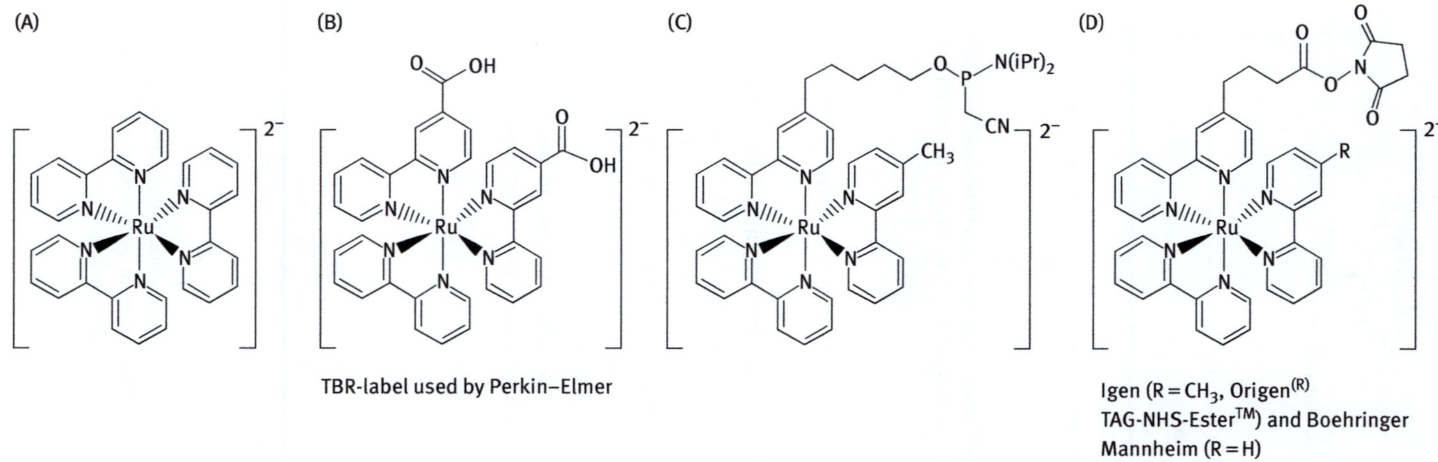

Fig. 7.3: (a) Structure of Tris(2,2′-bipyridyl)ruthenium (II), also known as $Ru(bpy)_3^{2+}$. (b)–(d) Structures of bifunctional $Ru(bpy)_3^{2+}$ derivatives (from Marquette and Blum [11]).

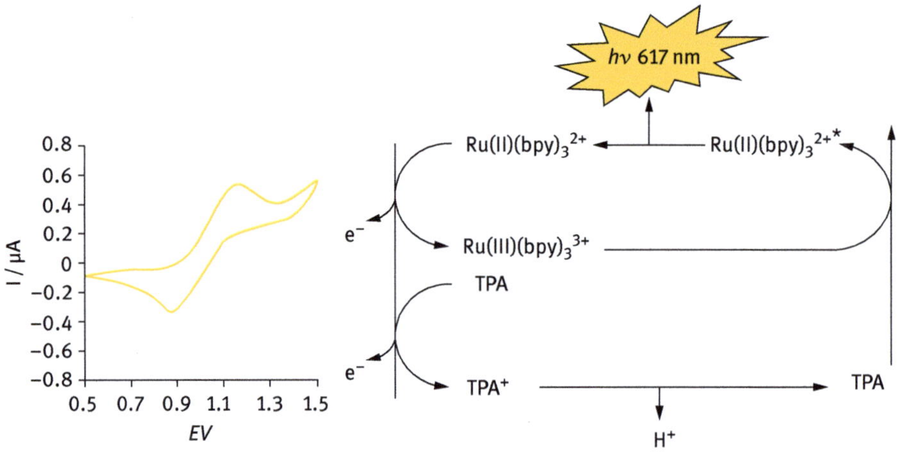

Fig. 7.4: Schematic representation of the electrochemiluminescent reaction of $Ru(bpy)_3^{2+}$ with TPA; inset: a typical cyclic voltammogram of $Ru(bpy)_3^{2+}$ in aqueous solution (from Marquette and Blum [11]).

related to light harvesting, applied in photoelectric devices to improve photoelectric conversion efficiency.

Zhang et al. reported that a strong and stable red irradiation at 634 nm could be stimulated electrochemically on the glassy carbon electrode (GCE) modified by zinc (II) meso-tetra(4-carboxyphenyl)porphyrin (ZnTCPP) in aqueous media [20]. The nature of ECL was thoroughly investigated, being exactly the production of $^1O_2$ from the electrochemical and chemical reactions between superoxide ion (O_2^-) and the intermediate anion of zinc porphyrin (Zn^+) (Fig. 7.5).

Porphyrins have a flexible and adjustable molecular structure in chemical modification. The introduction of molecular-recognition motifs (e.g., hydrogen bonding,

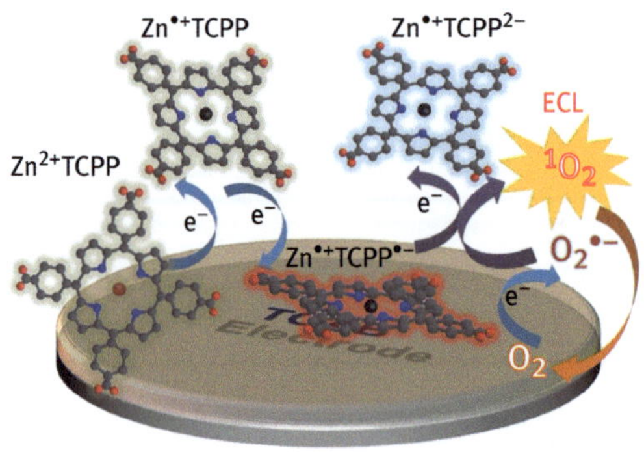

Fig. 7.5: Schematic illustration of the ECL mechanism of singlet oxygen based on ZnTCPP (from Zhang et al. [20]).

metal–ligand bonds, and π–π stacking) into the porphyrin building blocks will make them form new structures, such as fibers, cubes, sheets, simple micelles, and wheels. Shan and coworkers have developed several ECL sensors based on porphyrin nanocomposites. For example, Proto-porphyrin IX (CoPPIX) was combined with ultrathin carbon nitride nanosheets (C_3N_4) [21]. CoPPIX@C_3N_4 complex possessed much higher catalytic activity. Laponite nanosheets were utilized to manage the assembly and immobilization of protoporphyrin IX (PPIX).

They synthesized zirconium-based porphyrinic metal–organic frame-work (MOF) for the detection of a phosphoprotein [22]. The as-synthesized MOF-525-Zn serves as a three-in-one platform possessing oxygen nanocage, electron media, and bonding site in the ECL bioanalysis system, as shown in Fig. 7.6. Specifically, depending on the enrichment of oxygen molecules from the high porosity and tunable structures of MOFs, the active center ZnTCPP in MOF-525-Zn as electron media reacts with O_2 in the three-dimensional nanocage to produce $^1O_2$, resulting in the ECL signal. In addition, the amphiphilic surfactant TOAB facilitated O_2 to interact with electroactive ZnTCPP of MOF-525-Zn. Furthermore, with the coordination of the phosphate groups and inorganic Zr–O clusters as binding sites in MOF-525-Zn structures, MOF-525-Zn can be used as signal-amplifying probes for an ultrasensitive ECL PKA assay. This three-in-one approach not only exhibits great promise in clinic diagnostics and targeted therapy but also provides a novel promising platform for bioanalysis.

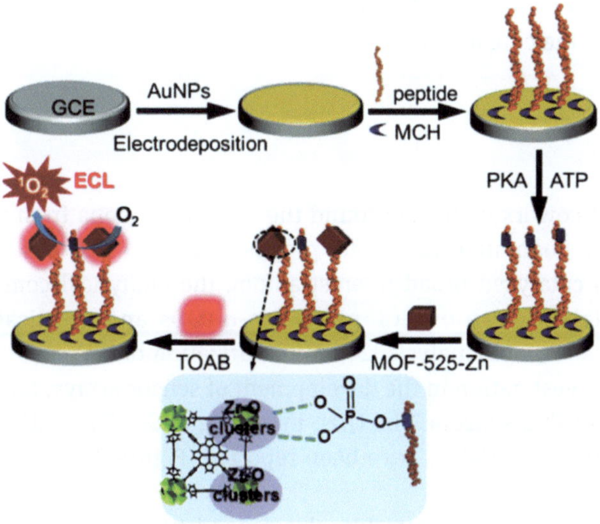

Fig. 7.6: Schematic representation of the MOF-525-Zn-based ECL biosensor for PKA assay (from Zhang et al. [22]).

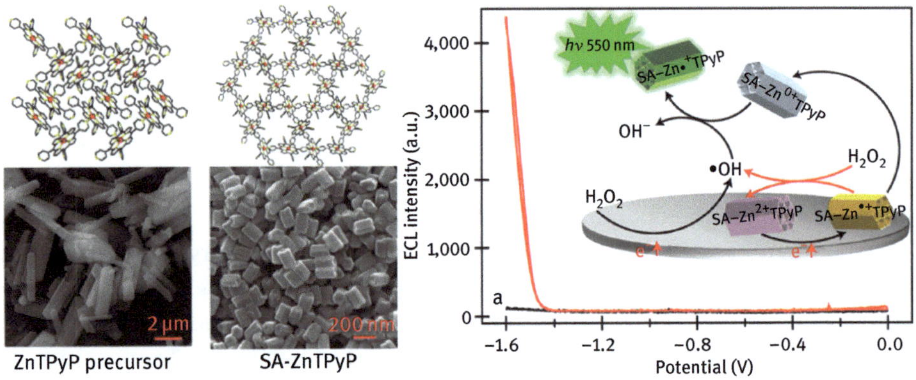

Fig. 7.7: Schematic representation of one-dimensional self-assembled porphyrin hexagonal nanoprisms and the enhanced ECL (from Cai et al. [23]).

They synthesized the one-dimensional nanostructure of zinc 5,10,15,20-tetra(4-pyridyl)-21H,23H-porphine (ZnTPyP) via a self-assembly technique [23]. Using sodium dodecyl sulfate as the "soft template," the self-assembled ZnTPyP (SA-ZnTPyP) had the morphology of hexagonal nanoprisms with a uniform size (diameter of 100 nm). The SA-ZnTPyP exhibited remarkably different spectral properties compared to those of the original ZnTPyP. The as-prepared SA- ZnTPyP was used to modify GCE, and the ECL behaviors of the SA-ZnTPyP/GCE were investigated (Fig. 7.7). The hydrophilic carbon dots (C-dots) could efficiently prevent the dissolution of SA-ZnTPyP in DMF containing 0.1 mol L^{-1} TBAP and, simultaneously, could accelerate electron transfer. Therefore, the enhanced ECL was realized by C-dots/SA-ZnTPyP/GCE by using H_2O_2 as the coreactant.

7.3.4 Quantum dots

In the early 2000s, Bard and coworkers further found the ECL phenomena from a series of quantum dots (QDs). Soon after, Ju's group proposed the first QDs-based ECL biosensor. These works catalyzed broad interest among the analytical community for both fundamental research on ECL properties of QDs and practical application of various QDs for ECL bioanalysis [24]. Miniaturization of biosensors is an attractive topic for the investigation in the development of sensor arrays, and this is also true in the field of ECL sensors. The QDs, including CdSe, CdS, CdTe, PbS, ZnSe, and Ag_2Se nanoparticles (NPs), have been reported. Figure 7.8 shows CdSe QDs apply in ECL.

In particular, high-fluorescence quantum yields, size-dependent luminescence, and stability against photobleaching make them a very attractive material for biosensing applications [26]. Numerous systems using QDs have been used as labels

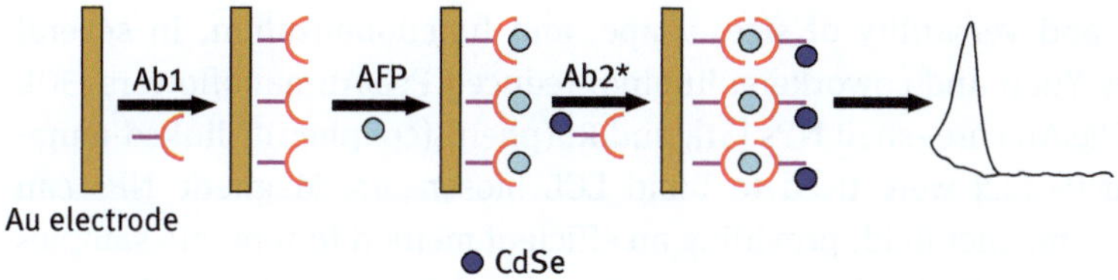

Fig. 7.8: Biosensor fabrication for the detection of AFP using CdSe QDs (from Liu et al. [25]).

within ECL detection systems as a result of their aforementioned-outlined advantages over more common emitters. It is well known that the size of QDs strongly affects the ECL behavior. The dependence of the ECL intensity on the size of QDs is related to the band gap of QDs, which increased with the decrease of the size of QDs. The detection of catechol derivatives was achieved by measuring the quenching of the ECL using the electro-oxidized products of catechols (dopamine and l adrenaline) [24]. Most of the ECL immunoassays of QDs are based on the quenching, inhibition, or enhancement of the ECL intensities via the well-studied coreactant ECL systems, which took $S_2O_8^{2-}$, H_2O_2, and SO_3^{2-} as the coreactants. In CdTe, QDs coated with a silica nanosphere label (Si/QD/Ab2) were attached onto the gold electrode surface through a subsequent "sandwich" immunoreaction [27].

Li et al. reported that the graphene QD ECL allows selective and sensitive cadmium ion detection using cysteine as the masking agent, indicating that graphene QD is a promising ECL luminophore for ECL detection [28, 29].

7.4 Enhancement strategies

7.4.1 Nanomaterials for signal amplification

The introduction of nanomaterials has deeply revolutionized the sensitivity and diversity of ECL biosensors. Nanomaterials with different sizes, shapes, chemical components, and unique properties have been adopted for different kinds of biosensing applications, playing an increasingly important role in the development of biosensors [29]. Nanomaterials can enhance the efficiency of both biological element part and transducer part in many different ways.

Nanomaterials with their enormous specific surface area provide robust carrier frameworks in which these ECL luminophores or certain biomolecules can be incorporated to give much stronger ECL signal for ultrasensitive biosensors. The most popular nanomaterials as carriers are metal NPs (especially gold nanoparticles (AuNPs)), magnetic NPs, and porous silica NPs. AuNPs are excellent carriers in various biological applications for great advantages such as ease of synthesis, great biocompatibility, and versatility of size, shape, and functionalization. In several works reported by Yuan and coworkers, luminol-reduced Pt@Au nanoflowers [30], luminol capped Pt@Au core–shell NPs [31], and $Ru(phen)_2(cpaphen)^{2+}$ linked-ampicillin capped AuNPs [32] were used to build ECL biosensors. Magnetic NPs can respond to external magnet field, providing an efficient method to separate samples from liquid suspension. By applying rotating functional magnetic force, the mass transfer between ECL reagents and magnetic NPs-linked enzymes could be enhanced, thus improving the ECL signal [33]. Due to high surface-to-volume ratio, good biocompatibility, and low cost, silica NPs have become another commonly used carrier for ECL biosensors. Silica NPs are widely employed for immobilizing a large loading of ECL luminophores by doping the luminophores into the NPs. Thus, silica NPs can protect luminophores from the surrounding environment and can increase the photostability and intensity of ECL signal.

To build a biosensing platform, one of the most important step is electrode modification. AuNPs are not only excellent carriers to improve the loading of ECL species but also wonderful electrode modification material that can promote the electron transfer and surface area of electrode. Recently, AuNPs-modified electrodes were applied in many different kinds of ECL biosensors, leading to effective improvement in ECL intensity and sensing sensitivity [34–36]. The remarkable electrical, chemical, mechanical, and structural properties of carbon nanotube (CNT) make it an attractive nanomaterial for electrode modification. CNTs can be easily derivatized with various functional groups for the attachment of biomolecules. A key aspect of CNTs applications in biosensors is their immobilization with suitable biomolecules. Another major aspect of CNTs applications is their ability to promote the electron transfer process. Similar to CNTs, graphene sheets are another popular electrode modification nanomaterials due to their strong mechanical strength, enormous specific surface area, excellent thermal conductivity, and electric conductivity. Graphene oxide (GO) contains a range of reactive oxygen functional groups, which give it great application value in biosensing.

Certain metal nanomaterials have catalytic effect on the reactions relevant to ECL emitting. Bringing in this kind of nanomaterials as catalysts can greatly improve the sensitivity of ECL biosensors. Yuan and coworkers reported a cholesterol biosensor based on luminol ECL catalyzed by AuNPs-modified electrode [37]. They found that AuNPs not only provided larger surface area but also formed the nanostructured interface to catalyze luminol ECL.

With the development of the modern biosensors, a single type of material could not fulfill the requirements for an ultrasensitive strategy of detection. Thus, multifunctional nanocomposites consisting of several materials with different performances are quite popular in biosensing applications. For instance, in a sandwich-type ECL immunosensor, the multifunctional nanocomposite of luminol capped gold-modified Fe_3O_4 (Luminol–AuNPs@Fe_3O_4) worked as ECL labels, catalysts, and magnetic separable carriers [38]. In the ratiometric ECL immunosensor designed by Ju's group, they used the luminol/Pd nanoclusters@graphene oxide nanocomposites as ECL probes, electrocatalyst, and signal amplification materials [39].

7.4.2 Resonance energy transfer (RET)

Energy transfer (ET) via dipole-induced interactions (Fçrster ET) or via exchange interactions (Dexter ET) is a promising technique to probe changes of the distance between a suitable matched donor and acceptor pair. In the past decades, several kinds of resonance ET have been reported in bioapplications on the basis of different excited modes of the donor luminescence, such as fluorescence (FRET), bioluminescence (BRET), and chemiluminesence (CRET). Recently, with the development of ECL technique combining the electrochemical and luminescent methods, the ECL-RET has attracted increasing interest in the trace analysis of nucleic acids, proteins, and small molecules, owing to its remarkable advantages of no excitation light source and no interference from the scattered light compared to the FRET. In the conventional ECL-RET biosensors, the researchers have focused on searching for perfect energy-overlapped donor–acceptor pairs to improve the sensitivity [40]. For example, Zhu et al [41]. used an efficient quenching ECL immunosensor based on ECL-RET and studied the sensitive analysis of prostate-specific antigen. In this protocol, nitrogen-doped graphene QDs, which could produce excellent ECL emission, were loaded onto $Ni(OH)_2$ with the three-dimensional hierarchical and stacked lamellar structure. Zhou et al. [42] provide a novel ECL-RET system using CdS:Eu nanocrystals as an ECL donor and Au nanorods as an ECL acceptor. Zhang et al. [43] designed an ECL biosensor for the determination of insulin using a novel ECL-RET strategy. In this strategy, carboxyl poly(9,9-dioctyfluorenyl-2,7-diyl) dots worked as the ECL donor and 3,4,9,10-perylenetetra-car-boxylic acid exploited as the ECL acceptor, and hydrogen peroxide (H_2O_2) was employed as the coreactant.

7.5 ECL for assay in biological matrices

ECL detections in immunosensors are mainly carried out with solid phase ECL assay heterogeneous formats, including direct, competitive, and sandwich immunoassays.

7.5.1 Direct interaction

In the direct probe format, first an electrode is modified with a species that serves as both a capture and a signal probe, as shown in Fig. 7.9 [44]. This is a species that

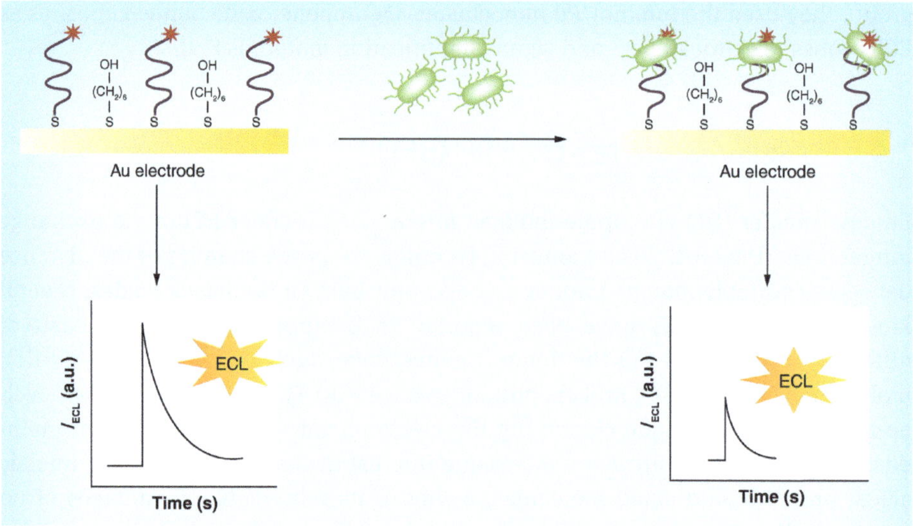

Fig. 7.9: Diagrams of direct probe electrogenerated chemiluminescence biosensors for the detection of bacteria (from Gross et al. [44]).

recognizes the analyte and has been labeled with the luminophore. When this electrode is placed in a solution of coreactant (e.g., TPrA) and the electrode properly biased, a strong ECL signal is observed. In the presence of the target (analyte), the target binds to the capture/signal probe and the ECL signal decreases. Generally, the decrease in ECL signal is proportional to the concentration of analyte.

Advantages: Cross-reactivity of secondary antibody (Ab2) is eliminated and is a quick methodology since only one antibody is used.

Disadvantages: There is lack of ECL (or coreactant) properties of the target analyte; labeling of every Abl is time consuming and expensive; no flexibility in choice of Abl label from one experiment to another; immunoreactivity of the Abl may be reduced as a result of labeling; and little signal amplification.

7.5.2 Competition assay

In a competitive format, the unlabeled analyte (usually the antigen) in the test sample is measured by its ability to compete with the labeled antigen in the immunoassay (Fig. 7.10). Typically, the detectable signal decreases with the increase of analyte concentration, which requires a high background signal toward zero analyte.

It employs an antibody and an analog of the analyte of interest, one immobilized on a solid phase and the other labeled with an ECL reporter. The analyte present in a

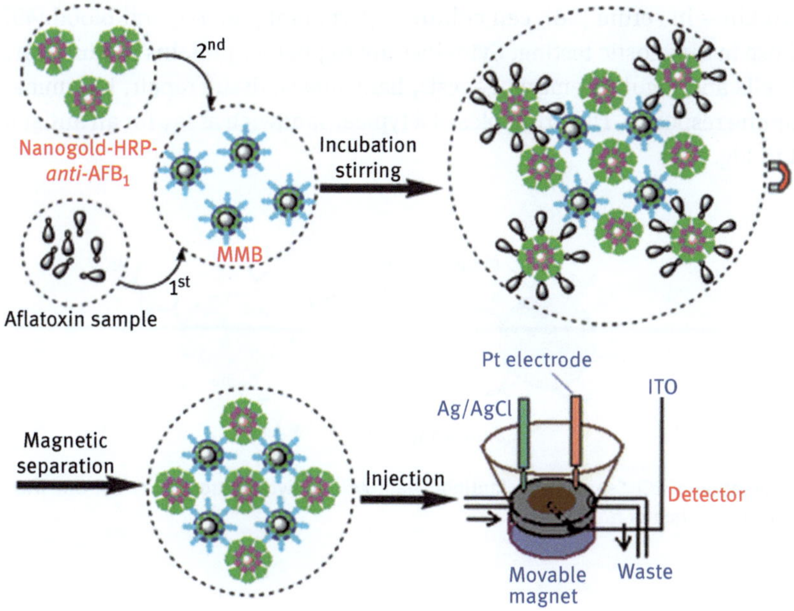

Fig. 7.10: Measurement process of the competitive immunoassay method (from Tang et al. [45]).

sample competes with the analog for binding to the antibody and decreases the accumulation of labels on the solid phase.

Advantage: Nonpurified primary antibodies may be used.

Disadvantages: Main drawbacks are limited sensitivity and working range; slow reaction kinetics; and lower precision, development of a negative endpoint.

7.5.3 Sandwich-type assay

The most widespread among them is the sandwich-type immunoassay. Generally, an electrochemical reaction occurs within a limited reaction layer, the so-called electrical double layer and/or the electrical diffuse layer at several nanometers distance from the surface of the WE. Consequently, only an extremely small quantity of the analyte that contacts with the surface of the WE may be utilized for the ECL detection. Accordingly, if all the ECL-labeled materials could be collected over the surface of the WE, much more ECL emission might be generated and a highly sensitive ECLIA could be realized.

ECL is a valuable tool in life science research laboratories to study basic physiology mechanisms and to detect, discover the causes for, and find potential cures for diseases [47–60]. For example, sandwich immunoassays have been used to detect numerous cytokines in serum [48], cell culture supernatant [46, 47], and blood [49] for potential use in diagnostic testing. Cytokines are regulatory proteins produced by white blood cells and regulate embryogenesis, hameotosis, tissue repair, inflammation, and immune response. The principles of a typical sandwich assay for an antigen are outlined in Fig. 7.11.

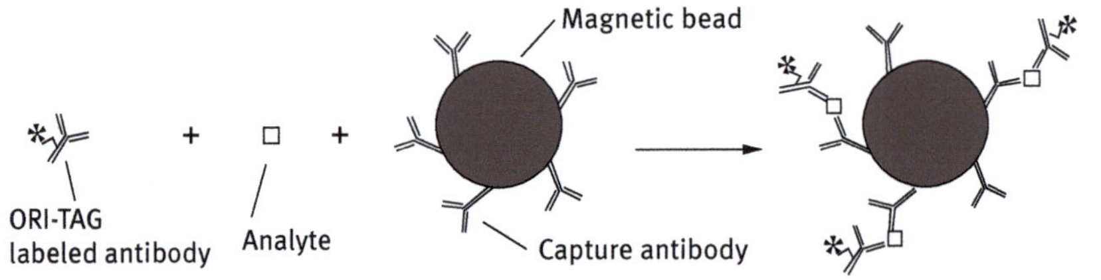

Fig. 7.11: Representation of ECL "sandwich" (antibody–antigen/analyte–antibody) assay. ORI-TAG refers to ECL label (from Richter [60]).

The sandwich-type formats give the highest level of sensitivity and specificity because of the use of a couple of match antibodies. The measurement of the labeled analyte (usually the antibody) is directly proportional to the amount of antigen present in the sample, thus resulting in an increase in the detectable signal with the increasing target analyte [59]. Therefore, sandwich-type assay is one of the most popular schemes in the immunosensings and immunoassays. Although the antigen–antibody reaction can cause the change of detectable signal to some extent, the change is comparatively small. High-affinity antibodies and appropriate labels are usually employed for the amplification of detectable signal.

7.6 Conjugation strategies

Conjugation strategies are important for ECL-based sensor. Especially in the case of QDs-based ECL, the limited functionalization of ligands and variable efficiency in the bioconjugation to antibodies are the main drawbacks of QDs. An illustration of some selected surface chemistries and conjugation strategies that are applied to QDs is shown in Fig. 7.12.

The approaches of conjugation QDs with Ab2 are as follows:
(1) direct conjugation of amino/carboxyl groups using active esters;
(2) direct conjugation to the QDs surface through the antibody thiol groups;
(3) indirect conjugation using avidin as bridge protein and bioti- nilated antibodies;

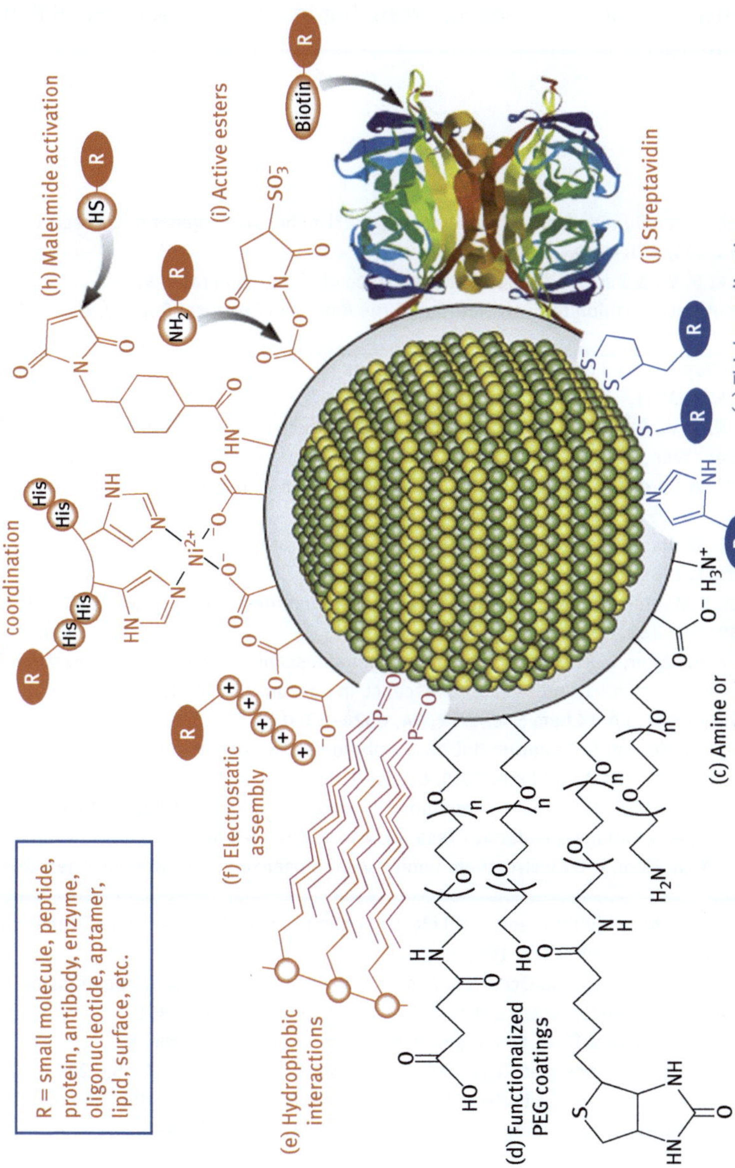

Fig. 7.12: An illustration of some selected surface chemistries and conjugation strategies that are applied to QDs. The grey periphery around the QD represents a general coating (from Algar et al. [61]).

(4) indirect conjugation using modified protein G as the bridge protein. Obviously, the employed QDs should have been previously modified to gain water solubility and the required functional groups, preferably carboxylate or amino groups.

It should be noticed that since QDs are typically poorer ECL emitters than the more commonly used ruthenium complexes or luminol, enhancement of ECL intensity, observed with QDs, is an important task. The enhancement can be achieved by reducing the barrier for electron injection, resulting in a lower onset potential and increased ECL intensity.

References

[1] Hercules, D. M. Chemiluminescence resulting from electrochemically generated species. Science, 1964, 145, 808–809.
[2] Santhanam, K. S. V., & Bard, A. J. Chemiluminescence of electrogenerated 9, 10-diphenylanthracene anion radical1. Journal of the American Chemical Society, 1965, 87, 139–140.
[3] Miao, W. Electrogenerated chemiluminescence and its biorelated applications. Chem Rev. 2008, 108, 2506–2553.
[4] Santhanam, K. S. V., & Bard, A. J. The electrochemical reduction of 9, 10-Diphenylanthracene. J Am Chem Soc. 1966, 88, 2669–2675.
[5] Faulkner, L. R., & Bard, A. J. Electroanalytical Chemistry. ed & A. J. Bard, Marcel Dekker, New York, 1977, 1–95.
[6] Faulkner, L. R., & Glass, R. S. Chemical and biological generation of excited states. Academic Press, New York, 1982.
[7] Maricle, D. L., & Maurer, A. Pre-annihilation electrochemiluminescence of rubrene. J Am Chem Soc. 1967, 89, 188–189.
[8] Bezman, R., & Faulkner, L. R. Mechanisms of chemiluminescent electron-transfer reactions. V. Absolute measurements of rubrene luminescence in benzonitrile and N, N-dimethylformamide. J Am Chem Soc. 1972, 94, 6324–6330.
[9] Zhou, H., Liu, J., & Zhang, S. Quantum dot-based photoelectric conversion for biosensing applications. TrAC, Trends Anal Chem. 2015, 67, 56–73.
[10] Roswell, D. F., & White, E. H. [36] The chemiluminescence of luminol and related hydrazides [M]//Methods in enzymology. Academic Press, New York, 1978,57, 409–423.
[11] Marquette, C. A., & Blum, L. J. Electro-chemiluminescent biosensing. Anal Bioanal Chem 2008, 390, 155–168.
[12] Lee, W. Y. Tris (2, 2′-bipyridyl) ruthenium (II) electrogenerated chemiluminescence in analytical science. Microchim Acta. 1997, 127, 19–39.
[13] Knight, A. W., & Greenway, G. M. Occurrence, mechanisms and analytical applications of electrogenerated chemiluminescence. A review. Analyst. 1994, 119, 879–890.
[14] Beideman, F. E., & Hercules, D. M. Electrogenerated chemiluminescence from 9, 10-diphenylanthracene cations reacting with radical anions. J Phys Chem. 1979, 83, 2203–2209.
[15] Kakhr, A., Mugnier, Y., & Laviron, E. Reduction electrochimique de la phenothiazine et du fluorobenzene a basse temperature. Electrochim Acta. 1983, 28, 1897–1898.

[16] Bard, A. J. (ed.). Electrogenerated chemiluminescence. CRC Press, Boca Raton, 2004.
[17] Tokel, N. E., & Bard, A. J. Electrogenerated chemiluminescence. IX. Electrochemistry and emission from systems containing tris (2, 2′-bipyridine) ruthenium (II) dichloride. J Am Chem Soc. 1972, 94, 2862–2863.
[18] Wilson, R., Akhavan-Tafti, H., DeSilva, R., & Schaap, A. P. Comparison between acridan ester, luminol, and ruthenium chelate electrochemiluminescence. Electroanalysis. 2001, 13, 1083–1092.
[19] Leland, J. K., & Powell, M. J. Electrogenerated chemiluminescence: an oxidative-reduction type ECL reaction sequence using tripropyl amine. J Electrochem Soc. 1990, 137, 3127–3131.
[20] Zhang, G. Y., Deng, S. Y., Zhang, X. J., & Shan, D. Cathodic electrochemiluminescence of singlet oxygen induced by the electroactive zinc porphyrin in aqueous media. Electrochim Acta. 2016, 190, 64–68.
[21] Deng, S. Y., Yuan, P. X., Ji, X. B., Shan, D., & Zhang, X. J. Carbon nitride nanosheet-supported porphyrin: a new biomimetic catalyst for highly efficient bioanalysis. ACS Appl Mater Interfaces. 2014, 7, 543–552.
[22] Zhang, G. Y., Cai, C., Cosnier, S., Zeng, H. B., Zhang, X. J., & Shan, D. Zirconium–metalloporphyrin frameworks as a three-in-one platform possessing oxygen nanocage, electron media, and bonding site for electrochemiluminescence protein kinase activity assay. Nanoscale. 2016, 8, 11649–11657.
[23] Cai, W. R., Zhang, G. Y., Lu, K. K., Zeng, H. B., Cosnier, S., Zhang, X. J., & Shan, D. Enhanced electrochemiluminescence of one-dimensional self-assembled porphyrin hexagonal nanoprisms. ACS Appl Mater Interfaces. 2017, 9, 20904–20912.
[24] Zou, G. Z., Ju, H. X. Electrogenerated chemiluminescence from a CdSe nanocrystal film and its sensing application in aqueous solution. Analytical chemistry, 2004, 76, 6871–6876.
[25] Liu, Q., Han, M., Bao, J., Jiang, X., & Dai, Z. CdSe quantum dots as labels for sensitive immunoassay of cancer biomarker proteins by electrogenerated chemiluminescence. Analyst. 2011, 136, 5197–5203.
[26] Bertoncello, P., Stewart, A. J., & Dennany, L. Analytical applications of nanomaterials in electrogenerated chemiluminescence. Anal Bioanal Chem. 2014, 406, 5573–5587.
[27] Muzyka, K. Current trends in the development of the electrochemiluminescent immunosensors. Biosens Bioelectron. 2014, 54, 393–407.
[28] Li, L. L., Ji, J., Fei, R., Wang, C. Z., Lu, Q., & Zhang, J. R. A facile microwave avenue to electrochemiluminescent two-color graphene quantum dots. Adv Funct Mater. 2012, 22, 2971–2979.
[29] Liu, Z., Qi, W., & Xu, G. Recent advances in electrochemiluminescence. Chem Soc Rev. 2015, 44, 3117–3142.
[30] Zhou, Y., Zhuo, Y., Liao, N., Chai, Y., & Yuan, R. Ultrasensitive immunoassay based on a pseudobienzyme amplifying system of choline oxidase and luminol-reduced Pt@ Au hybrid nanoflowers. Chem Commun. 2014, 50, 14627–14630.
[31] Liu, Y., Wang, H., Xiong, C., Yuan, Y., Chai, Y., & Yuan, R. A sensitive electrochemiluminescence immunosensor based on luminophore capped Pd@ Au core-shell nanoparticles as signal tracers and ferrocenyl compounds as signal enhancers. Biosens Bioelectron. 2016, 81, 334–340.
[32] Gui, G. F., Zhuo, Y., Chai, Y. Q., Xiang, Y., & Yuan, R. A novel ECL biosensor for β-lactamase detection: using Ru (II) linked-ampicillin complex as the recognition element. Biosens Bioelectron 2015, 70, 221–225.
[33] Weizmann, Y., Patolsky, F., Katz, E., & Willner, I. Amplified DNA sensing and immunosensing by the rotation of functional magnetic particles. J Am Chem Soc. 2003, 125, 3452–3454.

[34] Zhang, H. R., Wang, Y. Z., Wu, M. S., Feng, Q. M., Shi, H. W., Chen, H. Y., & Xu, J. J. Visual electrochemiluminescence detection of telomerase activity based on multifunctional Au nanoparticles modified with G-quadruplex deoxyribozyme and luminol. Chem Commun. 2014, 50, 12575–12577.
[35] Li, M., Kong, Q., Bian, Z., Ma, C., Ge, S., Zhang, Y., Yu, J. H., & Yan, M. Ultrasensitive detection of lead ion sensor based on gold nanodendrites modified electrode and electrochemiluminescent quenching of quantum dots by electrocatalytic silver/zinc oxide coupled structures. Biosens Bioelectron. 2015, 65, 176–182.
[36] Li, J., Ma, H., Wu, D., Li, X., Zhao, Y., Zhang, Y., Du, B., & Wei, Q. A label-free electrochemiluminescence immunosensor based on KNbO3–Au nanoparticles@ Bi2S3 for the detection of prostate specific antigen. Biosens Bioelectron. 2015, 74, 104–112.
[37] Zhang, M., Yuan, R., Chai, Y., Chen, S., Zhong, H., Wang, C., & Cheng, Y. A biosensor for cholesterol based on gold nanoparticles-catalyzed luminol electrogenerated chemiluminescence. Biosens Bioelectron. 2012, 32, 288–292.
[38] Lian, S., Huang, Z., Lin, Z., Chen, X., Oyama, M., & Chen, X. A highly selective melamine sensor relying on intensified electrochemiluminescence of the silica nanoparticles doped with [Ru (bpy) 3] 2+/molecularly imprinted polymer modified electrode. Sens Actuators B. 2016, 236, 614–620.
[39] Huang, Y., Lei, J., Cheng, Y., & Ju, H. Ratiometric electrochemiluminescent strategy regulated by electrocatalysis of palladium nanocluster for immunosensing. Biosensors and Bioelectronics, 2016, 77, 733–739.
[40] Liu, J. L., Zhao, M., Zhuo, Y., Chai, Y. Q., & Yuan, R. Highly efficient intramolecular electrochemiluminescence energy transfer for ultrasensitive bioanalysis of aflatoxin M1. Chem Eur J. 2017, 23, 1853–1859.
[41] Zhu, W., Wang, C., Li, X., Khan, M. S., Sun, X., Ma, H., Fan, D., & Wei, Q. Zinc-doping enhanced cadmium sulfide electrochemiluminescence behavior based on Au-Cu alloy nanocrystals quenching for insulin detection. Biosens Bioelectron. 2017, 97, 115–121.
[42] Zhou, H., Zhang, Y. Y., Liu, J., Xu, J. J., & Chen, H. Y. Electrochemiluminescence resonance energy transfer between CdS: Eu nanocrystals and Au nanorods for sensitive DNA detection. J Phys Chem C. 2012, 116, 17773–17780.
[43] Zhang, H., Zuo, F., Tan, X., Xu, S., Yuan, R., & Chen, S. A novel electrochemiluminescent biosensor based on resonance energy transfer between poly (9, 9-di-n-octylfluorenyl-2, 7-diyl) and 3, 4, 9, 10-perylenetetracar-boxylic acid for insulin detection. Biosens Bioelectron. 2018, 104, 65–71.
[44] Gross, E. M., Maddipati, S. S., & Snyder, S. M. A review of electrogenerated chemiluminescent biosensors for assays in biological matrices. Bioanalysis. 2016, 8, 2071–2089.
[45] Tang, D., Zhong, Z., Niessner, R., & Knopp, D. Multifunctional magnetic bead-based electrochemical immunoassay for the detection of aflatoxin B 1 in food. Analyst. 2009, 134, 1554–1560.
[46] Blohm, S., Kadey, S., McKeon, K., Perkins, S., & Sugasawara, R. Use of the ORIGEN electrochemiluminescence detection system for measuring tumor necrosis factor-α in tissue culture media. Biomed Prod. 1996, 21, 60.
[47] Shapiro, L., Heidenreich, K. A., Meintzer, M. K., & Dinarello, C. A. Role of p38 mitogen-activated protein kinase in HIV type 1 production in vitro. Proc Natl Acad Sci. 1998, 95, 7422–7426.
[48] Obenauer-Kutner, L. J., Jacobs, S. J., Kolz, K., Tobias, L. M., & Bordens, R. W. A highly sensitive electrochemiluminescence immunoassay for interferon alfa-2b in human serum. J Immunol Methods. 1997, 206, 25–33.

[49] Puren, A. J., Razeghi, P., Fantuzzi, G., & Dinarello, C. A. Interleukin-18 enhances lipopolysaccharide-induced interferon-γ production in human whole blood cultures. J Infect Dis. 1998, 178, 1830–1834.
[50] Gopalakrishnan, S. M., Warrior, U., Burns, D., & Groebe, D. R. Evaluation of electrochemiluminescent technology for inhibitors of granulocyte colony-stimulating factor receptor binding. J Biomol Screening. 2000, 5, 369–375.
[51] Weinreb, P. H., Yang, W. J., Violette, S. M., Couture, M., Kimball, K., Pepinsky, R. B., Lobb, R. P., & Josiah, S. A. Cell-free electrochemiluminescence assay for measuring β1-integrin–ligand interactions. Anal Biochem. 2002, 306, 305–313.
[52] Hughes, S. R., Khorkova, O., Goyal, S., Knaeblein, J., Heroux, J., Riedel, N. G., & Sahasrabudhe, S. α2-macroglobulin associates with β-amyloid peptide and prevents fibril formation. Proc Natl Acad Sci. 1998, 95, 3275–3280.
[53] Horiuchi, H., Lippé, R., McBride, H. M., Rubino, M., Woodman, P., Stenmark, H., Rybin, V., Wilm, M., Ashman, K., Mann, M., & Zerial, M. A novel Rab5 GDP/GTP exchange factor complexed to Rabaptin-5 links nucleotide exchange to effector recruitment and function. Cell. 1997, 90, 1149–1159.
[54] Mathew, A., Mathur, S. K., Jolly, C., Fox, S. G., Kim, S., & Morimoto, R. I. Stress-specific activation and repression of heat shock factors 1 and 2. Mol Cell Biol. 2001, 21, 7163–7171.
[55] Zhang, L., Schwartz, G., O'Donnell, M., et al. Development of a novel helicase assay using electrochemiluminescence. Anal Biochem. 2001, 293, 31–37.
[56] Zhang, L., Song, L., Terracina, G., Liu, Y., Pramanik, B., & Parker, E. Biochemical characterization of the γ-secretase activity that produces β-amyloid peptides. Biochemistry. 2001, 40, 5049–5055.
[57] De Baar, M. P., Van der Horn, K. H. M., Goudsmit, J., De Ronde, A., & De Wolf, F. Detection of human Immunodeficiency virus type 1 nucleocapsid protein p7 in vitro and in vivo. J Clin Microbiol. 1999, 37, 63–67.
[58] Khorkova, O. E., Patel, K., Heroux, J., & Sahasrabudhe, S. Modulation of amyloid precursor protein processing by compounds with various mechanisms of action: detection by liquid phase electrochemiluminescent system. J Neurosci Methods. 1998, 82, 159–166.
[59] Tang, D., & Ren, J. In situ amplified electrochemical immunoassay for carcinoembryonic antigen using horseradish peroxidase-encapsulated nanogold hollow microspheres as labels. Anal Chem. 2008, 80, 8064–8070.
[60] Richter, M. M. Electrochemiluminescence (ecl). Chem Rev. 2004, 104, 3003–3036.
[61] Algar, W. R., Tavares, A. J., & Krull, U. J. Beyond labels: a review of the application of quantum dots as integrated components of assays, bioprobes, and biosensors utilizing optical transduction. Anal Chim Acta. 2010, 673, 1–25.
[62] Ding, Z., Quinn, B. M., Haram, S. K., Pell, L. E., Korgel, B. A., & Bard, A. J. Electrochemistry and electrogenerated chemiluminescence from silicon nanocrystal quantum dots. Science, 2002, 296, 1293–1297.
[63] Myung, N., Bae, Y., & Bard, A. J. Effect of surface passivation on the electrogenerated chemiluminescence of CdSe/ZnSe nanocrystals. Nano Letters, 2003, 3, 1053–1055.
[64] Bae, Y., Myung, N., & Bard, A. J. Electrochemistry and electrogenerated chemiluminescence of CdTe nanoparticles. Nano Letters, 2004, 4, 1153–1161.
[65] Bae, Y., Lee, D. C., Rhogojina, E. V., Jurbergs, D. C., Korgel, B. A., & Bard, A. J. Electrochemistry and electrogenerated chemiluminescence of films of silicon nanoparticles in aqueous solution. Nanotechnology, 2006, 17, 3791.

Evgeny Katz

8 Signal-activated biomolecular release from alginate-modified electrodes

8.1 Introduction – signal-activated biomolecular release processes

Chemical systems based on stimuli-responsive materials designed to release loaded substances in response to external signals received considerable attention due to their potential applications in various areas [1–3], particularly for delivery of bioactive species (e.g., drugs) [4–7]. The releasing processes stimulated by various physical and chemical signals, such as light [8], magnetic field [9], temperature change [10], and pH variation [11], have been extensively studied. Drug release triggered by biomolecular signals (particularly represented by biomarkers signaling physiological dysfunctions) [12] are of special interest, being promising for the functional integration of releasing systems with biological processes [13, 14]. Such systems could be important for the development of closed-loop *sense-and-act* biomedical devices [15–18] in the general frame of a theranostic approach (a combination of diagnostics and therapy) [19]. In most of the presently designed substance-releasing systems activated by biochemical signals, the sensing component is represented by chemical receptors directly integrated with the releasing systems [20–22], significantly increasing the systems' complexity and limiting their versatility. The system design would be much easier and more flexible if the subsystems responsible for the signal-processing and signal-triggered drug release are realized in different units, essentially on different electrodes, communicating electrically. This chapter summarizes recent efforts in assembling bioelectronic systems for signal-stimulated substance release based on alginate polymers [23–30], as well as discusses future developments and possible applications of such systems.

8.2 Alginate polymer cross-linked with Fe^{3+} cations – a convenient matrix for molecular release stimulated by electrochemical signal

Alginate, a natural polymer, has attracted attention of researchers owing to its ease of availability, compatibility with hydrophobic as well as hydrophilic

Evgeny Katz, Department of Chemistry and Biomolecular Science, Clarkson University, Potsdam,NY

https://doi.org/10.1515/9783110570526-008

molecules, lack of toxicity, and attractive adhesive and mechanical properties [31]. Alginate gels are biodegradable in physiological conditions and are chemically erasable in basic (pH > 7) aqueous environments [31]. Owing to the biocompatibility of alginate polymers, they were used for the formation of membranes and thin films potentially useful in bioseparation and other biorelated applications [32]. Because of alginate ability to be ionically cross-linked with multivalent metal cations entrapping biomolecules into the biopolymer matrix, numerous reports have been published on encapsulation of proteins/enzymes [33–35], DNA [36, 37], cells [38, 39], and other biomolecular species [40, 41] in alginate hydrogels with the retention of their full biological activity. The alginate hydrogel is most notably used as films or microcapsules that can release components passively or in response to changed environmental conditions, through the controlled degradation of the assembly [42, 43]. Alternatively, the use of external stimuli allowing for triggering of release of encapsulated species on demand, irrespective of the environmental conditions, is less exploited for alginate in literature [44]. Among various stimuli used for triggering molecular release from signal-responsive matrices [45–68], the electrochemical trigger is particularly appealing because it enables a precise control over the dissolution process of polymer matrices sensitive to redox transformations.

This control can be achieved through cross-linking of alginate with iron cations, which possess distinctly different coordination chemistry of Fe^{2+} and Fe^{3+} cations. Fe^{2+} is a "soft" metal cation that tends to bind neutral ligands containing nitrogen and sulfur atoms, while Fe^{3+} cation is a typical example of a "hard" metal cation that preferentially binds oxygen atoms in negatively charged ligands such as carboxylate group [69]. The significant difference in binding of carboxylate groups by Fe^{2+} and Fe^{3+} is evident from stability constants of their citrate complexes, which have log K values 3.2 and 11.85, respectively [70]. Because binding of carboxylate groups by Fe^{2+} is substantially weaker than their binding by Fe^{3+}, it can be expected that "soft" Fe^{2+} cations will have a lower ability for cross-linking alginate in comparison to "hard" Fe^{3+} cations and interconversion between Fe^{2+} and Fe^{3+} will directly affect the alginate gel stability. As expected, Fe^{3+} cations produce cross-linked alginate resulting in the formation of hydrogel, while Fe^{2+} cations keep alginate in a soluble state in aqueous solution (Fig. 8.1).

Aiming at the electrochemically controlled formation and dissolution of alginate thin films resulting in the entrapment and release of biomolecules, respectively, Fe^{3+} cross-linking of alginate on an electrode surface has been studied [71]. Ionic iron can be oxidized and reduced electrochemically, thus controlling the alginate cross-linking electrochemically. The advantage of iron ions in comparison to Ca^{2+} (frequently used for alginate cross-linking [31]) is implementing of the electrochemical control over both process of the thin film formation and dissolution upon changing the iron ions oxidation state. The electrochemical fabrication

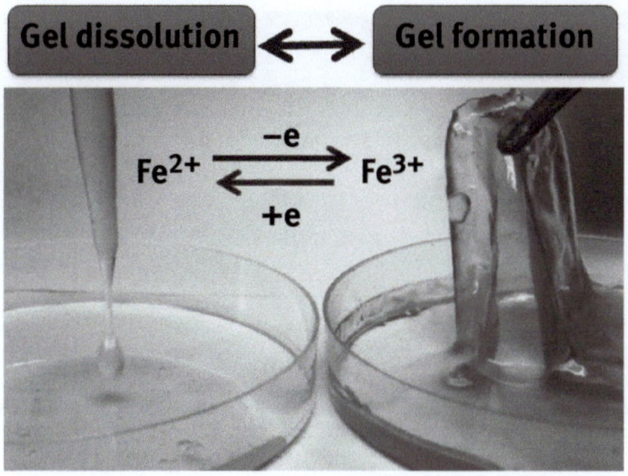

Fig. 8.1: The photograph represents reversible formation of alginate hydrogel in the presence of Fe^{3+} cations capable of alginate cross-linking and hydrogel dissolution when Fe^{3+} is converted to Fe^{2+} cations, which have much weaker interactions with alginate and do not cross-link the alginate molecules. This figure aims only at the schematic illustration of different interactions of alginate with iron ions in different oxidation states. Note that in the real systems described in the chapter the alginate matrix was produced on an electrode surface in the presence of electrochemically generated Fe^{3+} cations to yield a thin film, which was later dissolved upon electrochemical reduction of Fe^{3+} to Fe^{2+}. Adapted from Jin et al. [71] with permission.

method provides the possibility of encapsulating proteins and other biomolecules in alginate thin films as a model system for drug delivering. Applying oxidative potential to an electrode in the presence of soluble alginate and Fe^{2+} cations results in electrochemical oxidation of iron cations to yield Fe^{3+}, which immediately cross-link alginate, producing a thin film of the alginate hydrogel on the electrode surface. When other biomolecules (e.g., enzymes/proteins, DNA or drugs) [72] are present in the solution, the biomolecules are physically entrapped into the alginate hydrogel. The thickness of the alginate film and, therefore, the amount of the entrapped biomolecules can be controlled by the time period of the electrochemical deposition process (Figs. 8.2 and 8.3) [71]. In addition to the time control, spatial control for the alginate film deposition was possible by applying a scanning electrochemical microscope as a tool for the electrochemically produced alginate hydrogel patterns [71]. After the alginate film deposition is completed, the opposite process of its electrochemical dissolution is possible to release the biomolecules entrapped in the film during the film formation. A reductive potential can be applied to convert the cross-linking Fe^{3+} cations back to Fe^{2+}, which are not capable of the alginate cross-linking. As soon as the reductive potential is applied, the release of the entrapped biomolecules (e.g., lysozyme drug) starts (Fig. 8.4). Notably, uncontrolled leakage of the entrapped lysozyme molecules is negligible in comparison with the rate of the electrochemically stimulated release (Fig. 8.4, inset). The rate of the alginate film dissolution and, consequently, the rate of the release process are controlled by the value of the reductive potential applied – the

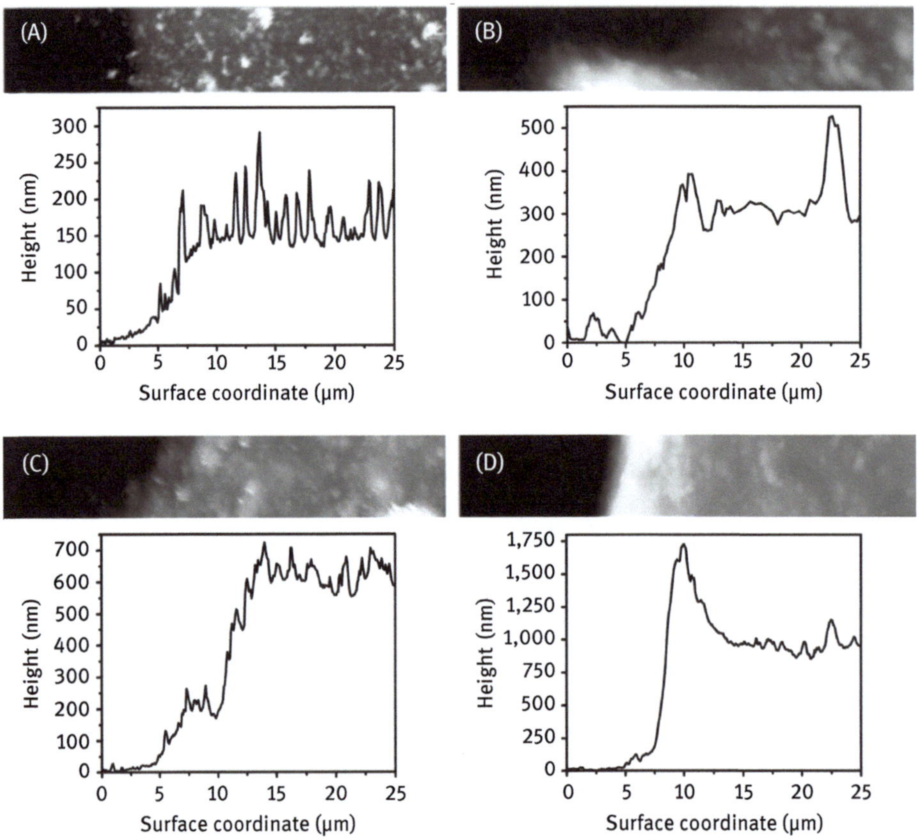

Fig. 8.2: The atomic force microscopy (AFM) topography images and the corresponding cross-sectional profiles of the alginate hydrogel films electrodeposited for 50 s (A), 100 s (B), 200 s (C), and 400 s (D) from the solution composed of alginate (1.5% w/w), $FeSO_4$ (35 mM), bovine serum albumin (used as a model for the entrapment release) (2.5 mg mL^{-1}) and 0.1 M Na_2SO_4 upon application of 0.8 V. The AFM images show the alginate film thickness increase with the increasing deposition time. Adapted from Jin et al. [71] with permission.

higher the applied negative potential, the faster is the release process. Overall, the electrochemical alginate hydrogel deposition-dissolution is a convenient approach to the electrochemically triggered biomolecular release. However, in the present realization, the electrochemical dissolution of the alginate hydrogel leading to the release process of the entrapped molecules is limited to the potential applied from an external electrical source. In the next step of designing "smart" signal-controlled release systems, the potential required to initiate the release process can be generated *in situ* through biocatalytic reactions.

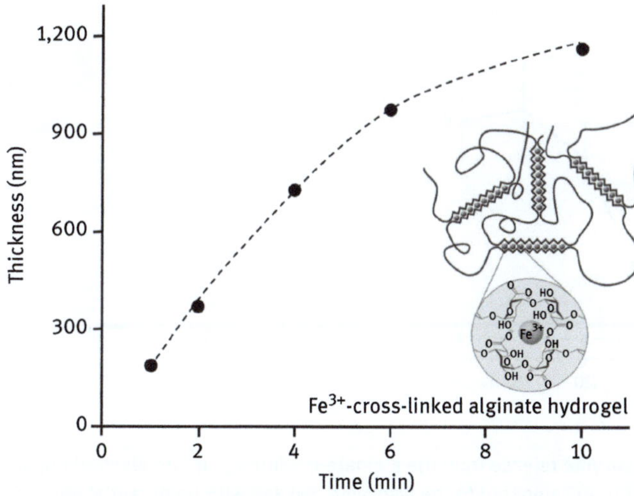

Fig. 8.3: The alginate hydrogel film thickness derived from the AFM images (see Fig. 8.2) as a function of the electrochemical deposition time period. The inset scheme shows the Fe^{3+} cross-linked alginate structure. Adapted from Jin et al. [71] with permission.

8.3 Self-operating release systems based on the alginate electrodes integrated with biosensing electrodes

The recently developed substance-releasing bioelectronic systems [23–30] are composed of two modified electrodes: one operating as a bioelectrocatalytic electrode activated by various biomolecular signals and producing reductive potential/current applied to the second electrode to activate release of substances loaded in a polymeric matrix, which is dissolved upon the reduction process. The polymeric matrix made of Fe^{3+}-cross-linked alginate was decomposed and dissolved upon reduction of cross-linking Fe^{3+} cations to the form of Fe^{2+} cations, which are not capable of cross-linking alginate [71, 72]. Thus, the reductive process stimulated by the biocatalytic electrode in the presence of biomolecular signals resulted in the polymer matrix dissolution and the concomitant release of loaded substances. The challenges of this approach are mostly in the design of the signal-processing electrode providing the reductive potential/current for the dissolution of the ferric-alginate matrix and in preventing the leakage of the loaded substances from the alginate matrix prior to its signal-stimulated dissolution. A few examples of experimental realizations of these systems are overviewed as follows, emphasizing the use of different biomolecular signals to stimulate the release process. The biomolecular signals activating the release process included the following: (i) reductive enzyme substrates (e.g., glucose), (ii) complex combinations of biomolecules

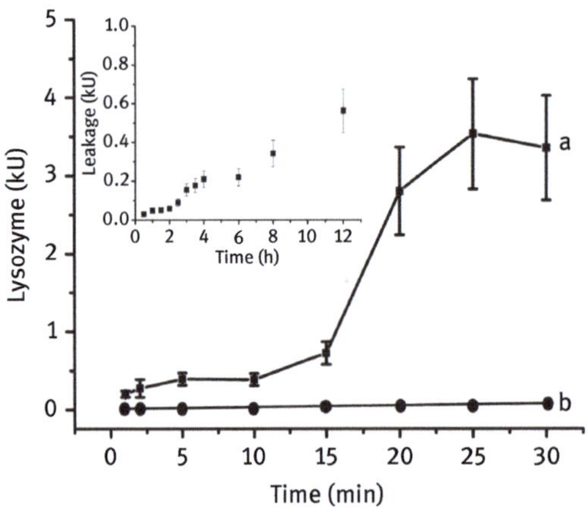

Fig. 8.4: The time-dependent lysozyme release from the alginate-modified graphite electrode upon application of −1.0 V (vs. Ag/AgCl) (a) compared to the lysozyme leakage with no potential applied (b). The inset shows the lysozyme leakage with no potential applied over a time period of 12 h. The experiments were performed in 0.1 M Na_2SO_4 solution, pH 6.0. The modified electrodes were prepared by the electrochemical deposition of the alginate film at +0.8 V for 60 s from the solution containing 35 mM Fe^{2+}, 15 mg mL^{-1} alginate and 5 mg mL^{-1} lysozyme in 0.1 M of Na_2SO_4 pH 6.0. Each point on the figure corresponds to utilization of a new electrode. Standard deviation was calculated based on the repetition of each experiment three times. The amount of the lysozyme released was measured (kilounits) according to the standard assay procedure (https://www.sigmaaldrich.com/technical-documents/protocols/biology/enzymatic-assay-of-lysozyme.html). Adapted from Jin et al. [72] with permission.

logically processed by biocomputing systems according to a predesigned program implemented in the system, (iii) various proteins and, finally, and (iv) biological cells (e.g., bacterial cells). The released substances included different drugs, drug-mimicking molecular species or nanoparticles (NPs), and even enzymes for activating biofuel cells. First, we will consider the signal-processing biocatalytic electrodes and then we will describe the substance-releasing electrodes and the use of their combinations for various biomedical and biotechnological applications.

In the simplest experimental realization, the biosensing electrode was functionalized with pyrroloquinoline quinone-dependent glucose dehydrogenase (PQQ-GDH; E.C. 1.1.5.2) biocatalytically oxidizing glucose. The material called "buckypaper" [73–75], made from compressed multiwalled carbon nanotubes (MWCNTs), was used as a conducting support for preparing the enzyme-modified electrode (Fig. 8.5A). PQQ-GDH was linked to the MWCNTs using a heterobifunctional cross-linker, 1-pyrenebutanoic acid succinimidyl ester (PBSE). Succinimidyl ester forms covalent amide bond with amino groups on protein lysine residues, while the other end of the linker, polyaromatic pyrenyl moieties, interacts with MWCNTs via π–π stacking (Fig. 8.5B). MWCNTs provided efficient direct nonmediated electron transfer from the PQQ-active center of the immobilized enzyme to the conducting support (Fig. 8.6A), thus resulting

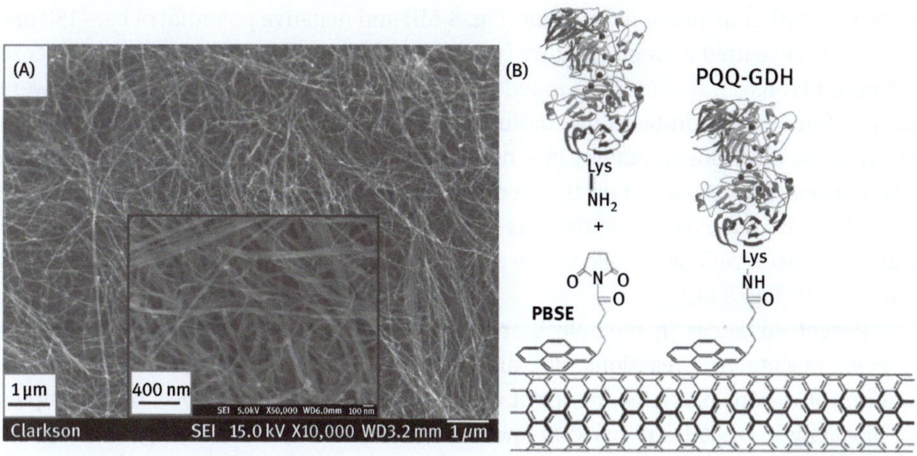

Fig. 8.5: (A) The scanning electron microscopy image of the buckypaper electrode. The inset shows the magnified image of MWCNTs. (B) Immobilization of the PQQ-GDH on MWCNTs with the help of the heterobifunctional linker PBSE, which provides covalent binding with amino groups of protein lysine residues through formation of amide bonds and also interacts with MWCNTs via π–π stacking of the polyaromatic pyrenyl moiety.

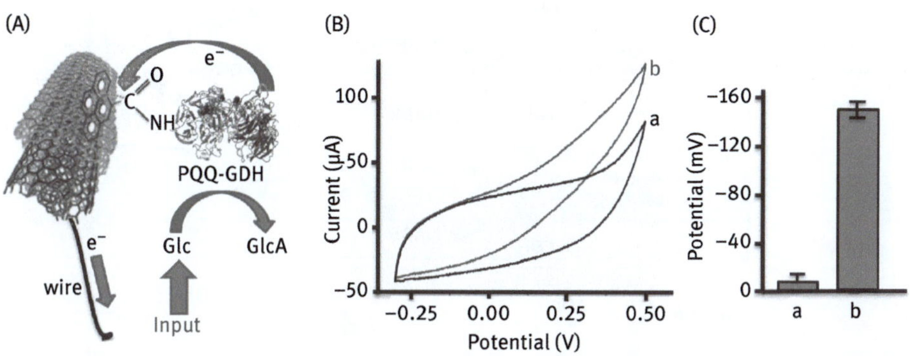

Fig. 8.6: (A) Biocatalytic oxidation of glucose at a PQQ-GDH-modified electrode made of compressed MWCNTs ("buckypaper"). The biomolecular signal represented by glucose resulted in the formation of a negative potential and anodic current on the modified electrode. (B) Cyclic voltammograms obtained with the PQQ-GDH-modified electrode in the absence (a) and presence (b) of 25 mM glucose, scan rate 1 mV s^{-1}. Background solution was composed of 50 mM 3-(*N*-morpholino)propanesulfonic acid buffer, pH 7.0, containing 100 mM Na_2SO_4 and 1 mM $CaCl_2$. Note electrocatalytic anodic current corresponding to the glucose oxidation at potentials more positive than −100 mV. (C) Potentials measured on the PQQ-GDH-modified electrode versus the reference electrode (open circuitry conditions) in the absence (a) and presence (b) of 25 mM glucose. Adapted from Katz et al. [30] with permission.

in the formation of an anodic current (Fig. 8.6B) and negative potential of ca. −150 mV when it is measured in open circuitry (vs. Ag|AgCl|KCl, 3 M) in the presence of glucose (25 mM; Fig. 8.6C). It should be noted that similar glucose-oxidizing electrodes were successfully used in implantable biofuel cells operating *in vivo* [76–78], thus demonstrating the electrode operation in a biological environment. The reductive potential and current were produced on the modified electrode only in the presence of glucose (note that the reductive potential was produced on the electrode in a broad range of glucose concentrations), while in the absence of glucose the electrode potential was ca. −10 mV (Fig. 8.6C).

Recent advances in biomolecular logic systems [79–83], particularly based on enzyme-biocatalyzed reactions [84], allowed electrode activation by complex combinations of various biomolecular input signals [85–87]. Figure 8.7 shows two examples of biocatalytic cascades mimicking operation of a three-input OR logic gate and three concatenated AND logic gates for processing biomolecular signals. The three-input OR logic gate [28] was realized with reactions biocatalyzed by three NAD^+-dependent

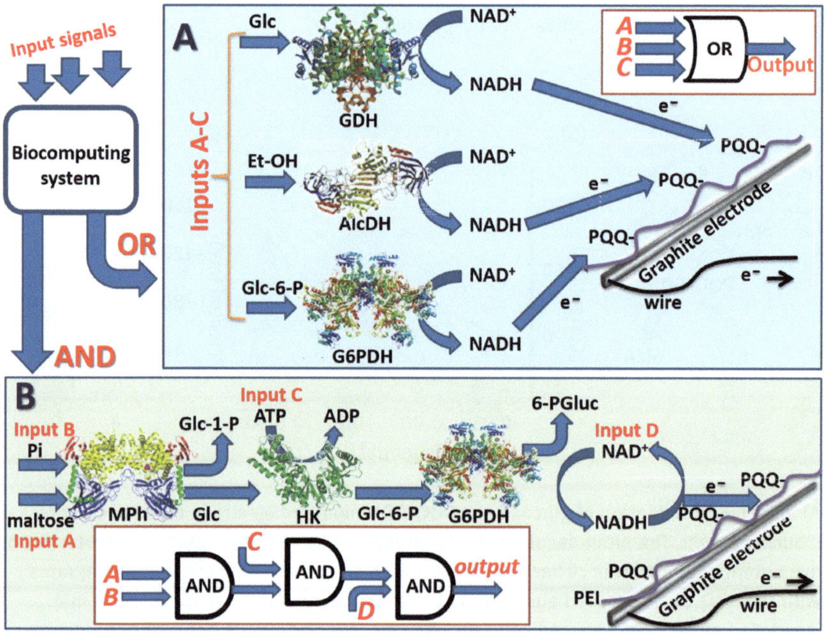

Fig. 8.7: Biocatalytic cascades logically processing multiple biomolecular input signals. (A) The biocatalytic reactions operating in parallel and mimicking three-input OR logic gate. (B) The biocatalytic cascade including three consecutive biocatalytic reactions and mimicking three concatenated AND logic gates. The biocatalytically produced NADH was electrocatalytically reoxidized back to NAD^+ at the PQQ-modified graphite electrode. The NADH oxidation generated a negative potential and anodic current on the modified electrode. (see the abbreviations in the list of the used abbreviations.) Adapted from Katz et al. [30] with permission.

enzymes, glucose dehydrogenase (GDH; E.C. 1.1.1.47), alcohol dehydrogenase (AlcDH; E.C. 1.1.1.1), and glucose-6-phosphate dehydrogenase (G6PGH; E.C. 1.1.1.49), operating in parallel (Fig. 8.7A). The biocatalytic reactions were activated by input signals *A*, *B*, and *C*, represented by the corresponding substrates: glucose (Glc), ethanol (Et-OH), and glucose-6-phosphate (Glc-6-P), respectively. Aiming at the simplest concept demonstration, logic value **0** for all input signals was defined as the absence of the substrates, while logic value **1** was defined as the experimentally convenient and optimized concentrations: Glc (2 mM), Et-OH (21 mM), and Glc-6-P (0.85 mM). All three reactions resulted in the generation of the reduced form of nicotinamide adenine dinucleotide (NADH) upon reduction of the β-nicotinamide adenine dinucleotide (NAD^+) cofactor and concomitant oxidation of the substrate inputs. The NADH formation was activated in the presence of any of the input signals applied in any combination (**0,0,1**; **0,1,0**; **1,0,0**; **1,1,0**; **1,0,1**; **0,1,1**; **1,1,1**). The only situation when NADH was not produced was the complete absence of all input signals (**0,0,0**). When NADH was produced in the biocatalytic reactions it was reoxidized back to NAD^+ at the electrode modified with PQQ, which is a well-known catalyst for NADH electrochemical oxidation [88] (PQQ immobilization was achieved by its covalent coupling to the adsorbed polyethyleneimine, PEI). The electrocatalytic process of NADH oxidation resulted in the formation of the reductive potential/current on the modified electrode.

On the other hand, the concatenated AND gates [28] were mimicked by the enzymatic cascade including three consecutive reactions activated by four input signals (Fig. 8.7B). The biocatalytic reaction of maltose phosphorylase (MPh, E.C. 2.4.1.8) was activated in the presence of maltose (Input *A*) and inorganic phosphate (Input *B*) resulting in glucose and glucose-1-phosphate byproduct formation. In the next reaction step, biocatalyzed by hexokinase (HK; E.C. 2.7.1.1), glucose was converted to Glc-6-P in the presence of adenosine 5′-triphosphate (ATP; Input *C*). Finally, Glc-6-P reduced NAD^+ (Input *D*) to NADH in the process biocatalyzed by G6PDH (E.C. 1.1.1.49). The NADH production was obtained only when all four input signals were applied at their logic **1** value (**1,1,1,1**): maltose (2 mM), sodium phosphate (2 mM), ATP (1 mM), and NAD^+ (2 mM) for inputs *A*, *B*, *C*, and *D*, respectively. The generated NADH was reoxidized and recycled to NAD^+ at the PQQ-modified electrode resulting in the formation of an anodic current and a negative potential of ca. −80 mV on the electrode. If any of the input signals were applied at the logic **0** value (meaning the physical absence of the corresponding species; 15 different input combinations), the biocatalytic cascade did not continue to the very end and NADH was not produced, thus inhibiting formation of the reductive potential on the electrode. It should be noted that the realized biocatalytic cascades and the input signals activating the processes do not have any specific biomedical meaning and only represent a convenient model where the final production of NADH and the potential formation on the electrode are controlled by the correct combination of multiple input signals. In the advanced system, the biocatalytic cascades can be activated by biomarkers signaling physiological conditions and their changes [86], thus resulting in the potential formation on the sensing electrode, reflecting biomedical conditions in real time.

The aforementioned systems demonstrated the electrode activation in the presence of small molecules operating as reductive substrates for enzymatic reactions. One step forward included electrode activation by biomolecular signals represented by proteins [26]. The electrode modified with a PQQ electrocatalyst for NADH oxidation was further functionalized with an antibody (polyclonal anti-rat IgG-antibody from rabbit) for immune recognition of a protein signal (Fig. 8.8). The electrode was ready to respond to the immune signal, which was represented in this model system by polyclonal anti-rabbit IgG-antibody from goat. When this antibody was added to the solution (36 μg mL^{-1}) where the modified electrode was immersed, it attached to the Fab fragment of the primary antibody, resulting in the immune complex (note that the primary antibody is from rabbit and the signal is the anti-rabbit antibody). After the formation of the complex, the electrode surface reacted with polyclonal anti-goat IgG-antibody from mouse conjugated with silica NPs, ca. 200 nm diameter, loaded with NAD^+-dependent GDH. The enzyme-loaded NPs were attached to the

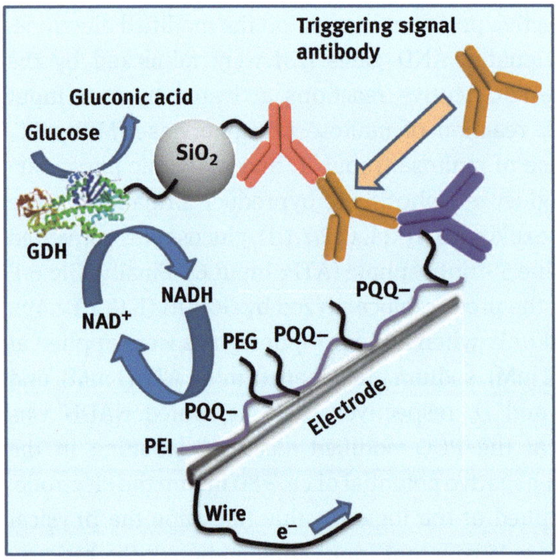

Fig. 8.8: The sensing electrode activation by assembling an immune complex on its surface in the presence of a protein signal. In this example, the signal was represented with an antibody specifically reacting with the complementary antibody attached to the electrode surface. The electrode activation was achieved by attaching a GDH-biocatalytic label to the affinity complex assembled on the electrode surface. The biocatalytic production of NADH resulted in the formation of the reductive potential/current on the PQQ-modified electrode. The PQQ electrocatalyst was covalently bound to PEI adsorbed on the graphite electrode surface. Note that the electrode surface was additionally functionalized with polyethylene glycol for reducing non-specific adsorption of proteins on the surface. Adapted from Katz et al. [30] with permission.

electrode surface through immune-complex formation of the anti-goat antibody with the Fab fragment of the signal antibody (note that the signal antibody was from goat). The GDH biocatalytic label attached to the electrode surface through the immune complex was able to reduce added NAD^+ to yield NADH in the presence of glucose. The produced NADH was electrocatalytically oxidized at the PQQ-modified electrode, producing the negative potential of ca. −80 mV and corresponding reductive current. It should be noted that the potential formation was obtained only when the whole complex was produced on the electrode surface that required the presence of the signal protein represented in this model system by anti-rabbit IgG antibody. In an advanced system, with the appropriate immune-recognition layer, the signal protein could be represented by any protein biomarker reporting various medical problems.

Aiming at increasing complexity of the biological signals further for the electrode activation and following the approach developed for the immune recognition of the protein signals described earlier, an electrode activated by the presence of bacterial cells was designed [23]. The sensing unit made of a graphite electrode modified with a mixed monolayer composed of PQQ and antibody (polyclonal anti-*E. coli* IgG-antibody from rabbit) was used to collect bacteria (*E. coli*) from the surrounding solution due to affinity interactions between the bacterial cells and the immobilized antibody. Then, the bacteria attached to the electrode surface through the affinity interaction with the immobilized antibody were reacted with silica nanoparticles (SiO_2-NPs; ca. 200 nm diameter) functionalized with the same bacteria-specific antibody and NAD^+-dependent GDH. The SiO_2-NPs were used as a dual platform for carrying the bioaffinity unit (antibody) attaching to the bacterial cells and the biocatalytic unit (GDH) that provided a biocatalytic redox reaction at the electrode surface. When the whole bioaffinity complex was assembled on the electrode surface, GDH attached to the complex was able to reduce NAD^+ in the presence of glucose in the proximity of the electrode surface. The biocatalytically produced NADH was reoxidized on the PQQ-modified electrode, thus yielding the reductive current and negative potential of ca. −80 mV [23]. This process was possible only in the presence of the bacterial cells, which were the signals activating the bioelectrocatalytic electrode.

The systems described earlier were activated by assembling a multicomponent ensemble capable of generating a reductive potential on the modified electrode. The assembling proceeded in the presence of a protein (an antibody in the given example) or microbial cells, both operating as activating signals. However, this approach suffers from the need of human-operator actions required to treat the modified surface with secondary antibody labeled with a biocatalytic tag to perform the potential-producing catalytic reaction. A different approach that does not require any human operation was developed using disassembling, rather than assembling process, for activating the sensing electrode [89]. The biosensing system was already preassembled on the electrode surface and was waiting for a biomolecular signal to start working. Figure 8.9A shows one of the designed configurations where a buckypaper electrode was primarily modified with PQQ-GDH, which is capable of

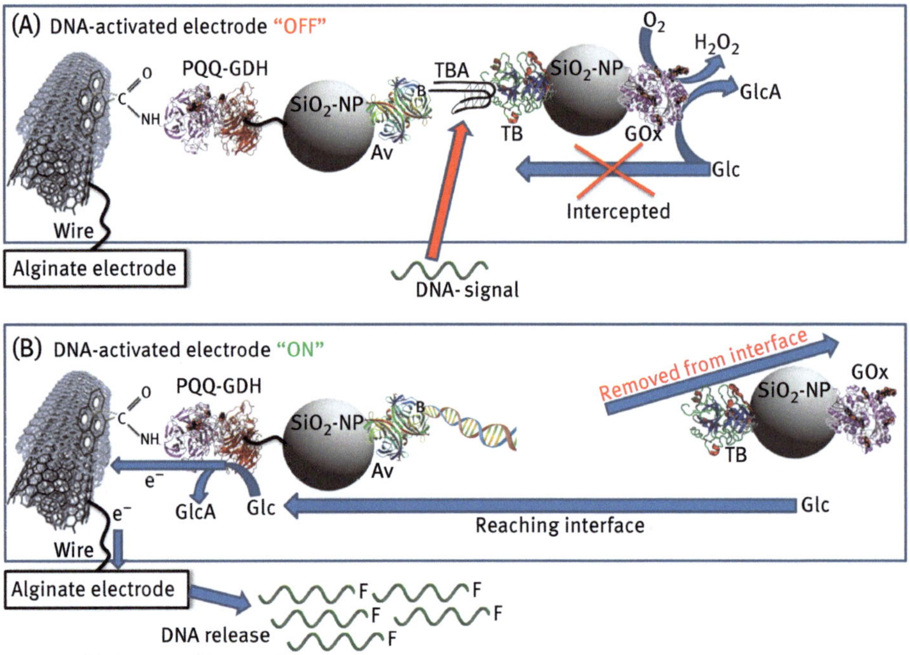

Fig. 8.9: (A) The biocatalytic-biosensing electrode in the "OFF" state due to interception of glucose by GOx immobilized on the SiO_2-NPs linked to the surface through the aptamer-thrombin complex. (B) The electrode in the "ON" state after removing thrombin-GOx-NPs by the DNA signal. Note that the biosening electrode is connected to an alginate-electrode for the signal-triggered DNA release. The amount of the released DNA is much greater than the amount of the DNA signal, thus resulting in amplification of the DNA signal. The sequence of the released DNA is not necessary the same as in the DNA signal. Adapted from Gamella et al. [89] with permission.

communicating with the electrode directly, generating a negative potential in the presence of glucose. However, this reaction was inhibited by the presence of glucose oxidase (GOx) bound to the external surface through a linker composed of thrombin and thrombin aptamer pair. NPs (SiO_2–NPs, 200 nm diameter) were used as a high-surface platform for immobilizing the linker and GOx, thus resulting in a high local concentration of GOx, which allowed affective interception of glucose, preventing it from reaching the internal layer with PQQ-GDH. Note that the glucose oxidation biocatalyzed by GOx does not result in any current at the modified electrode since the electrons received from glucose are used for O_2 reduction, resulting in H_2O_2 formation. Therefore, even in the presence of glucose in a solution, the electrode was in its mute, inactive state. The electrode was activated by applying a signal in the form of DNA (a short artificial oligonucleotide) complimentary to the thrombin aptamer. Hybridization of the DNA signal with the thrombin aptamer resulted in dissociation of the linker and removal of the GOx-functionalized SiO_2–NPs from the electrode surface (Fig. 8.9B). This resulted in glucose access to the internal layer where PQQ-GDH was activated, thus producing a negative potential. The activation process was possible only once since the preassembled sensing layer was decomposed irreversibly, but importantly it did not require any additional human-operator steps, thus allowing self-operation of the sensing system.

Another system based on the same disassembly approach was developed for activating the sensing electrode with ketone body (a biomarker of diabetic ketoacidosis) [90]. In this system, the linker between the catalytically active interface and the NPs holding GOx was composed of an immune complex of an isomeric form of the ketone body and the ketone body-specific antibody. The preassembled biomolecular film was in a nonactive OFF state because GOx consumed glucose preventing its penetration to the internal structure of the layer, thus keeping GDH in the mute state in the absence of the glucose substrate (Fig. 8.10A). The isomeric form of the ketone

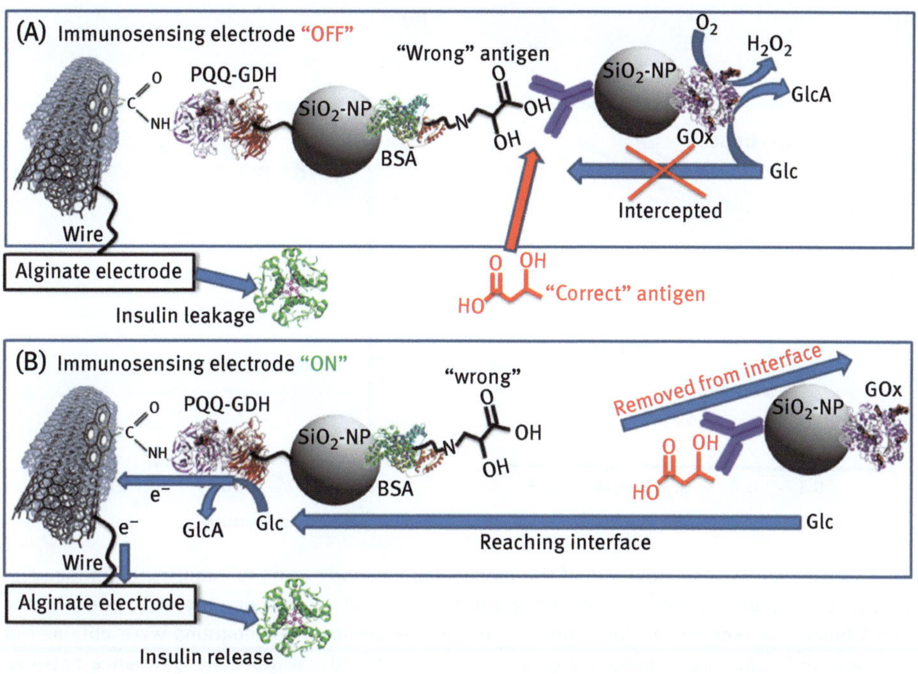

Fig. 8.10: (A) The biosensing electrode in the "OFF" state due to interception of glucose by GOx immobilized on the SiO_2-NPs linked to the surface through the immune interaction with the "wrong" antigen. Note that the immune complex is broken and the antibody-GOx-NPs are removed from the surface in the presence of 3-HBA biomarker ("correct" antigen). (B) The electrode in the "ON" state after removing antibody-GOx-NPs. Note that the biosening electrode is connected to an alginate electrode for the signal-triggered insulin release. Adapted from Gamella Carballo et al. [90] with permission.

body (4-amino-2-hydroxybutyric acid) was a "wrong" antigen for the used antibody with a low stability of the immune complex. The signal used to activate the biocatalytic electrode was the "correct" ketone body (3-hydroxybutyric acid, 3-HBA) with the structure corresponding to the antibody recognition sites. In the presence of the "correct" antigen, the previously formed immune complex was dissociated and the new stable complex of the antibody with the "correct" antigen was produced. This stable complex was removed from the electrode surface opening the biocatalytic interface for glucose and activating GDH (Fig. 8.10B). The electrocatalytic oxidation of glucose biocatalyzed by GDH was observed by cyclic voltammetry (Fig. 8.11A) and the bioelectrocatalytic process generated a negative potential on the modified electrode in its ON state (Fig. 8.11B). One more system based on the same concept was activated by pH changes splitting a pH-signal-cleavable linker composed of iminobiotin/avidin complex [24].

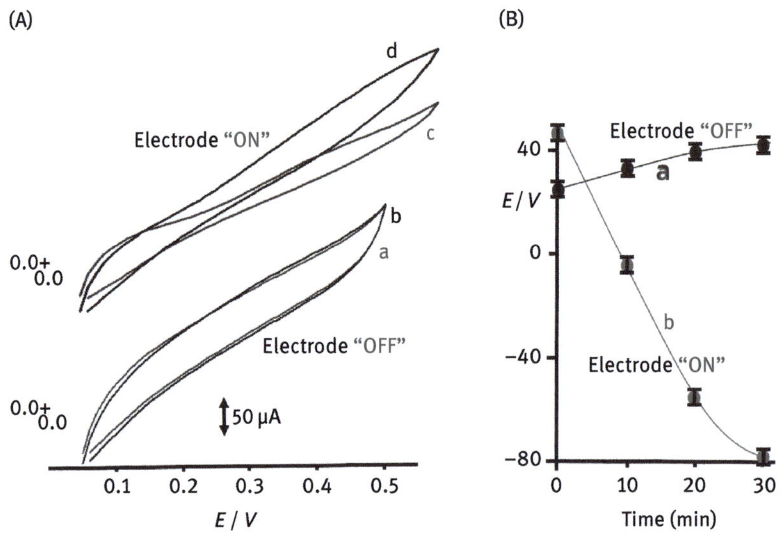

Fig. 8.11: (A) Cyclic voltammograms of the biosensing electrode in its full configuration (see Fig. 8.10A), curves a and b, and after removing antibody-GOx-NPs by reacting the electrode with the 3-HBA biomarker (see Fig. 8.10B), curves c and d. The cyclic voltammograms were obtained in the absence (a and c) and presence (b and d) of glucose 20 mM. Note the difference between curves c and d, which corresponds to the bioelectrocatalytic oxidation of glucose after removing the inhibiting GOx-NPs (the electrode in its ON state). The cyclic voltammograms in the absence and presence of glucose prior to removing GOx-NPs do not show any electrocatalytic current (the electrode is in its OFF state). Potential scan rate is 2 mV s^{-1}. (B) Potential generated on the biosensing electrode in the presence of 20 mM glucose (measured vs Ag|AgCl| KCl, 3 M, reference electrode) before (a) and after (b) removing antibody-GOx-NPs from the electrode surface. Adapted from Gamella Carballo et al. [90] with permission.

The negative potential and reductive current produced on the biocatalytic electrodes in the presence of biomolecular signals exemplified above were applied on another modified electrode coated with the Fe^{3+}-cross-linked alginate thin film containing loaded substances of various composition and complexity. It was already explained earlier that the reduction of Fe^{3+} cations to Fe^{2+} state results in the decomposition and dissolution of the alginate matrix, due to the reduced ability of Fe^{2+} cations to cross-link the alginate molecules [71]. Therefore, the electrochemically stimulated reduction process resulted in alginate matrix dissolution and release of the preloaded species from the dissolved thin film. Figure 8.12 schematically shows the ketone body-stimulated activation of the biosensing electrode (see also Fig. 8.10) and the concomitant decomposition/dissolution of the alginate film at the connected electrode resulting in the release of insulin entrapped in the alginate film. The intact molecular structure of the released insulin was confirmed by its mass spectrum (Fig. 8.13A). The rate of the signal-stimulated insulin release was significantly higher than its uncontrolled leakage (Fig. 8.13B).

It should be noted that the insulin release triggered by the ketone body diabetic ketoacidosis biomarker and DNA release activated by a DNA signal are only specific

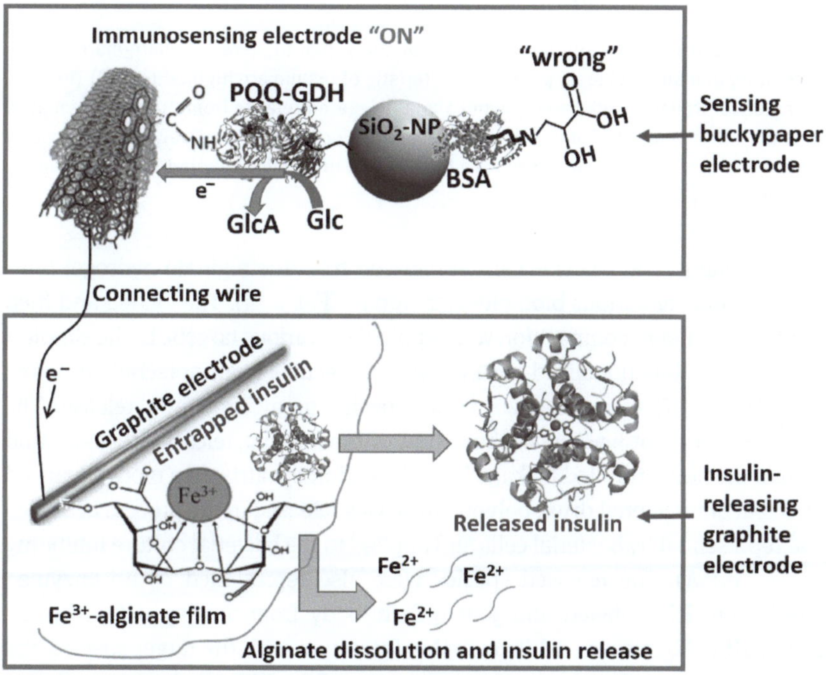

Fig. 8.12: Schematics of the two-electrode system operation (see also Figure 8.10). The activated immune-sensing electrode produces reductive current, which results in the reductive decomposition of the alginate film. Insulin physically entrapped in the alginate film is released upon the alginate film dissolution. (Adapted from ref. 90 with permission.)

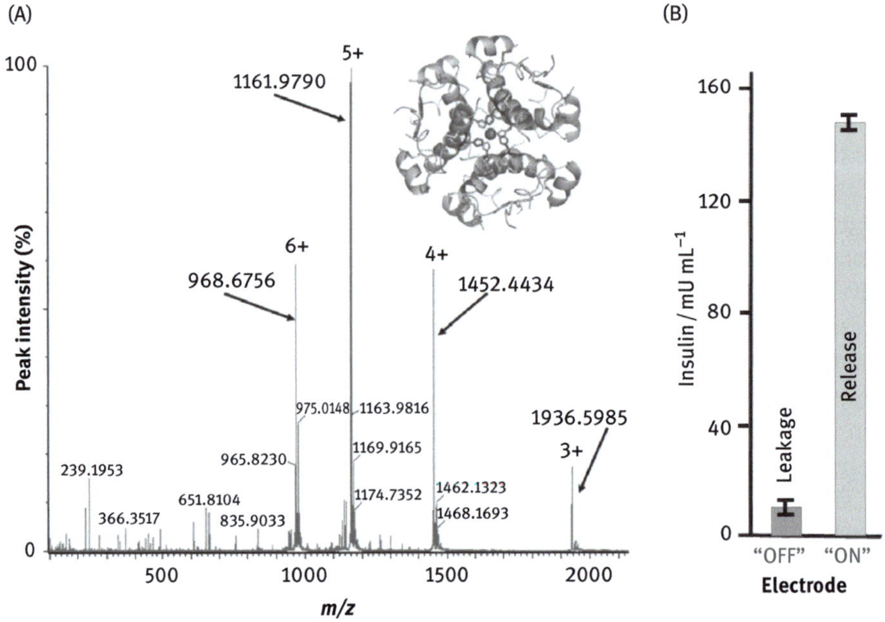

Fig. 8.13: (A) Mass spectrum of the solution produced by the signal-triggered alginate-matrix dissolution resulting in insulin release; peaks characteristic of insulin are highlighted. (B) The bar chart comparing uncontrolled insulin leakage and signal-triggered release from the alginate matrix corresponds to the "OFF" and "ON" states of the biosensing electrode, respectively. Insulin content was measured by the Bradford protein assay for a time period of 45 min. Adapted from Gamella Carballo et al. [90] with permission.

examples, while many other species can be released from the alginate hydrogel films decomposed by applying various biomolecular signals (Fig. 8.14). The loaded and then released species of different composition were applied for various targets. In the simplest example [29], Au-NPs functionalized with enzyme molecules (e.g., horseradish peroxidase (HRP), E.C.1.11.1.7) were used as model species mimicking drug release. The purpose of this experiment was the optimization of the loading/release processes and minimization of the noncontrolled leakage from the alginate matrix. In a more advanced system [23], a real antibacterial drug, polymyxin B, was released in the process activated by the signal represented by bacterial cells and applied to the bacterial culture inhibiting its growth (Fig. 8.15A). The released species were also represented by an enzyme–antibody conjugate (HRP-labeled anti-goat IgG-antibody from donkey; HRP-Ab; Fig. 8.14) [26]. After HRP-Ab release and its complex formation with the target species, the complex formation was analyzed by an immune assay of the HRP label (enzyme-linked immunosorbent assay), where HRP catalyzed biochemical transformations (Fig. 8.15B). The noncontrolled leakage of HRP-Ab from the alginate matrix (Fig. 8.15B, curve a) was negligible when compared to the signal-stimulated release (Fig. 8.15B, curve b). In this case, the released species were able to target specific biomolecules (e.g., complementary

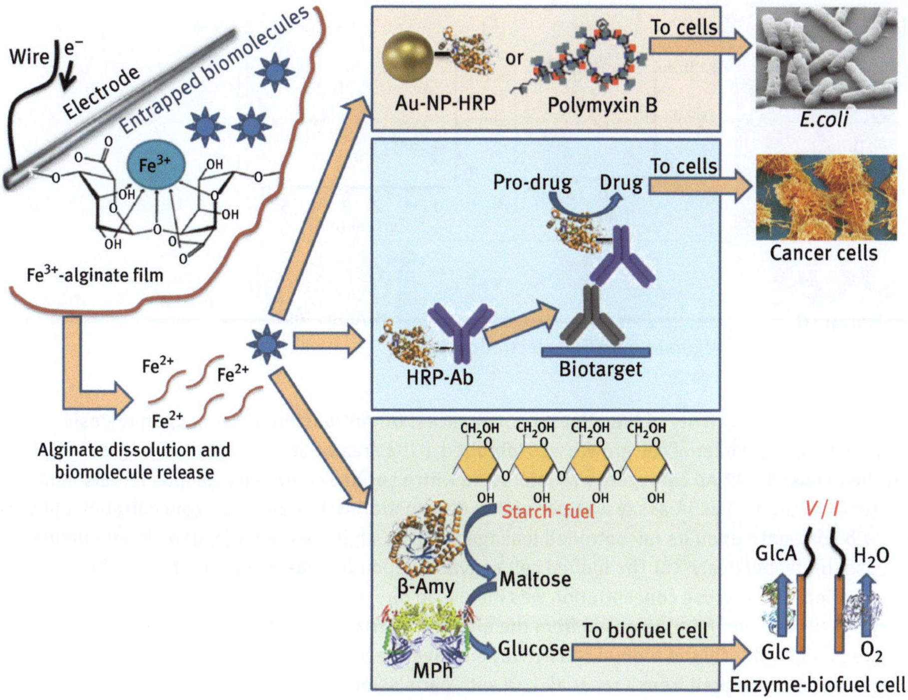

Fig. 8.14: Electrochemically stimulated reductive dissolution of Fe^{3+}-cross-linked alginate and concomitant release of the loaded substances, drugs, drug-mimicking species, and enzymes, aiming at different biomedical and biotechnological applications. (See the abbreviations in the list of the used abbreviations.) Adapted from Katz et al. [30] with permission.

polyclonal anti-rabbit IgG-antibody from goat), thus mimicking, for example, targeting cancer cells in future applications. The realized model system corresponds to the biomedical scenario where the "activator" (modeled with HRP-Ab) is released in response to a specific combination of biomarker signals and the "drug" is locally produced from the inactive "prodrug" with the help of the released "activator" (modeled with the reaction catalyzed by HRP; Fig. 8.15).

In addition to future biomedical applications of the signal-triggered release processes (mostly related to the signal-stimulated drug release), another system releasing enzymes, β-amylase (β-Amy; E.C. 3.2.1.2) and MPh (E.C. E.C. 2.4.1.8), was designed [27]. The released enzymes were decomposing starch and yielding glucose, which was used as a biomolecular "fuel" for a biofuel cell (Fig. 8.14). When the enzymes were released and after they decomposed starch yielding glucose, the current–voltage output produced by the biofuel cell was significantly increased corresponding to the increased glucose concentration (Fig. 8.15C). This example demonstrated the application of the releasing system for the activation of a biofuel cell triggered by biomolecular signals.

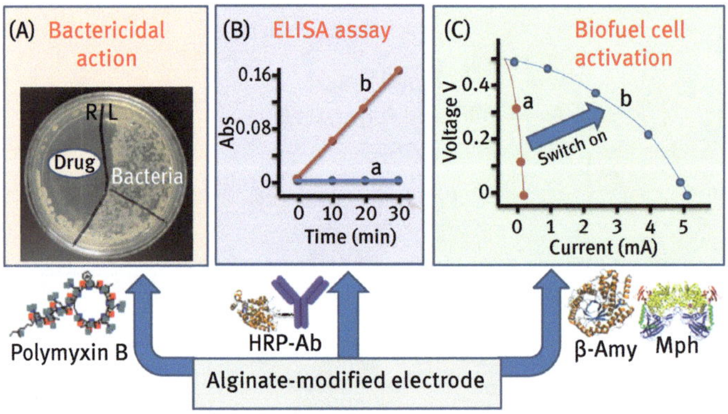

Fig. 8.15: Different activity of the released substances: (A) Antibacterial action upon releasing of polymyxin B. The bacterial cell growth was inhibited in the area where the released drug was applied. (B) The released HRP-Ab conjugate was interacted with a surface containing complementary immune-species and then an ELSA assay was performed demonstrating low and high concentrations of the HRP-Ab conjugate upon its uncontrolled leakage from the alginate matrix (a) and signal-stimulated release (b), respectively. (C) The biofuel cell activation upon increasing concentration of glucose in the solution. The glucose concentration was increased due to starch digestion in the presence of the enzymes (β-Amy and MPh) released from the alginate matrix. The plot shows polarization curves before (a) and after (b) the biofuel cell activation. (See the abbreviations in the list of the used abbreviations.) Adapted from Katz et al. [30] with permission.

8.4 Conclusion and perspectives

The systems described earlier illustrated the versatility of a bioelectronic approach to signal-triggered release of various substances. This approach provides inspiration for future research and possible practical applications. Indeed, the broad range of the applied biomolecular signals, also reaching the level of bacterial cells used as input signals, would allow a great variety of activating processes. While the biocatalytic/biosensing electrodes responded directly to small molecular species (e.g., glucose), the use of proteins and bacterial cells as input signals required additional processing steps similar to immune sensing, where secondary enzyme-labeled antibodies have to be attached to the primary affinity complex to activate the electrode. Obviously, the need of the enzyme-labeled antibody to trigger the release process is the drawback in the studied systems and the real drug-releasing systems operating *in vivo* under physiological conditions should respond to immune signals without additional treatment steps with human intervention. Thus, the presently designed systems can only be considered as the first prototype, only for concept demonstration, not yet ready for immediate practical application. This problem has been resolved by using pre-assembled systems being in the mute state until the inhibiting GOx species are removed from the surface by the chemical signal that splits the linker between the GOx and electrode surface, thus removing GOx and activating the biocatalytic electrode [89, 90].

The different biomolecular species and nano-objects released from the alginate matrix in the process triggered by the various signals could be used for many different biomedical and biotechnological applications. The ability of the modified electrodes to work in a biological environment [76–78] is promising for designing implantable bioelectronic devices [91] operating in biofluids and releasing biological substances in response to complex combinations of biomarkers. In future biomedical applications, a similar setup composed of two modified electrodes would allow separation of the potential-generating sensing electrode and drug-releasing electrode; they might be at different locations operating in different environments and under different conditions. Alternatively, Fe^{3+}-cross-linked alginate thin films containing loaded substances could be dissolved and the substances could be released by light signals [92] or by biocatalytic reactions performed inside the alginate films [93], bringing even more flexibility to future applications. Overall, this research direction could potentially contribute to future systems for autonomous sensing and actuating in the general frame of a theranostic approach. Another application of the systems discussed earlier belongs to the novel area of unconventional computing, particularly DNA computing, including DNA-signal amplification [89] and integrating enzyme- and DNA-based logic systems in one complex process [94, 95].

References

[1] Skirtach, A. G., Yashchenok, A. M., & Möhwald, H. Encapsulation, release and applications of LbL polyelectrolyte multilayer capsules. Chem Commun. 2011, 47, 12736–12746.

[2] Coll, C., Bernardos, A., Martínez-Máñez, R., & Sancenón, F. Gated silica mesoporous supports for controlled release and signaling applications. Acc Chem Res. 2013, 46, 339–349.

[3] Myrick, J. M., Vendra, V. K., & Krishnan, S. Self-assembled polysaccharide nanostructures for controlled-release applications. Nanotechnol Rev. 2014, 3, 319–346.

[4] Caldorera-Moore, M. E., Liechty, W. B., & Peppas, N. A. Responsive theranostic systems: integration of diagnostic imaging agents and responsive controlled release drug delivery carriers. Acc Chem Res. 2011, 44, 1061–1070.

[5] Vallet-Regí, M., Balas, F., & Arcos, D. Mesoporous materials for drug delivery. Angew Chem Int Ed. 2007, 46, 7548–7558.

[6] Moore, T., Chen, H. Y., Morrison, R., Wang, F. L., Anker, J. N., & Alexis, F. Nanotechnologies for noninvasive measurement of drug release. Molec Pharmaceutics. 2014, 11, 24–39.

[7] Qing, G. Y., Li, M. M., Deng, L. J., Lv, Z. Y., Ding, P., & Sun, T. L. Smart drug release systems based on stimuli-responsive polymers. Mini-Rev Med Chem. 2013, 13, 1369–1380.

[8] Niikura, K., Iyo, N., Matsuo, Y., Mitomo, H., & Ijiro, K. Sub-100 nm gold nanoparticle vesicles as a drug delivery carrier enabling rapid drug release upon light irradiation. ACS Appl Mater Interfaces. 2013, 5, 3900–3907.

[9] Bonini, M., Berti, D., & Baglioni, P. Nanostructures for magnetically triggered release of drugs and biomolecules. Cur Opin Colloid Interface Sci. 2013, 18, 459–467.

[10] Jiao, Y. F., Sun, Y. F., Chang, B. S., Lu, D. R., & Yang, W. L. Redox- and temperature-controlled drug release from hollow mesoporous silica nanoparticles. Chem Eur J. 2013, 19, 15410–15420.
[11] Michalak, M., Marek, A. A., Zawadiak, J., Kawalec, M., & Kurcok, P. Synthesis of PHB-based carrier for drug delivery systems with pH-controlled release. Eur Polym J. 2013, 49, 4149–4156.
[12] Munkhjargal, M., Matsuura, Y., Hatayama, K., et al. Glucose-sensing and glucose-driven "Organic Engine" with Co-immobilized enzyme membrane toward autonomous drug release systems for diabetes. Sens Actuat B. 2013, 188, 831–836.
[13] Bocharova, V., Zavalov, O., MacVittie, K., et al. A biochemical logic approach to biomarker-activated drug release. J Mater Chem. 2012, 22, 19709–19717.
[14] Mailloux, S., Zavalov, O., Guz, N., Katz, E., & Bocharova, V. Enzymatic filter for improved separation of output signals in enzyme logic systems towards 'Sense and Treat' medicine. Biomater Sci. 2014, 2, 184–191.
[15] Johnson, R. P., John, J. V., & Kim, I. Poly(L-histidine)-containing polymer bioconjugate hybrid materials as stimuli-responsive theranostic systems. J Appl Polym Sci. 2014, 131, art. # 40796.
[16] Yang, K., Feng, L. Z., Shi, X. Z., & Liu, Z. Nano-graphene in biomedicine: theranostic applications. Chem Soc Rev. 2013, 42, 530–547.
[17] Zhou, M., Zhou, N. D., Kuralay, F., et al. A self-powered "Sense-Act-Treat" system that is based on a biofuel cell and controlled by Boolean logic. Angew Chem Int Ed. 2012, 51, 2686–2689.
[18] Traitel, T., Goldbart, R., & Kost, J. Smart polymers for responsive drug-delivery systems. J Biomater Sci Polym Ed. 2008, 19, 755–767.
[19] Lim, E. K., Kim, T., Paik, S., Haam, S., Huh, Y. M., & Lee, K. Nanomaterials for theranostics: recent advances and future challenges. Chem Rev. 2015, 115, 327–394.
[20] Wang, L., Liu, M., Gao, C., Ma, L., & Cui, D. A pH-, thermo-, and glucose-, triple-responsive hydrogels: synthesis and controlled drug delivery. React Funct Polym. 2010, 70, 159–167.
[21] Zhao, Y., Trewyn, B. G., Slowing, I. I., & Lin, V. S. Y. Mesoporous silica nanoparticle-based double drug delivery system for glucose-responsive controlled release of insulin and cyclic AMP. J Am Chem Soc. 2009, 131, 8398–8400.
[22] Tokarev, I., Gopishetty, V., Zhou, J., et al. Stimuli-responsive hydrogel membranes coupled with biocatalytic processes. ACS Appl Mater Interfaces. 2009, 1, 532–536.
[23] Gamella, M., Guz, N., Mailloux, S., Pingarrón, J. M., & Katz, E. Antibacterial drug release electrochemically stimulated by the presence of bacterial cells – Theranostic approach. Electroanalysis. 2014, 26, 2552–2557.
[24] Gamella, M., Guz, N., Mailloux, S., Pingarrón, J. M., & Katz, E. Activation of a biocatalytic electrode by removing glucose oxidase from the surface – Application to signal triggered drug release. ACS Appl Mater Interfaces. 2014, 6, 13349–13354.
[25] Mailloux, S., Guz, N., Zakharchenko, A., Minko, S., & Katz, E. Majority and minority gates realized in enzyme-biocatalyzed systems integrated with logic networks and interfaced with bioelectronic systems. J Phys Chem B. 2014, 118, 6775–6784.
[26] Mailloux, S., Guz, N., Gamella Carballo, M., Pingarrón, J. M., & Katz, E. Model system for targeted drug release triggered by immune-specific signals. Anal Bioanal Chem. 2014, 406, 4825–4829.
[27] Mailloux, S., MacVittie, K., Privman, M., Guz, N., & Katz, E. Starch-powered biofuel cell activated by logically processed biomolecular signals. ChemElectroChem. 2014, 1, 1822–1827.
[28] Mailloux, S., Halámek, J., & Katz, E. A model system for targeted drug release triggered by biomolecular signals logically processed through enzyme logic networks. Analyst. 2014, 139, 982–986.
[29] Mailloux, S., Halámek, J., Halámková, L., Tokarev, A., Minko, S., & Katz, E. Biomolecular release triggered by glucose input – Bioelectronic coupling of sensing and actuating systems. Chem Commun. 2013, 49, 4755–4757.

[30] Katz, E., Pingarrón, J. M., Mailloux, S., et al. Substance release triggered by biomolecular signals in bioelectronic systems. J Phys Chem Lett. 2015, 6, 1340–1347.
[31] Donati, I., & Paoletti, S. Material properties of alginates. In: B. H. A. Rehm, ed. Alginates: biology and applications, Series: Microbiology monographs, Vol. 13, Springer, Dordrecht, 2009, 1–53.
[32] Tokarev, I., & Minko, S. Multiresponsive, hierarchically structured membranes: new, challenging, biomimetic materials for biosensors, controlled release, biochemical gates, and nanoreactors. Adv Mater. 2009, 21, 241–247.
[33] Chan, A. W., & Neufeld, R. J. Tuneable semi-synthetic network alginate for absorptive encapsulation and controlled release of protein therapeutics. Biomaterials. 2010, 31, 9040–9047.
[34] Pescosolido, L., Piro, T., Vermonden, T. et al. Biodegradable IPNs based on oxidized alginate and dextran-HEMA for controlled release of proteins. Carbohydr Polym. 2011, 86, 208–213.
[35] Barrias, C. C., Lamghari, M., Granja, P. L., Miranda, M. C. S., & Barbosa, M. A. Biological evaluation of calcium alginate microspheres as a vehicle for the localized delivery of a therapeutic enzyme. J Biomed. Mater Res A. 2005, 74A, 545–552.
[36] Krebs, M. D., Salter, E., Chen, E., Sutter, K. A., & Alsberg, E. Calcium phosphate-DNA nanoparticle gene delivery from alginate hydrogels induces in vivo osteogenesis. J Biomed Mater Res A. 2010, 92A, 1131–1138.
[37] Jiang, G., Min, S. H., Oh, E. J., & Hahn, S. K. DNA/PEI/alginate polyplex as an efficient in vivo gene delivery system. Biotechnol Bioprocess Eng. 2007, 12, 684–689.
[38] Dey, K., & Roy, P. Degradation of chloroform by immobilized cells of Bacillus sp in calcium alginate beads. Biotechnol Lett. 2011, 33, 1101–1105.
[39] Hoesli, C. A., Raghuram, K., Kiang, R. L. J., et al. Pancreatic cell immobilization in alginate beads produced by emulsion and internal gelation. Biotechnol Bioeng. 2011, 108, 424–434.
[40] Moebus, K., Siepmann, J., & Bodmeier, R. Alginate-poloxamer microparticles for controlled drug delivery to mucosal tissue. Eur J Pharm Biopharm. 2009, 72, 42–53.
[41] Sun, X., Shi, J., Zhang, Z., & Cao, S. Dual-responsive semi-interpenetrating network beads based on calcium alginate/poly(N-isopropylacrylamide)/poly(sodium acrylate) for sustained drug release. J Appl Polym Sci. 2011, 122, 729–737.
[42] Işıklan, N., İnal, M., Kurşun, F., & Ercan, G. pH Responsive itaconic acid grafted alginate microspheres for the controlled release of nifedipine. Carbohydr Polym. 2011, 84, 933–943.
[43] Xing, J., Deng, L., & Dong, A. Chitosan/alginate nanoparticles stabilized by poloxamer for the controlled release of 5-fluorouracil. J Appl Polym Sci. 2010, 117, 2354–2359.
[44] Castro, G. R., Kamdar, R. R., Panilaitis, B., & Kaplan, D. L. Triggered release of proteins from emulsan-alginate beads. J Contr Release. 2005, 109, 149–157.
[45] Svirskis, D., Travas-Sejdic, J., Rodgers, A., & Garg, S. Electrochemically controlled drug delivery based on intrinsically conducting polymers. J Contr Release. 2010, 146, 6–15.
[46] Boulmedais, F., Tang, C. S., Keller, B., & Vörös, J. Controlled electrodissolution of polyelectrolyte multilayers: a platform technology towards the surface-initiated delivery of drugs. Adv Funct Mater. 2006, 16, 63–70.
[47] George, P. M., LaVan, D. A., Burdick, J. A., Chen, C. Y., Liang, E., & Langer, R. Electrically controlled drug delivery from biotin-doped conductive polypyrrole. Adv Mater. 2006, 18, 577–581.
[48] Pan, D., Zhang, H., Fan, T., Chen, J., & Duan, X. Nearly monodispersed core–shell structural Fe_3O_4@DFUR–LDH submicro particles for magnetically controlled drug delivery and release. Chem Commun. 2011, 47, 908–910.
[49] Panczyk, T., Warzocha, T. P., & Camp, P. J. A magnetically controlled molecular nanocontainer as a drug delivery system: the effects of carbon nanotube and magnetic nanoparticle parameters from Monte Carlo simulations. J Phys Chem C. 2010, 114, 21299–21308.

[50] Choubey, J., & Bajpai, A. K. Investigation on magnetically controlled delivery of doxorubicin from superparamagnetic nanocarriers of gelatin crosslinked with genipin. J Mater Sci – Mater Med. 2010, 21, 1573–1586.
[51] Cai, K., Luo, Z., Hu, Y., et al. Magnetically triggered reversible controlled drug delivery from microfabricated polymeric multireservoir devices. Adv Mater. 2009, 21, 4045–4049.
[52] Hu, S. H., Tsai, C. H., Liao, C. F., Liu, D. M., & Chen, S. Y. Controlled rupture of magnetic polyelectrolyte microcapsules for drug delivery. Langmuir. 2008, 24, 11811–11818.
[53] Nž, K., Trewyn, B. G., & Lin, V. S. Y. Functionalized mesoporous silica nanoparticle-based visible light responsive controlled release delivery system. Chem Commun. 2011, 47, 2817–2819.
[54] Griffin, D. R., Patterson, J. T., & Kasko, A. M. Photodegradation as a mechanism for controlled drug delivery. Biotechnol Bioeng. 2010, 107, 1012–1019.
[55] Vivero-Escoto, J. L., Slowing, I. I., Wu, C. W., & Lin, V. S. Y. Photoinduced intracellular controlled release drug delivery in human cells by gold-capped mesoporous silica nanosphere. J Am Chem Soc. 2009, 131, 3462–3463.
[56] Lee, K. Y., Peters, M. C., & Mooney, D. J. Controlled drug delivery from polymers by mechanical signals. Adv Mater. 2001, 13, 837–839.
[57] Zhang, W., Gilstrap, K., Wu, L., et al. Synthesis and characterization of thermally responsive pluronic F127-chitosan nanocapsules for controlled release and intracellular delivery of small molecules. ACS Nano. 2010, 4, 6747–6759.
[58] Fong, W. K., Hanley, T. L., Thierry, B., Kirby, N., & Boyd, B. J. Plasmonic nanorods provide reversible control over nanostructure of self-assembled drug delivery materials. Langmuir. 2010, 26, 6136–6139.
[59] Chen, S., Li, Y., Guo, C., et al. Temperature-responsive magnetite/PEO-PPO-PEO block copolymer nanoparticles for controlled drug targeting delivery. Langmuir. 2007, 23, 12669–12676.
[60] Mathews, A. S., Cho, W. J., Kim, I., & Ha, C. S. Thermally responsive poly[*N*-isopropylacrylamide-*co*-2-hydroxyethylacrylate] colloidal crystals included in β-cyclodextrin for controlled drug delivery. J Appl Polym Sci. 2009, 113, 1680–1689.
[61 Kapoor, S., & Bhattacharyya, A. J. Ultrasound-triggered controlled drug delivery and biosensing using silica nanotubes. J Phys Chem C. 2009, 113, 7155–7163.
[62] Richard, D., Nguyen, I., Affolter, C., et al. Polyelectrolyte multilayer-mediated gene delivery for semaphorin signaling pathway control. Small. 2010, 6, 2405–2411.
[63] Asai, D., Kodama, K. B., Shoji, Y., et al. Drug delivery system based on responses to an HIV infectious signal. Med Chem. 2008, 4, 386–391.
[64] Yuan, L., Tang, Q., Yang, D., Zhang, J. Z., Zhang, F., & Hu, J. Preparation of pH-responsive mesoporous silica nanoparticles and their application in controlled drug delivery. J Phys Chem C. 2011, 115, 9926–9932.
[65] Negrini, R., & Mezzenga, R. pH-Responsive lyotropic liquid crystals for controlled drug delivery. Langmuir. 2011, 27, 5296–5303.
[66] Zhang, Z., Chen, L., Zhao, C., et al. Thermo- and pH-responsive HPC-g-AA/AA hydrogels for controlled drug delivery applications. Polymer. 2011, 52, 676–682.
[67] Aryal, S., Hu, C. M. J., & Zhang, L. Polymer-cisplatin conjugate nanoparticles for acid-responsive drug delivery. ACS Nano. 2010, 4, 251–258.
[68] Gordijo, C. R., Koulajian, K., Shuhendler, A. J., et al. Nanotechnology-enabled closed loop insulin delivery device: in vitro and in vivo evaluation of glucose-regulated insulin release for diabetes control. Adv Funct Mater. 2011, 21, 73–82.
[69] Pearson, R. G. Hard and soft acids and bases. J Am Chem Soc. 1963, 85, 3533–3539.
[70] Furia, T. E., CRC Handbook of Food Additives, Vol 2, 2nd ed., CRC Press, Cleveland, 1972.

[71] Jin, Z., Güven, G., Bocharova, V., et al. Electrochemically controlled drug-mimicking protein release from iron-alginate thin-films associated with an electrode. ACS Appl Mater Interfaces. 2012, 4, 466–475.
[72] Jin, Z., Harvey, A. M., Mailloux, S., et al. Electrochemically stimulated release of lysozyme from alginate matrix cross-linked with iron cations. J Mater Chem. 2012, 22, 19523–19528.
[73] Strack, G., Babanova, S., Farrington, K. E., Luckarift, H. R., Atanassov, P., & Johnson, G. R. Enzyme-modified buckypaper for bioelectrocatalysis. J Electrochem Soc. 2013, 160, G3178–82.
[74] Holzinger, M., Le Goff, A., & Cosnier, S. Carbon nanotube/enzyme biofuel cells. Electrochim Acta. 2012, 82, 179–190.
[75] Zebda, A., Gondran, C., Le Goff, A., Holzinger, M., Cinquin, P., & Cosnier, S. Mediatorless High-power glucose biofuel cells based on compressed carbon nanotube-enzyme electrodes. Nature Commun. 2011, 2, art. #370.
[76] Halámková, L., Halámek, J., Bocharova, V., Szczupak, A., Alfonta, L., & Katz, E. Implanted biofuel cell operating in a living snail. J Am Chem Soc. 2012, 134, 5040–5043.
[77] Szczupak, A., Halámek, J., Halámková, L., Bocharova, V., Alfonta, L., & Katz, E. Living battery – biofuel cells operating in vivo in clams. Energy Environ Sci. 2012, 5, 8891–8895.
[78] MacVittie, K., Halámek, J., Halámková, L., et al. From “cyborg” lobsters to a pacemaker powered by implantable biofuel cells. Energy Environ Sci. 2013, 6, 81–86.
[79] Katz, E., ed. Biomolecular computing – From logic systems to smart sensors and actuators. Willey-VCH, Weinheim, 2012.
[80] Benenson, Y. Biocomputers: from test tubes to live cells. Mol BioSyst. 2009, 5, 675–685.
[81] Benenson, Y. Biomolecular computing systems: principles, progress and potential. Nat Rev Genet. 2012, 13, 455–468.
[82] Stojanovic, M. N., Stefanovic, D., & Rudchenko, S. Exercises in molecular computing. Acc Chem Res. 2014, 47, 1845–1852.
[83] Stojanovic, M. N., & Stefanovic, D. Chemistry at a higher level of abstraction. J Comput Theor Nanosci. 2011, 8, 434–440.
[84] Katz, E., & Privman, V. Enzyme-based logic systems for information processing. Chem Soc Rev. 2010, 39, 1835–1857.
[85] Katz, E., Bocharova, V., & Privman, M. Electronic interfaces switchable by logically processed multiple biochemical and physiological signals. J Mater Chem 2012, 22, 8171–8178.
[86] Privman, M., Tam, T. K., Bocharova, V., Halámek, J., Wang, J., & Katz, E. Responsive interface switchable by logically processed physiological signals: towards “smart” actuators for signal amplification and drug delivery. ACS Appl Mater Interfaces. 2011, 3, 1620–1623.
[87] Privman, M., Tam, T. K., Pita, M., & Katz, E. Switchable electrode controlled by enzyme logic network system: approaching physiologically regulated bioelectronics. J Am Chem Soc. 2009, 131, 1314–1321.
[88] Katz, E., Lötzbeyer, T., Schlereth, D. D., Schuhmann, W., & Schmidt, H. L. Electrocatalytic oxidation of reduced nicotinamide coenzymes at gold and platinum electrode surfaces modified with a monolayer of pyrroloquinoline quinone: effect of Ca^{2+} cations. J Electroanal Chem. 1994, 373, 189–200.
[89] Gamella, M., Guz, N., & Katz, E. DNA release from a bioelectronic interface stimulated by a DNA signal: amplification of DNA signals. Electroanalysis. 2016, 28, 2692–2696.
[90] Gamella Carballo, M., Guz, N., Pingarrón, J. M., Aslebagh, R., Darie, C. C., & Katz, E. Bioelectronic system for insulin release triggered by ketone body mimicking diabetic ketoacidosis in vitro. Chem Commun. 2015, 51, 7618–7621.
[91] Katz, E., ed. Implantable bioelectronics – Devices, materials and applications. Wiley-VCH, Weinheim, Germany, 2014.

[92] Narayanan, R. P., Melman, G., Letourneau, N. J., Mendelson, N. L., & Melman, A. Photodegradable iron(III) cross-linked alginate gels. Biomacromolecules. 2012, 13, 2465–2471.
[93] Gamella, M., Privman, M., Bakshi, S., Melman, A., & Katz, E. DNA release from Fe^{3+}-cross-linked alginate films triggered by logically processed biomolecular signals – Integration of biomolecular computing and actuation. ChemPhysChem. 2017, 18, 1811–1821.
[94] Guz, N., Fedotova, T. A., Fratto, B. E., Schlesinger, O., Alfonta, L., Kolpashchikov, D., & Katz, E. Bioelectronic interface connecting reversible logic gates based on enzyme and DNA reactions. ChemPhysChem. 2016, 17, 2247–2255.
[95] Mailloux, S., Gerasimova, Y. V., Guz, N., Kolpashchikov, D. M., & Katz, E. Bridging the two worlds: a universal interface between enzymatic and DNA computing systems. Angew Chem Int Ed. 2015, 54, 6562–6566.

List of abbreviations

2-HBA	4-amino-2-hydroxybutyric acid
3-HBA	3-hydroxybutyric acid
6-PGluc	6-phosphogluconic acid
ADP	adenosine 5'-diphosphate
AFM	atomic force microscopy
AlcDH	alcohol dehydrogenase (enzyme)
ATP	adenosine 5'-triphosphate
BSA	bovine serum albumin
ELISA	enzyme-linked immunosorbent assay
Et-OH	ethanol
G6PGH	glucose-6-phosphate dehydrogenase (enzyme)
GDH	glucose dehydrogenase (NAD^+-dependent enzyme)
Glc	glucose
Glc-1-P	glucose-1-phosphate
Glc-6-P	glucose-6-phosphate
GlcA	gluconic acid (product of glucose oxidation)
HK	hexokinase (enzyme)
HRP	horseradish peroxidase (enzyme)
HRP-Ab	horseradish peroxidase-labeled anti-goat IgG-antibody
MOPS	3-(*N*-morpholino)propanesulfonic acid (buffer)
MPh	maltose phosphorylase (enzyme)
MWCNTs	multi-walled carbon nanotubes
NAD^+	β-nicotinamide adenine dinucleotide (oxidized form)
NADH	β-nicotinamide adenine dinucleotide (reduced form)
NPs	nanoparticles
PBSE	1-pyrenebutanoic acid succinimidyl ester
PEG	polyethylene glycol
PEI	polyethyleneimine
Pi	inorganic phosphate
PQQ	pyrroloquinoline quinone
PQQ-GDH	pyrroloquinoline quinone-dependent glucose dehydrogenase (enzyme)
SECM	scanning electrochemical microscope
β-Amy	β-amylase (enzyme)

You Yu, Shaojun Dong

9 Self-powered electrochemical biosensors

9.1 Introduction

The electrochemical biosensor is an analytical biodevice, which is designed for the detection of a specific analyte by the electrochemical methods [1, 2]. Electrochemical biosensors have been the subject of basic analytical research for nearly 50 years [3, 4]. Numerous electrochemical biosensors have been provided across different fields such as individual healthcare [5], food analysis [6], environmental protect [7, 8], and even space industry [9], for the high sensitivity, rapid response, outstanding selectivity, and extensive university [10].

A typical principle of an electrochemical biosensor can be divided into three individual processes: First, the sensitive biological element, which could be a biologically derived material or biomimetic component, for instance, microorganisms, organelles, cell receptors, enzymes, antibodies, and nucleic acids, can interact, bind, or recognize with the corresponding analyte, and can change itself or the regarding analyte [11–13]. Under external electric power, the interaction of the characteristic analyte with the biological element can cause the signal generation to easily identify and quantify in an electrochemical method. At last, the signal reflects in electronic display, buzzer, or even bluetooth communication to a cell phone [14, 15]. Through the standard curves, the respective analyte can be qualitatively and quantitatively detected. Thus, a representative electrochemical biosensor that is also considered to have three components contains not only the core biological sensing element but also other accessories such as the power sources and the signaling electronics. Complex assemble procedures of the traditional electrochemical biosensors have limited the further improvements along with increasing requirement of smaller volume, lower request for familiar researcher, and wider applicative range [16, 17]. A miniature electrochemical biosensor with simple fabricate processes is highly expected.

For now, the extreme challenge associated with miniaturization and portability of sensors is the power supply. Power supplies [18], such as batteries [19] and solar cells [20], are difficult to miniaturize. Moreover, with external power supplies, the structure of sensor should require a special design that allows for the replacement or recharging, which goes against the requirements of neither an implantable sensor nor a long-term sensor [21, 22]. Therefore, major efforts are focusing on constructing a self-powered electrochemical biosensor by merging

You Yu, State Key Laboratory of Electroanalytical Chemistry, Changchun Institute of Applied Chemistry, Chinese Academy of Science, Changchun, Jilin, 130022, P.R. China
Shaojun Dong, Shaojun Dong, State Key Laboratory of Electroanalytical Chemistry, Changchun Institute of Applied Chemistry, Chinese Academy of Science, Changchun, Jilin, 130022, P.R. China

https://doi.org/10.1515/9783110570526-009

the biosensing transducer and the power supply, which could be a traditional metal-based battery or fuel cell. For higher integrated construction, the self-powered biosensor can even contain the display characters by the version signal of the biosensor itself. Mostly, major electrochemical galvanic cells utilized in self-powered biosensors are biofuel cells (BFCs) since 2001 [23]. Additionally, the sensor field is primarily made of electroanalytical methods, which typically operate in three-electrode modes with work electrodes, counter electrodes, and reference electrodes. Instead, the self-powered biosensor operates in two-electrode modes (anodes and cathodes) and reduces the system from three-electrode system into two electrodes, which is well beneficial for popularization and applications.

BFCs are a special category of FCs whose biocatalysts employ microorganisms or enzymes instead of precious noble metal catalysts [24, 25]. BFCs are usually recognized as an interesting branch of energy conversion technologies, which can convert bioenergy or other energy like photoenergy and chemical energy into electricity. Compared to traditional FCs, BFCs can operate well in mild conditions such as ambient temperature, neutral pH, and nontoxic substrates. BFCs have already been used as practical power sources in vivo for microelectronics including micro pumps, pacemakers, and so on [16, 26–29].

Researchers familiar with the field of electrochemical biosensors are attracted by the commercial acceptance of the glucose biosensor for testing the blood glucose level among diabetic patients at first [30–32]. In fact, there is a high coincidence of the BFC substrates and the corresponding biosensor analytes such as glucose, lactate, and ethanol [33–35]. With the characters of low cost, high efficiency, and high selectivity, BFCs have a huge potential to be the self-powered sensor by detecting the variation of the parameters itself affected by the target analyte.

BFC's performance is characterized in terms of several parameters, such as the open circuit potential (OCV), short circuit current (SCC), and the power output. The OCV provides measurements of the maximum voltage output associated with BFCs. It is also termed as the resting potential or zero-current potential, because it defines a potential at which there is no net current flow and, consequently, at which no work is done. In an ideal case, the OCV is determined by the difference between the thermodynamic onset catalytic potentials of the two electrodes (indicated in red in Fig. 9.1A), and then decreased by all the nonstandard conditions in the majority of cases. The SCC achieved ideally with zero resistance in the entire system to get the maximum cell current, which usually relate to the limit capacity of anode electron donation or cathode electron acceptance (Fig. 9.1B). The SCC changed along with the current output densities of either the anode or the cathode. The power output of the BFC depends on the current achieved at cell voltage range. Useful power is achieved at current and voltage values that are a compromise between the limiting cases of open circuit and short circuit. The highest fuel cell current is usually delivered at low cell voltages.

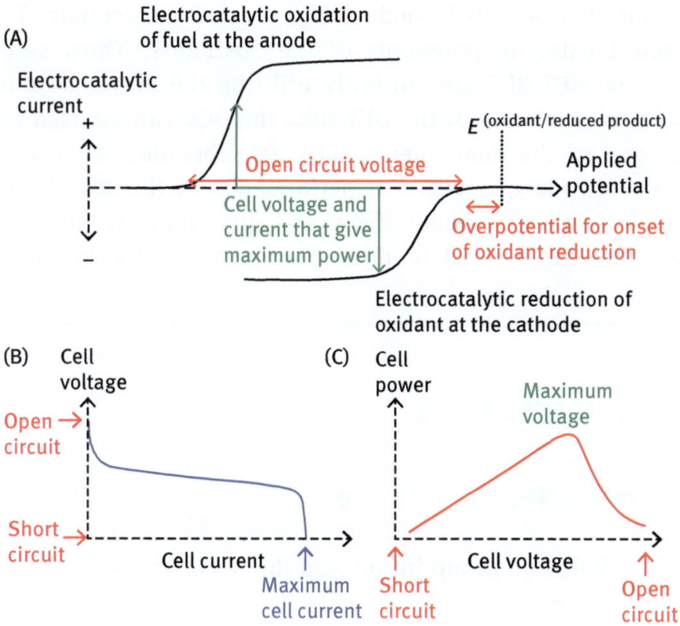

Fig. 9.1: Voltage and current response for a pair of fuel cell electrodes tested separately (A) or operating together in a fuel cell (B and C). Features that determine fuel cell performance are highlighted. Reprinted from Cracknell et al. [24] with permission by the American Chemical Society.

With regard to a regular BFC, the mode of electron transfer is either direct electron transfer (DET) or mediated electron transfer (MET) in BFCs, depending on which the electrons flow from/to the electrode, the enzyme itself or the mediator. Figure 9.2 shows a BFC with the MET configuration; a mediator shuttles electrons between the enzymes and the electrodes. On the contrary, electrons are transferred directly from the electrode to the enzyme without the

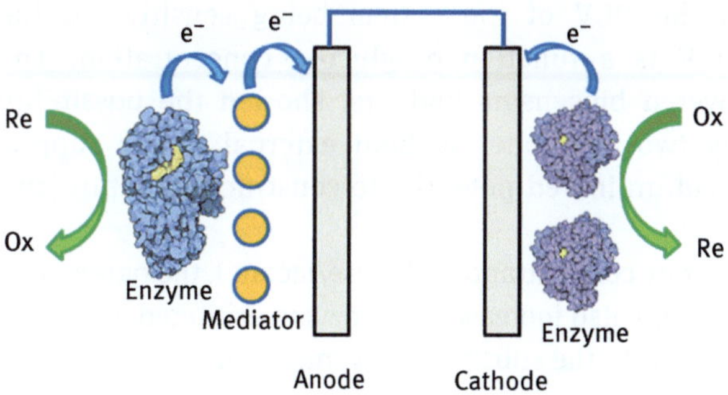

Fig. 9.2: Schematic of typical BFC with a DET biocathode and a MET bioanode.

mediators in DET mode. For an MET mode BFC, the OCV is certain by the thermodynamic onset catalytic potentials of the mediators. Thus, self-powered sensors based on MET-BFC are currently utilizing the power output and the SCC to measure the analyte. For the DET-BFC, the OCV can be used as the measurements. If one of the components such as substrates, enzymes, mediator, or even the electrode changes, the performance of the BFC itself will be different. Therefore, we can qualitatively and quantitatively monitor which and how much the target analyte does change according to these differences.

9.2 Positive enhanced effect

9.2.1 Positive substrate effect on enzyme

In 2001, for the first time, Willner's group introduced the novel concept of self-powered biosensor devices based on chemical-to-electrochemical energy transformations occurring in BFC elements [23]. Two similar self-powered biosensor devices were represented based on BFCs, which consisted of two kinds of enzyme-functionalized Au electrodes as bioanode. For the first sensor, a glucose oxidase (GOx) electrode (Fig. 9.3A) was assembled by the reconstitution of apo-GOx on an aminoethyl flavine adenine dinucleotide phosphate (amino-FAD; Fig. 9.3) covalently linked to a pyrroloquinolino quinone (PQQ; Fig. 9.3) monolayer. In the second system, a lactate dehydrogenase (LDH)-layered electrode (Fig. 9.3B) was generated by a PQQ monolayer linked to an Au electrode-coupled aminoethyl-functionalized NAD^+ (Fig. 9.3). The affinity complex formed between LDH and the PQQ-NAD^+ monolayer assembly was cross-linked with glutaric dialdehyde to yield the integrated electrically contacted LDH-functionalized electrode. In both systems, the biocathodes employed a glutaric dialdehyde–cross-linked Cyt c/Cox monolayer assembled on an Au electrode to catalyze the oxygen reduction reaction (Fig. 9.3C). The biocatalytic oxidation of the glucose biofuel with the OCV of the system being sensitive to the substrate level; thus, the OCV is a function of glucose concentration. The introduction of such self-powered biosensors had first showed the possibility to fabricate a sensor of only two electrodes without external power supply. Obviously this novel design had unlimited potential to construct miniature and implantable devices.

With an increase in the substrate concentration (glucose/lactate), the parameters of this electrochemical biosensor did also increase. The enzyme exhibited a positive relationship of the onset potential with the substrate concentration.

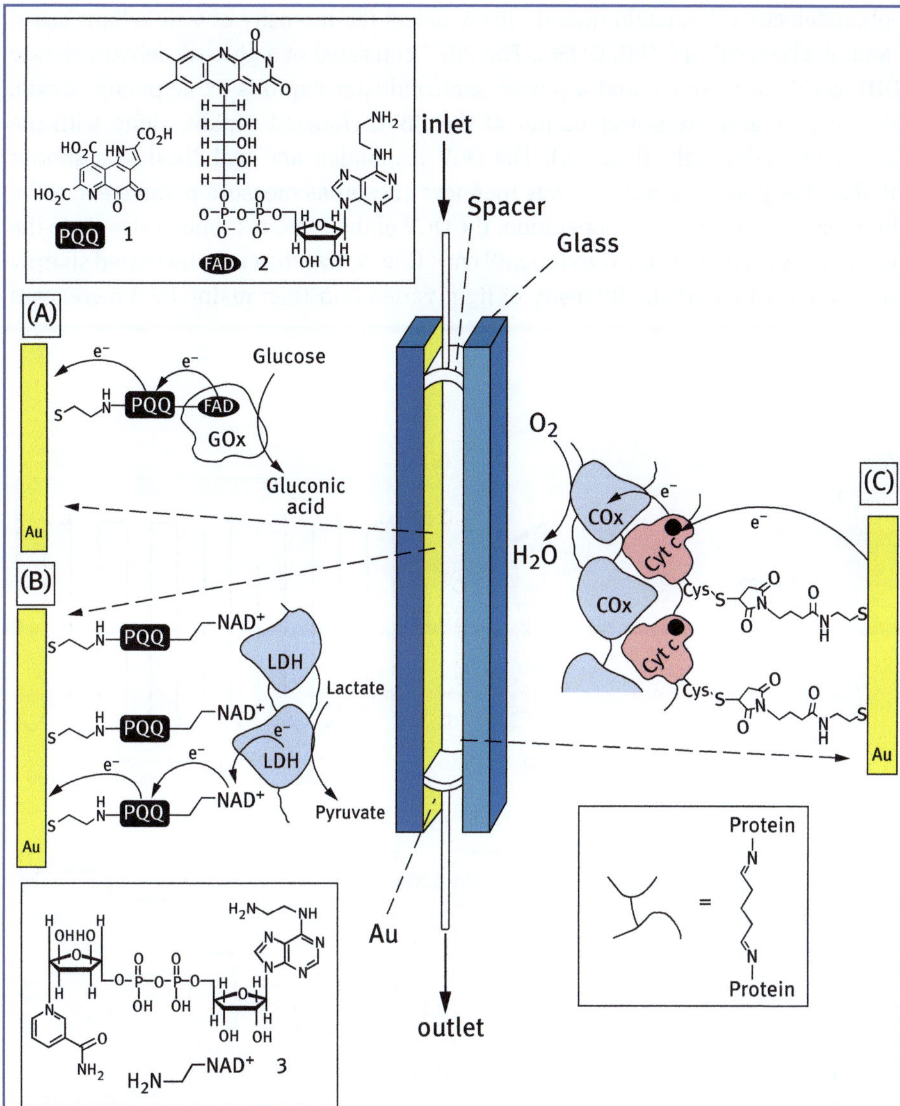

Fig. 9.3: Configuration of self-powered BFC-based biosensors is composed of (A) PQQ-FAD/GOx-functionalized bioanode utilizing glucose analyte as a fuel or (B) PQQ-NAD^+/LDH-functionalized bioanode utilizing lactate analyte as a fuel, and (C) Cyt c/COX-functionalized biocathode utilizing O_2 as an oxidizer in combination with both bioanodes (A or B). Reprinted from Katz et al. [23] with permission by the American Chemical Society.

9.2.2 Positive surroundings effect on photocathode

Beyond the substrate in the solution, surrounding light intensity can also be detected by the electrochemical biosensors. Recently, Dong's group constructed a self-powered biophotodetectors that could quantitatively detect the intensity of visible light based on a photochemical BFC (PBFC) [36]. The PBFC consisted of a glucose dehydrogenase (GDH)-modified bioanode and a p-type semiconductor cuprous oxide photocathode. Both the OCV and the power output of the PBFC changed rapidly along with the intensity of visible light (Fig. 9.4). The OCV responded and stabilized in a shorter time than the power output and was therefore chosen as measure parameters of the light intensity. By continuous operation, the OCV of the PBFC responded steeply to the light intensity changes from 57 to 180 mW cm^{-2} (Fig. 9.4D); the curve increased sharply in a very short time as the intensity of light varied and then maintained a constant

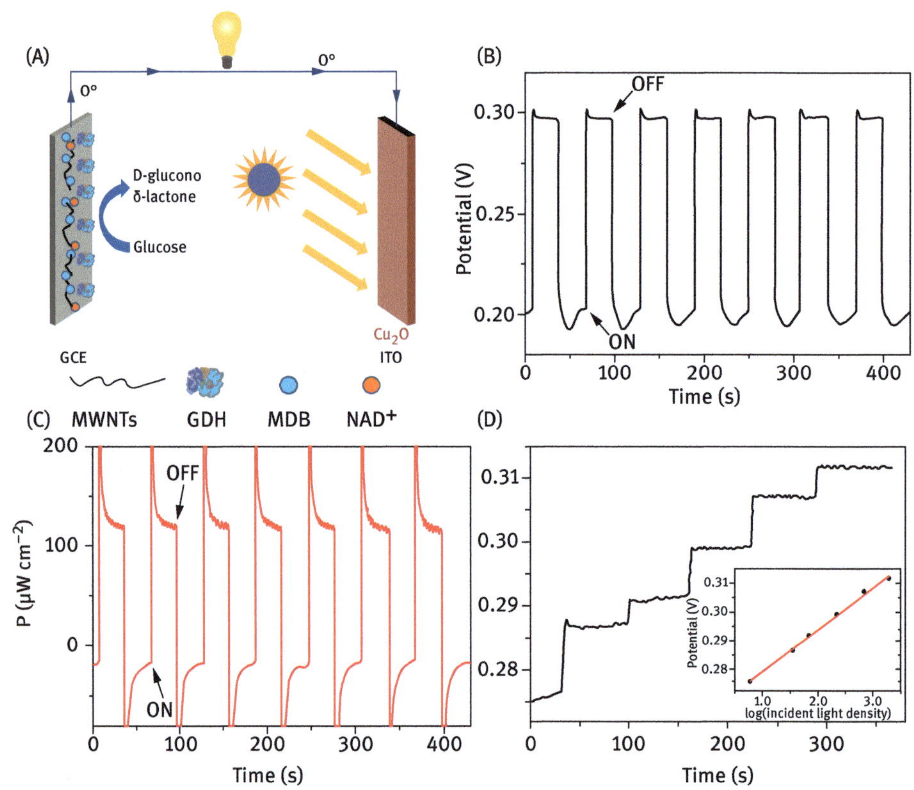

Fig. 9.4: (A) A schematic diagram of the operating principle of the illumination compensation device. (B) OCV in response to 100 mW cm^{-2} visible light with on/off cycles of 30 s. (C) Power output cycles of the PBFC response to visible light with on/off cycles of 30 s at 0.22 V in 0.1 M phosphate buffer solution (PBS) containing 30 mM glucose and 5 mM NAD$^+$. (D) OCV of the PBFC response to continuous changes in light intensity of 57, 81, 93, 117, 147, and 180 mW cm^{-2} in 0.1 M PBS containing 30 mM glucose and 5 mM NAD$^+$. Inset shows the linear function of the plot of OCV versus the logarithm of light intensity. Reprinted from Yu et al. [36] with permission by the Royal Society of Chemistry.

value under a constant intensity of light. It indicated a positive linear relationship between OCV and the logarithm of the light intensity.

Similar to the biophotodetector, a novel PBFC has been assembled using a lactate oxidase–modified bioanode and a photocathode covered by organic semiconductor polyterthiophene (pTTh), which can not only detect the illuminance by the OCV but also accurately and simultaneously monitor the perspiration lactate by the SCC (Fig. 9.5A) [37]. As mentioned earlier, there is a

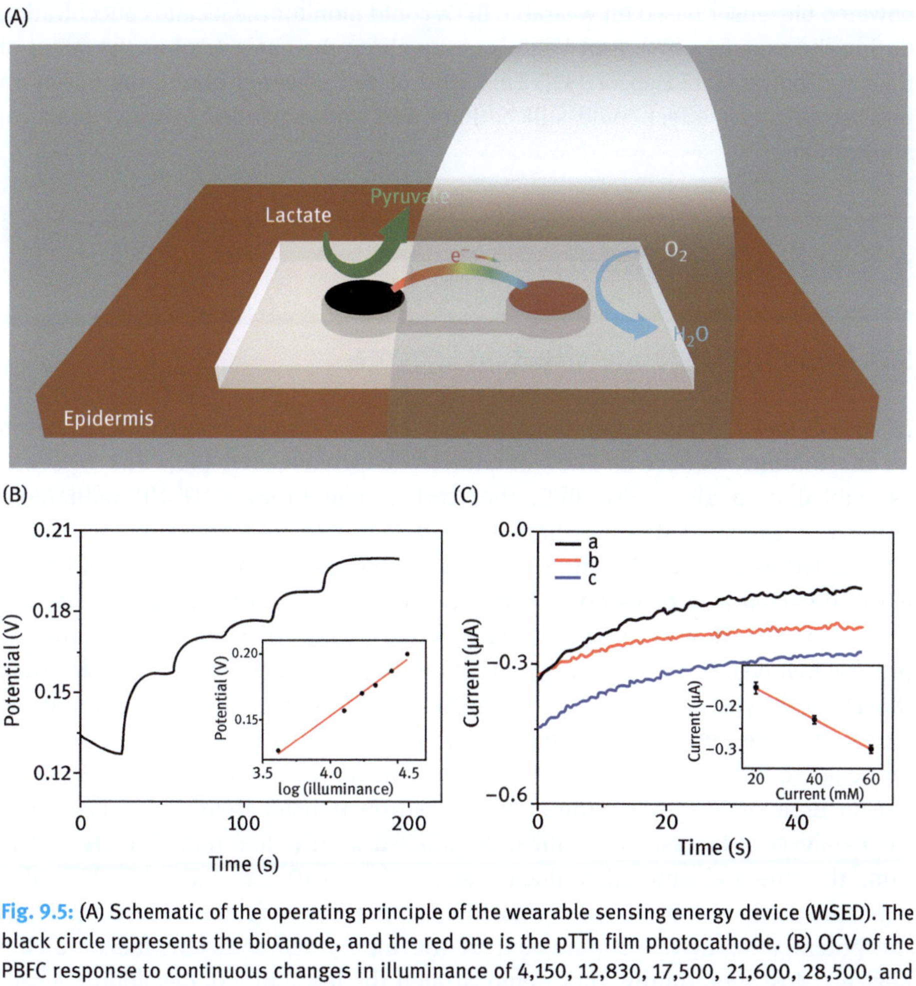

Fig. 9.5: (A) Schematic of the operating principle of the wearable sensing energy device (WSED). The black circle represents the bioanode, and the red one is the pTTh film photocathode. (B) OCV of the PBFC response to continuous changes in illuminance of 4,150, 12,830, 17,500, 21,600, 28,500, and 37,600 Lux. The inset shows the linear function of the OCV versus the logarithm of illuminance. The above were all in 0.1 M PBS (pH 6.0) containing 30 mM lactate. (C) Current–time recording of short-circuit current for increasing lactate concentration from 20 to 60 mM measured at 28,500 Lux. The inset shows the linear function of the plot of short-circuit current versus the concentration of lactate. Reprinted from Yu et al. [37] with permission by the Royal Society of Chemistry.

positive linear relationship between OCV of the BFC and the logarithm of the ambient illuminance by assembling with the organic semiconductor photocathode (Fig. 9.5B). After identifying the real-time illuminance, SCC was chosen to detect the lactate concentration (Fig. 9.5C). Inspiringly, a linear relationship has been found to exist between the lactate concentration and the SCC. The self-powered BFCs could monitor the lactate concentrations and respond to the varied illuminance at the same time.

As a simple method of self-powered sensor assembly, lots of efforts have been made based on substrate effect. For instance, Wang's group built the epidermal self-powered biosensor based on wearable BFCs could monitor the lactate concentration during the exercises [38]. And Vagin's group chose the cholesterol as the target of their self-powered biosensor [39]. This kind of self-powered biosensors based on positive substrate effect could still be improved and expanded to wider practical applications.

9.3 Negative inhibitive effect

9.3.1 Negative analyte effect on enzyme

In 2010, Dong's group reported the first microchip self-powered biosensor based on the inhibitive effect on the miniaturized enzymatic BFC [40]. The BFC was assembled of a glucose/O_2 BFC, the (carbon nanotubes (CNTs)/thionine/gold nanoparticles (AuNPs)) GDH film modified indium-doped tin oxide (ITO) electrode acted as a bioanode, and the (CNTs/polylysine /laccase) film-modified ITO electrode acted as a biocathode (Fig. 9.6) [41]. The direct catalyzed four-electron reduction of O_2 to water was realized in laccase-based biocathode. The catalysis process involves laccase at the type 1 (T1) Cu site, internal electron transfer from the T1 Cu to type 2/type 3 (T2/T3) trinuclear Cu cluster, and O_2 reduction at the T2/T3 site. The binding of cyanide (CN^-) mainly took place onto the T2 Cu and its negative effect on oxygen reduction was attributed to a perturbed T2/T3 Cu cluster unfavorable toward the internal electron transfer from T1 Cu [42]. Thus, the catalysis of laccase is significantly affected by CN^- addition. For CN^- detection, this method showed a linear range of 3.0×10^{-7} to 5.0×10^{-4} M and a detection limit of 1.0×10^{-7} M. This BFC integration with other electronics has the potential to fit the requirement of the totally self-sufficient lab on-a-chip devices. Most importantly, they demonstrated the feasibility of developing a self-powered biosensor based on the inhibitive effect on the miniaturized enzyme BFC for the first time.

Then, Dong's group developed another self-powered biosensor based on the inhibition effect for trace Hg^{2+} detection [43]. A one-compartment miniature

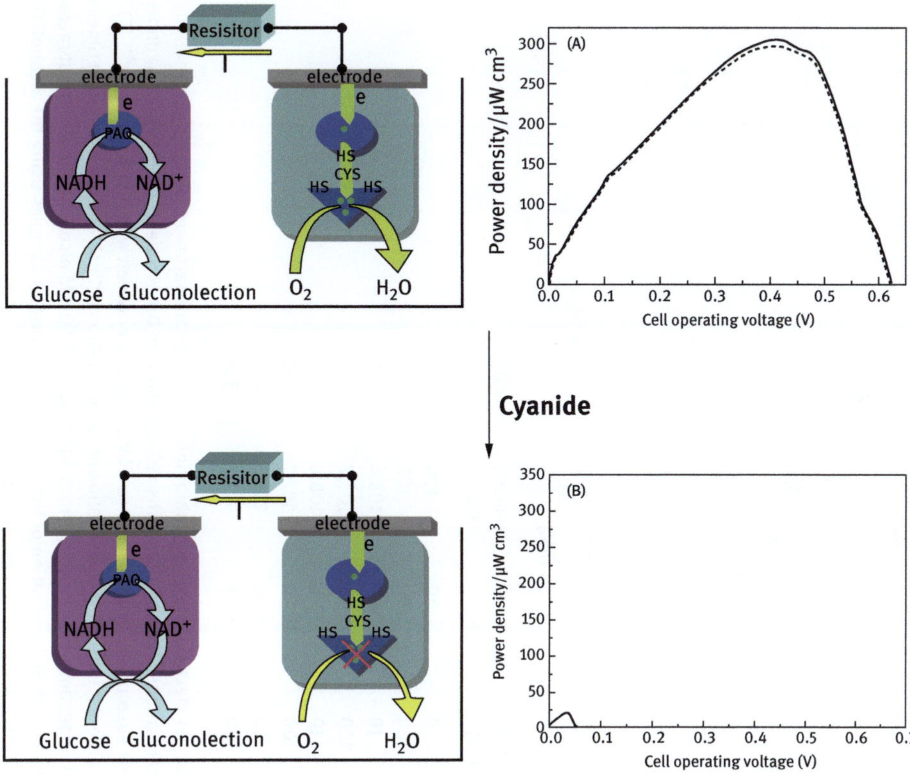

Fig. 9.6: Bioelectrocatalysis mechanism of the cyanide inhibitive effect at the laccase immobilized electrode. (A) Polarization of the uninhibited control and regeneration with 30 min fresh PBS washing of the glucose/O_2 BFC. (B) Polarization of 1 mM cyanide inhibition of the glucose/O_2 BFC. The electrolytes are 10 mM NADH, 40 mM glucose, and 0.2 M PBS (pH 6.5) solution. Reprinted from Deng et al. [40] with permission by the American Chemical Society.

alcohol/O_2 BFC was used, where the single-walled carbon nanohorns modified carbon fiber microelectrodes were used as the substrates for both bioanode and biocathode, and alcohol dehydrogenase (ADH) and bilirubin oxidase (BOD) acted as biocatalysts, respectively (Fig. 9.7A). The assembled alcohol/O_2 BFC showed an OCV of 636 mV and maximum power density of 137 μW cm^{-2}. The existence of Hg^{2+} in the cell solution inhibited the catalytic abilities of the biocatalysts (i.e., BOD and ADH), and thus, led to a remarkable decrease in the OCV of the BFC (Fig. 9.7B). This method showed a linear range of 1,500 nM for an Hg^{2+} array with a detection limit of 1 nM. Besides this self-powered Hg^{2+} sensor shows good selectivity against other relevant metal ions (Fig. 9.7C) and good recovery in different real water sample (Fig. 9.7D).

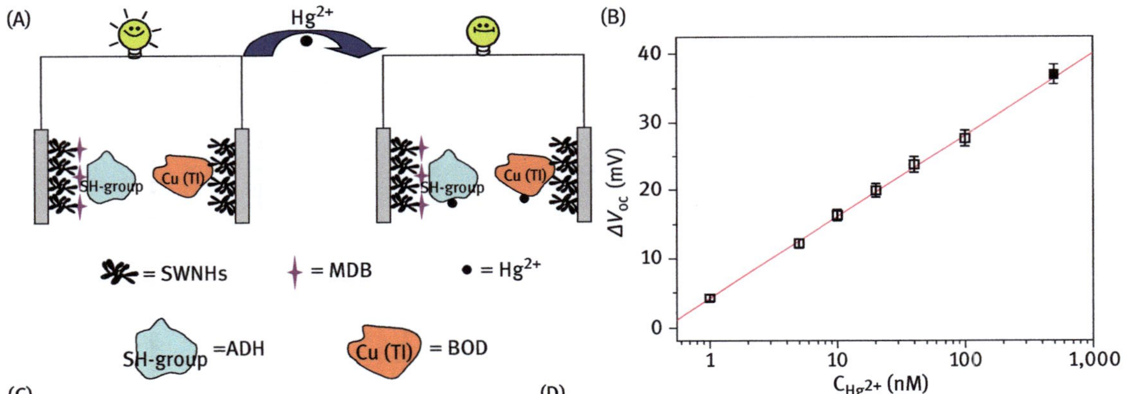

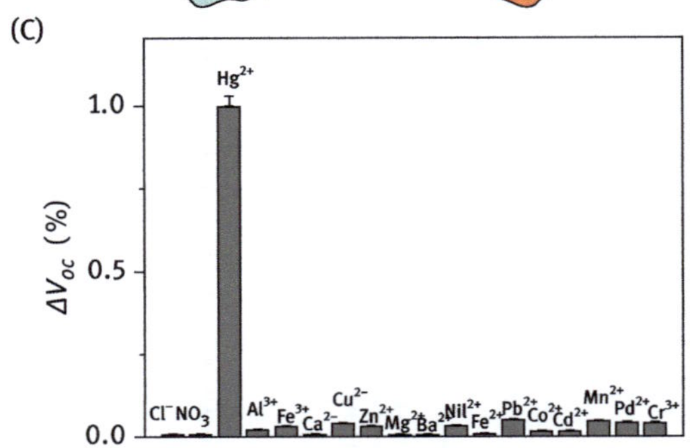

(D)

Water samples	Added (nM)	Measured (nM)	Recovery (%)	RSD (%, $n = 3$)
Mini-Q	10	10	100	3.3
	100	100	100	2.7
Tap water	10	10.4	103.6	3.2
	100	103.8	103.8	3.1
Ground water	10	10.8	108.0	2.3
	100	107.2	107.2	2.9
Lake water	10	10.6	105.5	3.5
	100	99.17	99.17	3.7

Fig. 9.7: (A) Configuration of self-powered Hg^{2+} sensor and bioelectrocatalysis mechanism of Hg^{2+} inhibition of the one-compartment miniature alcohol/O_2 BFC. (B) Linear relationship between ΔV_{oc} and the logarithm of Hg^{2+} concentrations. The solution of the cell operation is 0.1 M quiescent PBS (pH 7.4) under ambient air containing 5 mM NAD^+ and 50 mM alcohol. ΔV_{oc} represents the absolute value of difference OCV of the BFC between with and without Hg^{2+}. (C) Normalized ΔV_{oc} of the self-powered sensor in response to Hg^{2+} and other different ions. The concentration of each metal ion is 1 μM, and the incubation time is 20 min. The solution of the cell operation is 0.1 M quiescent PBS (pH 7.4) under ambient air containing 5 mM NAD^+ and 50 mM alcohol with or without metal ions. ΔV_{oc} represents the absolute value of difference OCV of the BFC between. (D) Determination of Hg^{2+} content in real water samples using the alcohol/O_2 type self-powered sensor. Reprinted from Wen et al. [43] with permission by the American Chemical Society.

9.3.2 Negative inhibitive effect on enzyme reaction

In 2012, Dong's group proposed a self-powered acetaldehyde sensor based on the regulation of BFC performance caused by kinetics suppression [44]. The core component of this self-powered biosensor is a membraneless ethanol/air BFC. Carbon nanotubes' covalently binding with amine-terminated ionic liquid (CNTs-IL-NH_2) serves as the electrode substrates, and ADH and BOD are immobilized on the bioanode and biocathode, respectively. ADH is a zinc metalloenzyme catalyzing the reversible oxidation of primary short-chain alcohols to their corresponding carbonyl compounds using NAD^+ or $NADP^+$ as a cofactor.

$$CH_3CH_2OH + NAD^+ \rightarrow CH_3CHO + NADH + H^+$$

When acetaldehyde is added, the kinetics of enzyme-catalytic ethanol electro-oxidation was suppressed (Fig. 9.8). The sensor is able to detect an acetaldehyde concentration of 5–200 μM, with a detection limit of 1 μM. This BFC has the excellent selectivity from other carbonyl compounds.

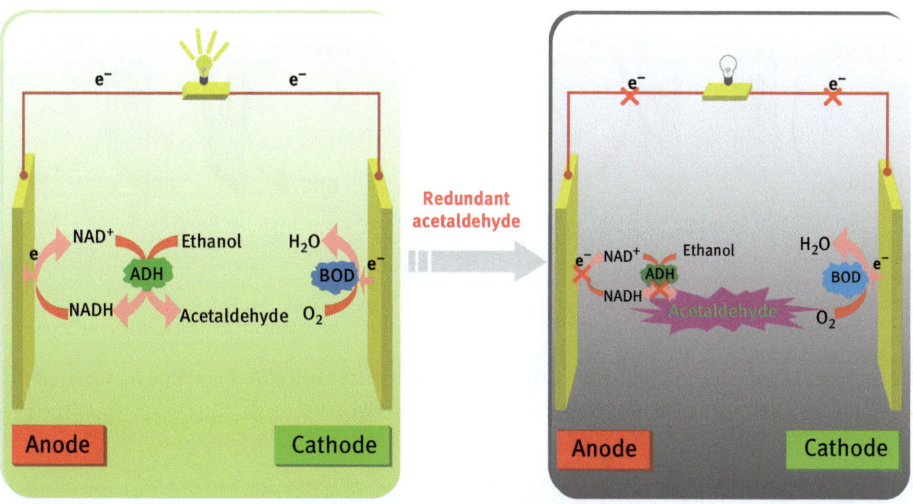

Fig. 9.8: Principle for the acetaldehyde sensor based on the ethanol/air BFC. Reprinted from Zhang et al. [44] with permission by the American Chemical Society.

9.3.3 Activation of negative inhibitive effect

In 2011, Minteer's group provided a self-powered enzymatic ethylenediaminetetraacetic acid (EDTA) sensor based on the inhibition and subsequent activation of GOx-based BFC [45]. It was fabricated by coupling the (poly(vinylpyridine)Os(bipyridyl)

$2Cl^{2+/3+}$)/GOx bioanode with a platinum gas diffusion electrode. The self-powered sensor was "turned off" by using a fuel solution containing 100 mM glucose and 10 mM $CuCl_2$ and activated by adding a slight excess of EDTA into the fuel solution (relative to the amount of Cu^{2+} (Fig. 9.9)). The amount of EDTA was detected by using different concentrations of Cu^{2+} to inhibit the bioanode and using an appropriate amount of EDTA to reactivate the enzyme. This work represents a new approach for self-powered enzymatic sensing through sensing signal turned on, rather than turned off. Since the stability of many enzymatic biofuels decreases with the operating time, this approach should be advantageous in the presence of the analyte from a dormant state to an operational state, which limits the number of enzyme turnovers that take place while the sensor is at rest. Importantly, it is the first time when a nonredox-active molecule is detected.

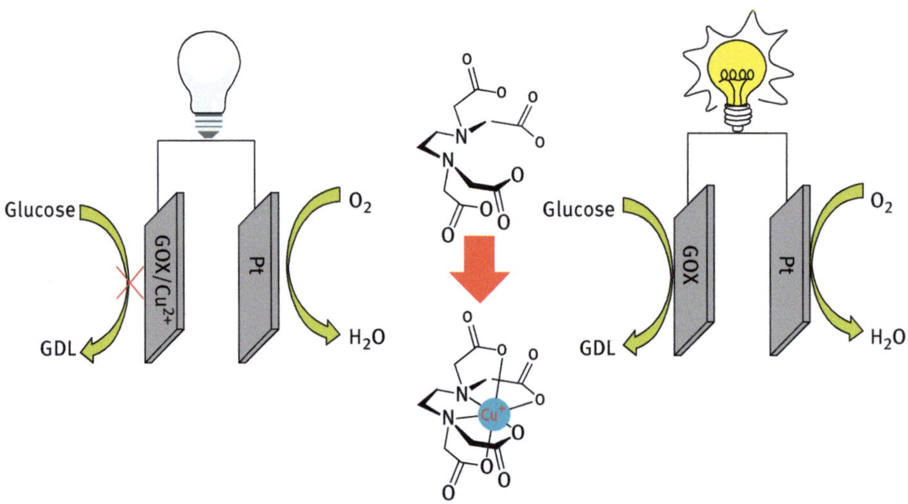

Fig. 9.9: Activation of a Cu^{2+}-inhibited GOx self-powered sensor by EDTA. Reprinted from Meredith et al. [45] with permission by the American Chemical Society.

9.3.4 Aptamer inhibitive effect

In 2010, Dong's group described the first example of controlled power release of BFCs by aptamer-based biochemical signals processed according to the Boolean logic operations "programmed" into biocomputing systems [46]. The BFC logically controlled by an aptamer system was composed of two on-chip patterned ITO electrodes modified with an enzyme/aptamer-based self-assembled multilayer consisting of poly(diallyldimethylammonium chloride)-wrapped CNTs, enzyme (glucose oxidase (all used as GOD) for the bioanode and BOD for the biocathode), aptamer, and bovine serum albumin (BSA) (Fig. 9.10).

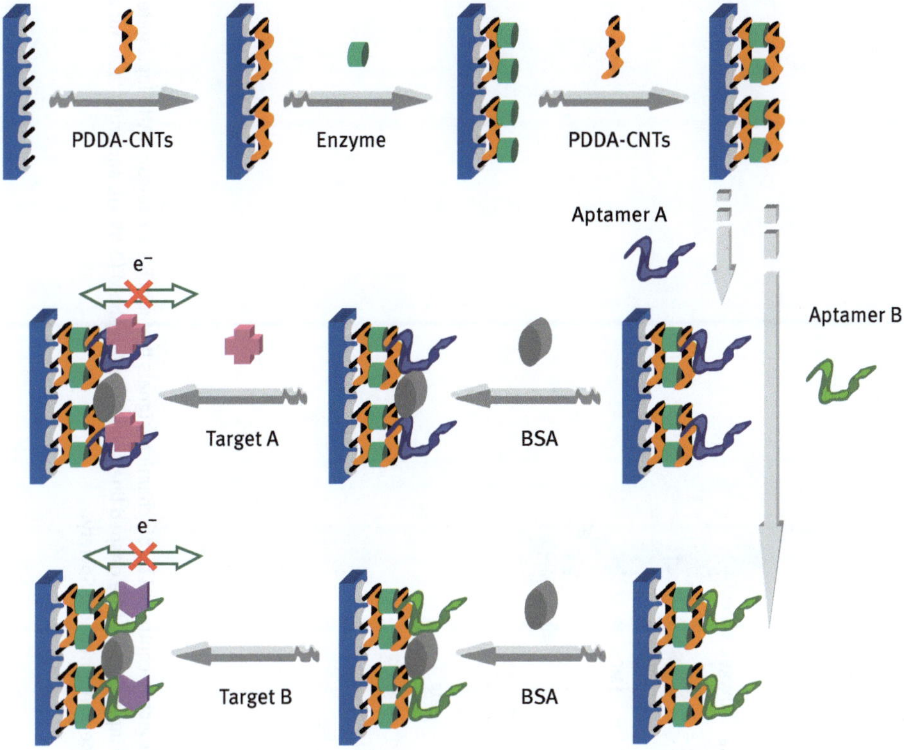

Fig. 9.10: A schematic illustration of the fabrication procedure for the bioanode and biocathode of the BFC logically controlled by aptamer-based biochemical signals (enzymes: GOD for the bioanode and BOD for the biocathode). Reprinted from Zhou et al. [46] with permission by the American Chemical Society.

There is another strategy for aptamer inhibitive effect self-powered biosensor. By assembly of the as-prepared GOD/aptamer-based bioanode and BOD/aptamer-based biocathode to form a microchip-based compartmentless glucose/O_2 BFC, the fabricated BFC operates in an air-saturated base solution (i.e., air-saturated pH 7.4 Tris-HCl buffer (20 mM Tris-HCl, 140 mM NaCl, 5 mM KCl, 5 mM $MgCl_2$)) containing 20 mM glucose as the fuel and 0.5 M ferrocene monocarboxylic acid as the redox mediator. In the presence of both thrombin and lysozyme, aptamer-target recognition occurred at both the bioanode (between thrombin-binding aptamer and thrombin) and biocathode (between lysozyme-binding aptamer and lysozyme); the OCV of the BFC outputs was executed according to the truth table for a NAND gate (Fig. 9.11). This work first introduced aptamer logic systems into BFCs to control the power release of the BFCs. Such integration may not only give us an avenue to control BFC power released by aptamer-based biocomputing systems but also indicate an interesting concept of "mutual benefits" between aptamers and BFCs, that is, utilizing aptamer-based biochemical signals in the logic operation for controlling the BFCs' power release

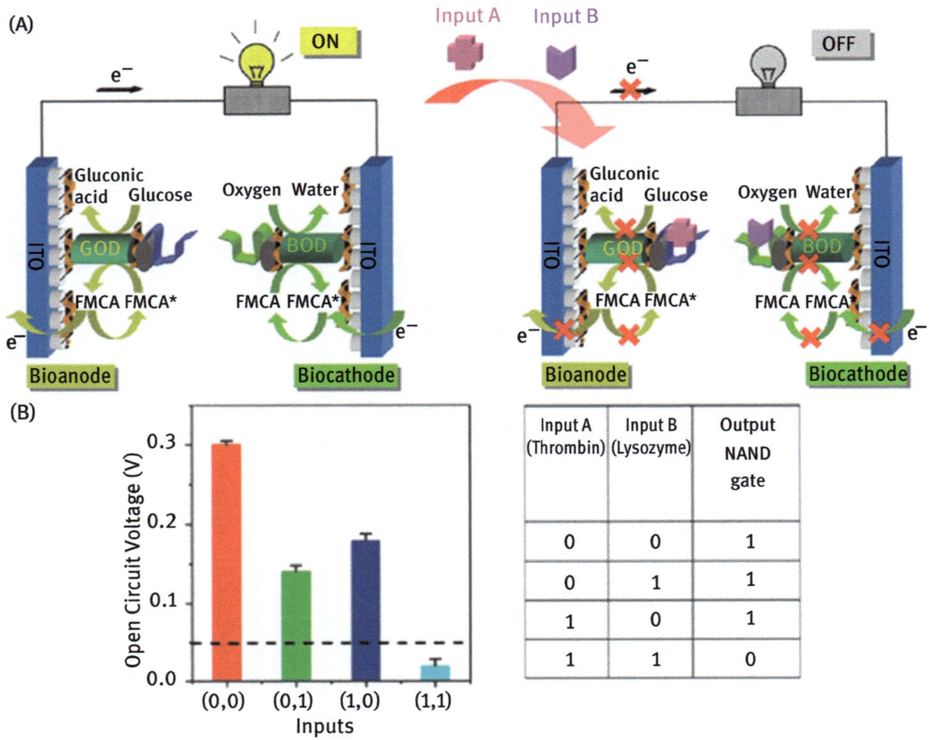

Input A (Thrombin)	Input B (Lysozyme)	Output NAND gate
0	0	1
0	1	1
1	0	1
1	1	0

Fig. 9.11: (A) A schematic illustration of the assembled aptamer-based BFC logically controlled by biochemical signals that mimics a Boolean NAND logic gate. (B) A bar diagram showing the OCV of the BFC for different combinations of input signals (the dashed line shows the threshold (0.05 V)), and truth table for a NAND logic gate. Reprinted from Zhou et al. [46] with permission by the American Chemical Society.

and applying the BFCs as self-powered and intelligent biosensors for logic aptasensing. In contrast to traditional biosensors, the fabricated logic aptasensors are self-powered, smart, and able to sense whether both specific targets are present by using the built-in Boolean logic.

9.4 Self-powered display signal

9.4.1 Self-powered electrical fluorescence biosensor

BFCs can be designed as self-powered electrical signal supply to control the fluorescence switch operation effectively. Dong's group first introduced BFCs into the fluorescence switch system to control the fluorescence intensity through bioelectrocatalysis and generate electricity [47]. The bioanode was constructed by immobilizing GDH on the carbon nanotubes–ionic liquid–NAD^+ nanocomposites and oxidized glucose with the onset potential of −0.11 V versus Ag/AgCl. BOD immobilized on CNTs was used as the biocathode with the onset potential of 0.59 V versus Ag/AgCl toward the reduction of oxygen. And then, a Prussian Blue (PB) electrode was inset between the bioanode and biocathode, which was not only used as an electrode in the system to generate electricity but also acted as an electrical signal switcher to effectively control the fluorescence switch simultaneously (Fig. 9.12). Therefore, the self-powered fluorescence system was capable of switching by adjusting the absorbance change of the PB component with the aid of the related fluorescence quenching mechanisms.

9.4.2 Self-powered electrochemiluminescence biosensor

In 2013, Wang's group proposed a self-powered 3D microfluidic electrochemiluminescence (ECL) biosensing platform [48]. First, a noble metal-free primary battery (C|$FeCl_3$ |NaCl|$AlCl_3$|Al) was employed in the origami device. The OCV of each origami battery was 1.53 ± 0.15 V and could maintain the output after 15 days (storage in air at room temperature). Moreover, the power density reached about 0.52 ± 0.026 mW cm^{-2}, which could even push the LED light. And then, this primary battery was tried to drive the ECL reaction with 1 mM luminol and 1 mM H_2O_2 in 0.1 M PBS (pH 9.0). Based on the effective detection of H_2O_2, this self-powered ECL sensing device was improved to monitor the glucose with the solutions pretreated by GOx. This self-powered glucose biosensor had a linear range from 0.1 to 3 mM ($R = 0.9954$), with the detection limit of 0.1 mM (Fig. 9.13). This work creatively utilized a primary battery to drive the ECL biosensors toward glucose, and provided practical significance in low-cost and disposable biosensing devices.

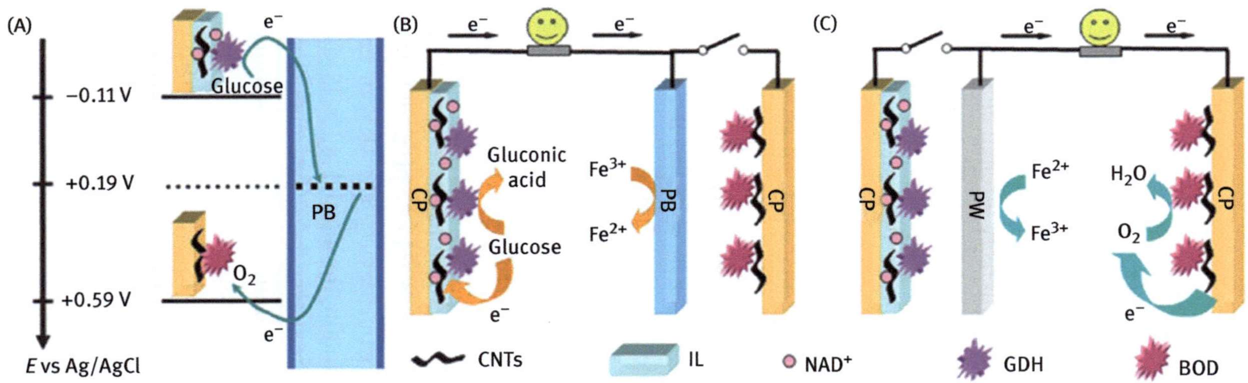

Fig. 9.12: (A) Thermodynamics of electron transfer. Electrons would spontaneously flow from the negative potential to the more positive potential. (B) A schematic diagram of bioreduction of PB by the bioanode. When the PB electrode was connected with the bioanode, GDH would biocatalyze the oxidation of glucose and the generated electrons transport to the PB electrode via the external circuit to reduce PB. (C) A schematic diagram of biooxidation of Prussian White by the biocathode. When the PW electrode was connected with the biocathode, BOD would biocatalyze the reduction of oxygen and compel PW oxidation to generate electrons. Reprinted from Bai et al. [47] with permission by the Royal Society of Chemistry.

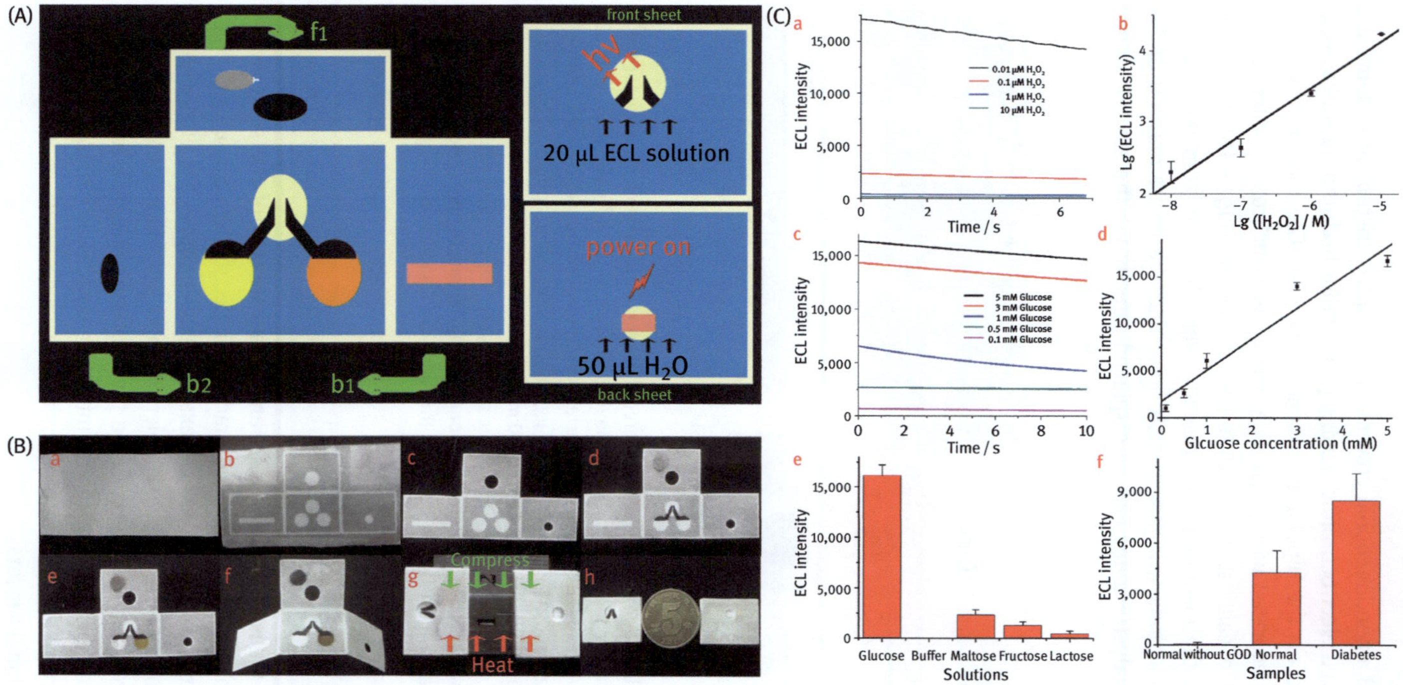

Fig. 9.13: (A) Schematic diagram of the working principle of m-s-OECLD. f: front sheet; b1: the sheet folded to the back first; b2: the back sheet. (B) Fabrication process of one battery m-s-OECLD. a: filter paper; b: paper patterning; c: size cutting; d: electrodes integrating; e: solution adding and baking; f: paper folding; g: chip assembling; h: accomplished chips. (C)a-b: ECL responses of different H_2O_2 concentrations: 0.01 mM, 0.1 mM, 1 Mm, and 10 mM. c-d: ECL responses of different glucose concentrations: 0.1 mM, 0.5 mM, 1 mM, 3 mM, and 5 mM. e: Selectivity of glucose detection. f: Detection of glucose in human blood samples. Reprinted from Zhang et al. [48] with permission by the Royal Society of Chemistry.

9.4.3 Self-powered bipolar electrochromic biosensor

Wang's group provided a nondirectional power source to drive the bipolar systems and developed a self-powered electrochromic sensing platform for high-throughput screening applications based on the bipolar electrochemistry [49]. The bipolar electrodes consisted of one end decorated with PB, which was an electrochromic material, and one end covered with a catalyst or biocatalyst such as GDH (Fig. 9.14). When the electrode potentials of the electrochromic reaction were different from the catalytic reactions, these two reactions will power each other and the target reaction or the performance of the catalyst/enzyme can be displayed directly via the electrochromic reaction.

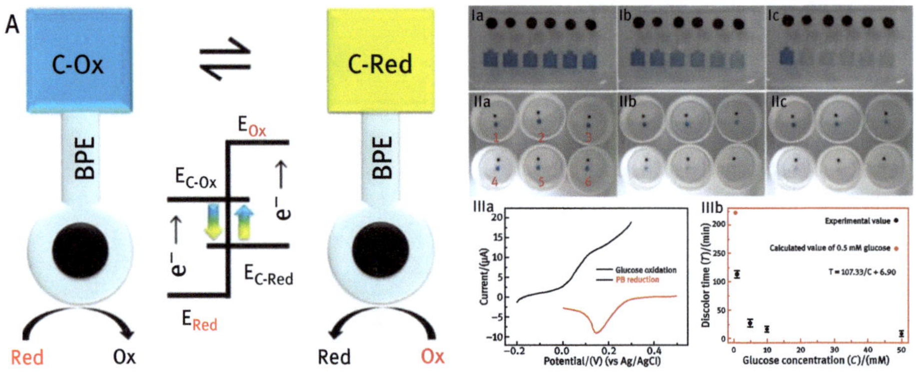

Fig. 9.14: (A) Schematic illustration on the self-powered bipolar electrochromic strategy for the displaying analysis of catalytic redox reaction. Displaying analysis of the different amount of GDH enzyme and its substrate (glucose). (I) SP-BP-EC-E array with the anodes decorated with different amounts of GDH (left to right, 0, 0.4, 0.8, 1.2, 16, 2.0 μg) at (a) 0 s, (b) 180 s, and (c) 1,020 s. (II) The displaying different concentrations (0, 0.5, 1, 5, 10, 50 mM) of glucose with 4 μg of CNT and 2.0 μg of GDH. (III) Electrochemical mechanism validation of the self-powered bipolar biosensing strategy (a) and the time-quantified mode of the SP-BP-EC-E responding to different glucose concentrations (b). Reprinted from Zhang et al. [49] with permission by the American Chemical Society.

9.5 Conclusion and outlooks

It is really important to list the advantages and disadvantages of the self-powered electrochemical biosensors compared to the traditional electrochemical sensors. For a self-powered biosensor, the circuitry and the external power supply are finally not the essential accessories any more, leading to simpler assemble processes by a two-electrode design instead of a three-electrode design. And the biosensor with only two electrodes also has smaller volume for implantable and miniature devices. Additionally, the biosensor has higher sensitivity and, by employing the biocatalysts, specificity when compared to the traditional sensors.

However, the self-powered electrochemical biosensors still face some problems for further practical applications and wider suitable ranges. First, the concern analytes are usually limited by the categories of biocatalysts, such as glucose, alcohol, lactate, and other the small biomolecules. The range that the self-powered biosensor can monitor is still not enough even by expanding the substrates to cholesterol or alkane. Thus, the directions would point to search for more biocatalysts and utilize the cathode characters to monitor other information instead of just focusing on the biomolecules. If so, the self-powered biosensors could be adaptable in many conditions. Secondly, the power densities of the BFCs are low and it cannot directly drive the signal electrics to integrate the whole components into a single sensor. By improving the power density of the BFC, a totally integrated self-powered electrochemical biosensor can be expected.

In summary, future research will focus on addressing the stability improving the power density and shortening response time of self-powered biosensors. Self-powered electrochemical biosensors with remarkable and unique performances are expected to exhibit new opportunities in diverse practical areas.

References

[1] Arechederra, R. L., & Minteer, S. D. Self-powered sensors. Anal Bioanal Chem. 2011, 400, 1605–1611.

[2] Grattieri, M., & Minteer, S. D. Self-powered biosensors. ACS Sensors. 2018, 3, 44–53.

[3] Zhou, M., & Dong, S. Bioelectrochemical interface engineering: toward the fabrication of electrochemical biosensors, biofuel cells, and self-powered logic biosensors. Acc Chem Res. 2011, 44, 1232–1243.

[4] Zhou, M. Recent progress on the development of biofuel cells for self-powered electrochemical biosensing and logic biosensing: a review. Electroanalysis. 2015, 27, 1786–1810.

[5] Windmiller, J. R., & Wang, J. Wearable electrochemical sensors and biosensors: a review. Electroanalysis. 2013, 25, 29–46.

[6] Wang, N., Zhang, N., & Wang, M. Wireless sensors in agriculture and food industry: recent development and future perspective. Comput Electron Agric. 2006, 50, 1–14.

[7] Wang, Z. L. Toward self-powered sensor networks. Nano Today. 2010, 5, 512–514.

[8] Lee, M., Bae, J., Lee, J., Lee, C.-S., Hong, S., & Wang, Z. L. Self-powered environmental sensor system driven by nanogenerators. Energy Environ Sci. 2011, 4, 3359–3363.

[9] Zhou, M., & Wang, J. Biofuel cells for self-powered electrochemical biosensing and logic biosensing: a review. Electroanalysis. 2012, 24, 197–209.

[10] Bullen, R. A., Arnot, T. C., Lakeman, J. B., & Walsh, F. C. Biofuel cells and their development. Biosens Bioelectron. 2006, 21, 2015–2045.

[11] Davis, F., & Higson, S. P. Biofuel cells–recent advances and applications. Biosens Bioelectron. 2007, 22, 1224–1235.

[12] Rasmussen, M., Abdellaoui, S., & Minteer, S. D. Enzymatic biofuel cells: 30 years of critical advancements. Biosens Bioelectron. 2016, 76, 91–102.

[13] Kim, J., Jia, H., & Wang, P. Challenges in biocatalysis for enzyme-based biofuel cells. Biotechnol Adv. 2006, 24, 296–308.

[14] Calabrese Barton, S., Gallaway, J., & Atanassov, P. Enzymatic biofuel cells for implantable and microscale devices. Chem Rev. 2004, 104, 4867–4886.
[15] de Poulpiquet, A., Ciaccafava, A., & Lojou, E. New trends in enzyme immobilization at nanostructured interfaces for efficient electrocatalysis in biofuel cells. Electrochim Acta. 2014, 126, 104–114.
[16] Szczupak, A., Halámek, J., Halámková, L., Bocharova, V., Alfonta, L., & Katz, E. Living battery – biofuel cells operating in vivo in clams. Energy Environ Sci. 2012, 5, 8891–8895.
[17] MacVittie, K., Halámek, J., Halámková, L., Southcott, M., Jemison, W. D., Lobel, R., & Katz, E. From "cyborg" lobsters to a pacemaker powered by implantable biofuel cells. Energy Environ Sci. 2013, 6, 81–86.
[18] Tarascon, J. M. Key challenges in future Li-battery research. Philos Trans Series A, Mathematical, physical, and engineering sciences. 2010, 368, 3227–3241.
[19] Ceder, G. Opportunities and challenges for first-principles materials design and applications to Li battery materials. MRS Bulletin. 2011, 35, 693–701.
[20] O'Regan, B., & Grätzel, M. A low-cost, high-efficiency solar cell based on dye-sensitized colloidal TiO_2 films. Nature. 1991, 353, 737–740.
[21] Chen, T., Barton, S. C., Binyamin, G., Gao, Z., Zhang, Y., Kim, H.-H., & Heller, A. A miniature biofuel cell. J Am Chem Soc. 2001, 123, 8630–8631.
[22] Falk, M., Andoralov, V., Silow, M., Toscano, M. D., & Shleev, S. Miniature biofuel cell as a potential power source for glucose-sensing contact lenses. Anal Chem. 2013, 85, 6342–6348.
[23] Katz, E., Bückmann, A. F., & Willner, I. Self-powered enzyme-based biosensors. J Am Chem Soc. 2001, 123, 10752–10753.
[24] Cracknell, J. A., Vincent, K. A., & Armstrong, F. A. Enzymes as working or inspirational electrocatalysts for fuel cells and electrolysis. Chem Rev. 2008, 108, 2439–2461.
[25] Mano, N., Mao, F., & Heller, A. A miniature biofuel cell operating in a physiological buffer. J Am Chem Soc. 2002, 124, 12962–12963.
[26] Halamkova, L., Halamek, J., Bocharova, V., Szczupak, A., Alfonta, L., & Katz, E. Implanted biofuel cell operating in a living snail. J Am Chem Soc. 2012, 134, 5040–5043.
[27] Rasmussen, M., Ritzmann, R. E., Lee, I., Pollack, A. J., & Scherson, D. An implantable biofuel cell for a live insect. J Am Chem Soc. 2012, 134, 1458–1460.
[28] Castorena-Gonzalez, J. A., Foote, C., MacVittie, K., Halámek, J., Halámková, L., Martinez-Lemus, L. A., & Katz, E. Biofuel cell operating in vivo in rat. Electroanalysis. 2013, 25, 1579–1584.
[29] Katz, E., & MacVittie, K. Implanted biofuel cells operating in vivo-methods, applications and perspectives-feature article. Energy Environ Sci. 2013, 6, 2791–2803.
[30] Yu, J., Liu, S., & Ju, H. Glucose sensor for flow injection analysis of serum glucose based on immobilization of glucose oxidase in titania sol–gel membrane. Biosens Bioelectron. 2003, 19, 401–409.
[31] Wang, S. G., Zhang, Q., Wang, R., & Yoon, S. F. A novel multi-walled carbon nanotube-based biosensor for glucose detection. Biochem Biophys Res Commun. 2003, 311, 572–576.
[32] Crouch, E., Cowell, D. C., Hoskins, S., Pittson, R. W., & Hart, J. P. A novel, disposable, screen-printed amperometric biosensor for glucose in serum fabricated using a water-based carbon ink. Biosens Bioelectron. 2005, 21, 712–718.
[33] Sokic-Lazic, D., de Andrade, A. R., & Minteer, S. D. Utilization of enzyme cascades for complete oxidation of lactate in an enzymatic biofuel cell. Electrochim Acta. 2011, 56, 10772–10775.
[34] Topcagic, S., & Minteer, S. D. Development of a membraneless ethanol/oxygen biofuel cell. Electrochim Acta. 2006, 51, 2168–2172.
[35] Deng, L., Shang, L., Wen, D., Zhai, J., & Dong, S. A membraneless biofuel cell powered by ethanol and alcoholic beverage. Biosens Bioelectron. 2010, 26, 70–73.

[36] Yu, Y., Han, Y., Xu, M., Zhang, L., & Dong, S. Automatic illumination compensation device based on a photoelectrochemical biofuel cell driven by visible light. Nanoscale. 2016, 8, 9004–9008.
[37] Yu, Y., Zhai, J., Xia, Y., & Dong, S. Single wearable sensing energy device based on photoelectric biofuel cells for simultaneous analysis of perspiration and illuminance. Nanoscale. 2017, 9, 11846–11850.
[38] Jia, W., Valdes-Ramirez, G., Bandodkar, A. J., Windmiller, J. R., & Wang, J. Epidermal biofuel cells: energy harvesting from human perspiration. Angew Chem Int Ed. 2013, 52, 7233–7236.
[39] Sekretaryova, A. N., Beni, V., Eriksson, M., Karyakin, A. A., Turner, A. P., & Vagin, M. Y. Cholesterol self-powered biosensor. Anal Chem. 2014, 86, 9540–9547.
[40] Deng, L., Chen, C., Zhou, M., Guo, S., Wang, E., & Dong, S. Integrated self-powered microchip biosensor for endogenous biological cyanide. Anal Chem. 2010, 82, 4283–4287.
[41] Deng, L., Shang, L., Wang, Y., Wang, T., Chen, H., & Dong, S. Multilayer structured carbon nanotubes/poly-l-lysine/laccase composite cathode for glucose/O_2 biofuel cell. Electrochem Commun. 2008, 10, 1012–1015.
[42] Shleev, S., Tkac, J., Christenson, A., Ruzgas, T., Yaropolov, A. I., Whittaker, J. W., & Gorton, L. Direct electron transfer between copper-containing proteins and electrodes. Biosens Bioelectron. 2005, 20, 2517–2554.
[43] Wen, D., Deng, L., Guo, S., & Dong, S. Self-powered sensor for trace Hg^{2+} detection. Anal Chem. 2011, 83, 3968–3972.
[44] Zhang, L., Zhou, M., & Dong, S. A self-powered acetaldehyde sensor based on biofuel cell. Anal Chem. 2012, 84, 10345–10349.
[45] Meredith, M. T., & Minteer, S. D. Inhibition and activation of glucose oxidase bioanodes for use in a self-powered EDTA sensor. Anal Chem. 2011, 83, 5436–5441.
[46] Zhou, M., Du, Y., Chen, C., Li, B., Wen, D., Dong, S., & Wang, E. Aptamer-controlled biofuel cells in logic systems and used as self-powered and intelligent logic aptasensors. J Am Chem Soc. 2010, 132, 2172–2174.
[47] Bai, L., Jin, L., Han, L., & Dong, S. Self-powered fluorescence controlled switch systems based on biofuel cells. Energy Environ Sci. 2013, 6, 3015–3021.
[48] Zhang, X., Li, J., Chen, C., Lou, B., Zhang, L., & Wang, E. A self-powered microfluidic origami electrochemiluminescence biosensing platform. Chem Commun. 2013, 49, 3866–3868.
[49] Zhang, X., Zhang, L., Zhai, Q., Gu, W., Li, J., & Wang, E. Self-powered bipolar electrochromic electrode arrays for direct displaying applications. Anal Chem. 2016, 88, 2543–2547.

Carlo Santoro, Dmitry Pankratov, Ioannis Ieropoulos, Francesca Soavi

10 Supercapacitors in bioelectrochemical systems

10.1 Bioelectrochemical systems

10.1.1 Introduction

Bioelectrochemical systems (BESs) are devices composed of two electrodes, the anode and the cathode, a separator and the electrolyte [1, 2]. At the anode, the oxidation reaction occurs while at the cathode reduction reaction takes place. Different from classical abiotic electrochemical systems such as fuel cells, batteries, and supercapacitors, BESs have biotic components and features in at least one of the two electrodes [1, 2]. It may happen that both of the electrodes interact with the biotic matter. The biotic components are generally microorganisms and enzymes that categorize fuel cells called microbial fuel cells (MFCs) [1, 2] and enzymatic fuel cells (EFCs) [3]. The latter are also known by the general term, that is, biofuel cells (BFCs). BESs dealing with microorganisms can also be exploited for the following: (i) generating hydrogen at the cathode under an external voltage supply in a microbial electrolysis cell (MEC); (ii) reducing carbon dioxide to methane, acetate, or other valuable compounds in a microbial electrosynthesis cell (MEC); (iii) desalinate water while producing electricity (MDC) [2].

In acidic or alkaline fuel cells, extreme pHs are required to accelerate the reactions and increase the efficiency of the overall system. Moreover, often the temperature is increased for increasing the reaction kinetics [4, 5]. H^+ and OH^- are reactants within the redox process and therefore high concentration of those reactants can be achieved at low and high pHs respectively. In BESs in which a biological component is present, neutral or circumneutral conditions are required to operate in order for enzymes or bacteria to work efficiently. BESs can work in ubiquitous media containing organics that bacteria or enzymes can use as fuel for the electrochemical device to produce useful electricity [2, 3]. Because of the presence of biological and

Carlo Santoro, Ioannis Ieropoulos, Bristol BioEnergy Centre, Bristol Robotics Laboratory, University of The West of England, Bristol, UK

Dmitry Pankratov, Department of Chemistry, Technical University of Denmark, Kongens Lyngby, Denmark

Francesca Soavi, Department of Chemistry "G. Ciamician", Alma Mater Studiorum – University of Bologna, Italy

https://doi.org/10.1515/9783110570526-010

enzymatic life, BESs are forced to operate in conditions closer to room temperature in order to preserve their activities [2, 3].

10.1.2 Enzymatic fuel cells

EFCs exploit oxidoreductases to convert chemical energy into electric power. Depending on the BES design, fuel, and oxidant, individual enzymes [6] or enzymatic cascades [7] can be utilized to catalyze fuel oxidation and/or oxidant reduction, or single-enzyme approach can be used, where the same enzyme is immobilized on the anodic and cathodic sides [8–10]. High activity of immobilized enzymes allows achieving diffusional limited rates of redox reactions, whereas enzymatic substrate specificity may result in simplified design of EFC, since membrane is not necessary required for anolyte and catholyte separation to avoid parasitic crossover processes.

About 1,800 redox enzymes are known to date [11] and any of them may potentially be employed as a biocatalyst in EFC; however, their practical application in many cases is limited because of hampered electrochemical communication between the active center of enzyme and electrode surface (especially an important factor in direct electron transfer [DET] approach) with the availability of substrate(s) and stability of the bioelement in an immobilized state. Particular advantages of EFCs over conventional fuel cells coupled with broad possibilities of EFC miniaturization to the nanoscale level make these kind of BESs an attractive power source aimed to operate in physiological media in vitro [12, 13], ex vivo [14], and in vivo [15–17].

Carbohydrate/O_2 EFC (mainly glucose/O_2) became the most popular type of this kind of power sources, beginning with pioneering reports by Yahiro et al. [18] and continuing till date [19], including all examples of supercapacitive EFCs (sEFCs). Half-cell potentials of two-electron glucose oxidation to gluconolactone and four-electron reduction of oxygen equal to −0.36 V and 0.82 V, respectively (pH 7, 25 °C, vs. the standard hydrogen electrode [SHE]) determine the thermodynamic maximum voltage of a glucose/oxygen EFC, operating at neutral pH at 1.18 V. This value can be increased up to 1.24 V for complete 24-electron oxidation of glucose, whereas the maximum open circuit voltage (OCV) practically achieved for the glucose/O_2 EFC so far is close to 1 V [20].

A great variety of enzymes were employed in EFC, intended for different substrates and operational conditions. The most popular anodic enzymes in glucose oxidizing bioanodes are glucose oxidase (GOx) [21], cellobiose dehydrogenase (CDH) [22], glucose dehydrogenase (GDH) [23], pyranose dehydrogenase [24], and so on. Fructose dehydrogenase was widely used in fructose/O_2 EFC, whereas NiFe and FeFe hydrogenases are common anodic bioelements in H_2/O_2 EFCs [25]. Alcohol dehydrogenases were efficiently employed for oxidation of ethanol and other alcohols [26]. Cathodic bioelements for EFCs are mainly presented by a broad family of blue multicopper oxidases (laccases, bilirubin oxidases [BOx]). Oxygen-reducing enzymes were thoroughly evaluated in a recent review by Mano and de Poulpiquet [27]. Apart from O_2-reducing enzymes, several heme-containing enzymes were utilized to catalyze reduction of peroxides [28], which can also be used for the development of EFCs.

10.1.3 Microbial fuel cells and microbial desalination cells

In MFCs, the oxidation reaction of organics occurs at the anode and the reduction of an oxidant occurs at the cathode to complete the cell redox reaction. Among the reductants used for the anode oxidation reaction, a plethora of organics were investigated as fuel [29, 30]. Anodic fuels studied are based on simple or complex organic molecules investigated individually or combined into liquid electrolytes [29, 30]. The catalysts in this case are of biotic type. They are electroactive bacteria (single species or organized in consortia) that through their anaerobic respiration utilize the solid anode electrode as the terminal electron acceptor [31]. In this way, the electrons produced from the break down of organics are transferred to the anodic electrode. The electrons move through an external circuit and generate electric current. Among several oxidants available for the cathode reduction, the most utilized is oxygen [32]. The latter is readily and highly available at no cost in the atmosphere and therefore suitable for MFC applications. Oxygen-reduction reaction (ORR) has its own issues when operated in a neutral media as described above, therefore a catalyst (inorganic is preferred) to accelerate the reaction is needed [33, 34]. Dual aspects such as organic/pollutant removal and power generation are achieved within the system during operations [1, 2].

Another interesting bioelectrochemical technology recently introduced is the microbial desalination cell (MDC) that utilizes the same anodic and cathodic processes as the MFC but a different cell design. Two membrane separators are included and an additional compartment (desalination cell) between anode and cathode compartment is added [35, 36]. Salt solution flows through the desalination cell and ions diffuse from the central compartment to the other two compartments through anion and cation exchange membranes. Osmosis also occurs with the transport of water from the lateral compartment to the desalination cell. Therefore, MDCs are able to degrade organics, generate electricity, and reduce the salinity content of concentrated salt water.

10.1.4 Supercapacitors

Electrochemical capacitors, also called supercapacitors or ultracapacitors, are playing a key role as energy storage and conversion systems for applications requiring high peak-to-average power demand [37, 38].

Many types of supercapacitors making use of electrodes and electrolytes with different nature and composition have been demonstrated [39–41]. The most common supercapacitors are the electrochemical double layer capacitors (EDLCs) that use high surface area carbon electrodes that store/deliver charge by an electrostatic process [41, 42]. EDLC cells feature two high surface area carbon electrodes separated by a porous membrane, the separator, and ionically connected with the electrolyte. Specifically, on application of an external voltage bias, electrodes are polarized with opposite sign. In parallel, ions from the electrolyte migrate and diffuse to counterbalance the surface electrode charge. This gives rise to the formation of the so-called electrical double layer [41, 42]. This electrostatic process is faster (seconds to minutes time scale) and more reversible than other electrochemical (redox) processes, which provides EDLCs the inherent characteristic of fast charge/discharge, high-energy storing efficiency, and highly prolonged life cycle [41, 42]. A life cycle longer than 10^5 charge/discharge cycles and coulombic efficiencies higher than 99.5% are feasible with EDLCs. According to the double-layer Helmholtz theory, the polarized electrode/electrolyte interface can be modeled as an electrical capacitor. The corresponding capacitance $C_{\text{electrode}}$ is related to the electrode of surface area (S), the vacuum permittivity ($\varepsilon_0 = 8.854 \times 10^{-12}\ \text{F m}^{-1}$), the relative permittivity of the electrolyte solvent ε_r, and the thickness of the double layer (d, corresponding to the Debye length) by eq. (10.1):

$$C_{electrode} = \frac{S \times \varepsilon_0 \times \varepsilon_r}{d} \tag{10.1}$$

The very high specific surface area of the carbon electrodes used in EDLCs (up to 1,500–2000 $\text{m}^2\ \text{g}^{-1}$), and the very short length of d, in the order of the molecule size bring to specific electrode capacitances in the order of 100 F g^{-1} and above, depending on the chemical nature of the electrolyte that impacts on the value of ε_r [41, 42]. In order to get high $C_{\text{electrode}}$ values great efforts are devoted for designing nanostructured carbons high surface area and porosity tailored to provide easy access and fast diffusion of electrolyte ions to the polarized surface (in the nanometer range). Examples are activated carbons, carbons obtained by template methods, carbide-derived carbon nanotubes (CNTs) and graphenes [39–41].

In EDLCs, each electrode/electrolyte interface represents a capacitor so that the complete cell can be modeled with two capacitors and a resistance (the equivalent series resistance, ESR) connected in series [41]. Therefore, EDLC cell capacitance (C_{EDLC}) results from a series combination of the negative and positive electrode capacitances, (C_-, C_+), that is:

$$\frac{1}{C_{EDLC}} = \frac{1}{C_-} + \frac{1}{C_+} \tag{10.2}$$

Energy and power performance of EDLCs depend on C_{EDLC}, ESR, and on the highest voltage that can be reached during charge that in turn is related to the electrochemical stability of the electrolyte. For aqueous electrolytes, this is constrained below 1.5–2 V and for organic electrolytes can be as high as 3–3.5 V [41].

The capacitance, in Farads, corresponds to the amount of charge (Q) that can be stored over a potential difference of 1 V between the electrodes, and for ideal systems is a constant, that is:

$$C_{EDLC} = \frac{dQ}{dV} \tag{10.3}$$

It turns out that during a galvanostatic charge/discharge at the current i, the EDLC cell voltage linearly increases/decreases over time with a slope (s):

$$s = \frac{dV}{dt} = \frac{i}{C_{EDLC}} \tag{10.4}$$

The energy that can be stored/delivered is:

$$E = i \int V dt \tag{10.5}$$

and the highest value that can be obtained at negligible ESR is E_{max}, that is:

$$E_{max} = \frac{1}{2} C_{EDLC} \times V_{max}{}^2 \tag{10.6}$$

In practical cases, ESR is not negligible, therefore an ohmic drop occurs at the beginning of the discharge that lowers the cell voltage by a value that depends on current and ESR. Specifically, the highest feasible voltage becomes:

$$V = V_{max} - i \times ESR \tag{10.7}$$

The linear relation between the stored charge and voltage, and the fast response of EDLCs, make them extremely attractive for integration with several energy harvesters systems, including those with low voltage output like MFCs [43]. EDLCs store and deliver-on-demand the energy harvested from the ambient with the effect of improving power generation quality [43]. Indeed, they can buffer discontinuities at a small time-scale and to boost up power output. EDLCs can, indeed, store energy at any voltage (provided that is lower than V_{max}) and current rate provided by the MFC, in the form of charge separation at the electrode/electrolyte interfaces and deliver it back by a very fast electrostatic discharge [44]. The maximum power (P_{max}) delivered by EDLCs, that corresponds to the power delivered to an external load with impedance equal to the EDLC ESR is not affected by the energy harvester performance and is:

$$P_{max} = \frac{1}{4} \frac{V_{max}{}^2}{ESR} \tag{10.8}$$

The practical power (P) that is the power delivered through any load, is obtained by:

$$P = \frac{i \int V dt}{\Delta t} \tag{10.9}$$

where Δt is the discharge duration.

10.1.5 Bioelectrochemical systems combined with external supercapacitors

Enzymatic and MFCs produce relatively low electrochemical output and the harvesting for real applications result to be quite complicated. EFCs and MFCs and generally BESs operate in aqueous media and therefore they are limited in their voltage operation. In order to increase the energy and power output within the system, simple connection in series and parallel can be done [45, 46]. In the first case, the connection in series enhances the voltage but this is possible only when the different cell units do not share the same electrolyte or electrode short-circuits or shunt current can occur. The connection in parallel can occur also in the case of same electrolyte shared and allow increasing the current output. Both voltage and current are important parameters to be accounted when energy harvesting is considered [45, 46].

Enzymatic and MFCs produce relatively low energy and power output and therefore in order to harvest the current/power efficiently, BESs are connected with external supercapacitors [47]. The results of this hybridization have brought practical applications and useful harvesting of the electricity produced [47]. EFCs are usually associated with biosensing applications [48, 49] while MFCs can also be associated with biosensing [50, 51] even if other applications were investigated. In the case of marine environments, sediment MFCs coupled with external electronic circuitry containing supercapacitors were able to power sensors successfully [52–54].

Other examples of successful applications are given by the several energetically autonomous robots powered with MFCs coupled with external supercapacitors and/or batteries reported in literature. Gastrorobot was first introduced in 2000 at the University of South Florida and was powered by MFCs operating as an artificial stomach with *Escherichia coli*, HNQ mediator, and glucose as organic feedstock, and the metabolized chemicals were then inserted into a chemical fuel cell [55]. The power/energy produced was then used for charging external Ni–Cd batteries, which then powered the motor of the train. The EcoBots family (EcoBot I, EcoBot II, and EcoBot III) [56–58] is another successful example of integration of MFCs that produced energy/power that is harvested into external energy storage units. In EcoBot I and II, the external energy storage was composed by electrolytic capacitors, which were capable of delivering pulsed discharges allowing the phototactic robot to progress toward a light source [56, 57]. EcoBot III design was more complex since it was able to ingest, digest, and egest food by moving to the feeding base [58]. Similarly, Row-bot composed by on-board MFCs was a floating robot capable of rowing within a polluted environment and rowing while feeding itself through opening its "mouth" [59].

In these robots, the power produced by the MFCs was because of the oxidation of organics by electroactive bacteria and it was stored temporarily in external capacitors. The discharge of capacitors occurred when a voltage level was achieved. The integration of MFCs with supercapacitors allowed delivery of higher level of energy compared to the one produced continuously by MFCs despite the energy is delivered in a pulsated manner.

10.2 Supercapacitive bioelectrochemical systems

10.2.1 Introduction

In the previous sections, EFCs and MFCs were briefly described and also supercapacitors were concisely introduced. The combination of the two diverse types of electrochemical technology, separated with external supercapacitors and power management board, has brought the possibility of harvesting such a small amount of energy/power for practical applications. Only recently, electrode electrostatic and faradaic of both enzymatic and MFCs were exploited and used as internal supercapacitors electrodes. Therefore a new field named biosupercapacitor emerged.

10.2.2 Supercapacitive enzymatic fuel cells

Primary attempts to utilize inherent electrode capacitance as active functional element in EFC design were performed in parallel and independently by research groups of Prof. S. Cosnier and Prof. S. Shleev, followed by patent applications with priority dates of March 07, 2013, and April 10, 2013, respectively [60, 61]. Later, initial research papers regarding this subject were published. Based on the previously reported principle of dual-function electrodes [62], Pankratov et al. fabricated membrane-, mediator-, and cofactor-less sEFC (defined as self-charging biosupercapacitor), where energy converting and capacitive parts were deposited on the opposite sides of graphite foil [63]. Energy converting part of this glucose/O_2 sEFC was based on BOx and CDH for the cathode and anode, respectively. Enzymes were immobilized on the surface of gold nanoparticles, modified with a mixture of thiols in case of the anode. Nanocomposite polyaniline/CNT was utilized as a main charge-storing component in both bioelectrodes. A maximum OCV value obtained in air saturated phosphate buffer saline (PBS) containing 50 mM glucose was 0.40 V, indicating a fully charged state. Peak power density was registered when 500 Ω load was applied between the bioelectrodes, it was 1.2 mW cm^{-2}, dropping to 0.02 mW cm^{-2} (close to continuous power output from non-capacitive EFC of the same design) after 50 min of discharge process. The example presented above does not display any significant improvement of the power conversion efficiency, when capacitive component was introduced in the sEFC design, but can be considered as a proof-of-principle demonstration of a new experimental dual-function platform to overcome bottleneck for the power supply from conventional EFCs. Direct immobilization of BOx on the surface of polyaniline/CNT nanocomposite significantly improves OCP value and stability of the biocathode both in continuous and pulse modes, and, therefore, may lead to further improvement of the sEFC performance [64].

Coupling CNT-based laccase biocathode with Zn anode demonstrates opportunity to use hybrid biobatteries in a pulse mode. When an external load of 2 kΩ was introduced between the electrodes, fast and reproducible recovery of the OCV value in the self-charge process could be observed [67].

According to the classification of sEFCs proposed previously [68], hybrid biodevice described above can be defined as pseudo-sEFC in analogy with conventional capacitors, taking into account the main charge-storing principle. The very first example of double-layer sEFC was presented by Agnes et al., where bioelectrodes were based on laccase and GOx /catalase, entrapped into a compressed porous CNT matrix [65]. This device could deliver up to 16 mW at 0.5 V in air saturated PBS containing 200 mM glucose with remarkable operational stability, when continuous

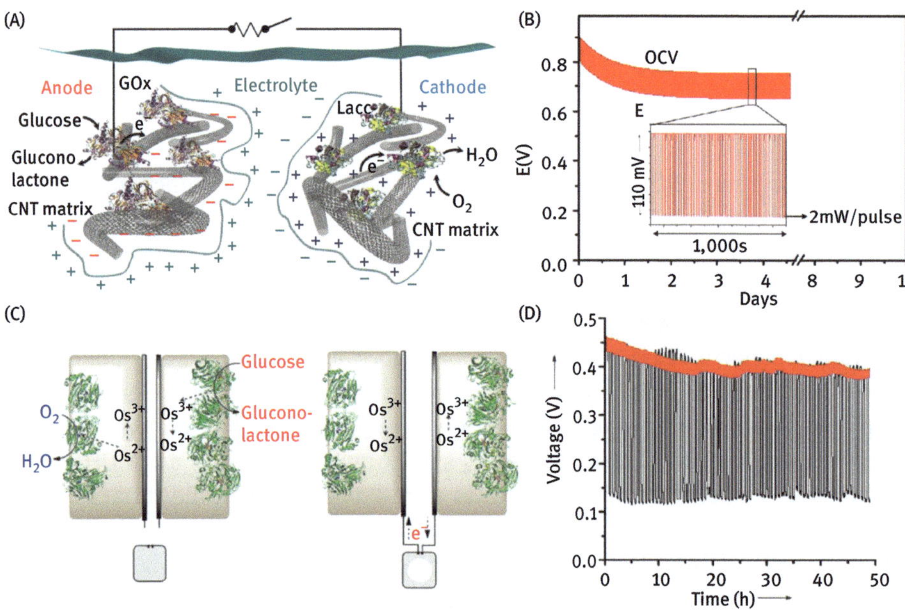

Fig. 10.1: Schematic representation and performance of representative glucose/oxygen supercapacitive enzymatic fuel cells. (A) Scheme of the electrochemical double layer sEFC based on glucose oxidase and laccase. (B) Operational stability test of the sEFC under 40,000 discharge pulses of 3 mA for 0.01 s followed by self-charging for 10 s. (C) Scheme of the Nernstian pseudo-sEFC employed glucose dehydrogenase and bilirubin oxidase. (D) Operational stability test of the sEFC under short-term discharging by applying a constant current of 5 μA for 1 s followed by self-charging for 100 s (red curve) and by applying a constant load of 22 kΩ for 20 min followed by self-charging for 20 min (black curve). (A, B) Adapted from Agnes et al. [65], Copyright (2014) from The Royal Society of Chemistry. (C, D) Adapted from Pankratov et al. [66] with permission from John Wiley and Sons.

self-charge/discharge pulses of 2 mW were applied for 10 ms every 10 s for at least 5 days (40 000 cycles) (Fig. 10.1A and B).

Narvaez Villarubia et al. have retrofitted conventional sEFC construct by building a paper-based capillary-driven microfluidic system, which allows a continuous biofuel supply [69]. Dual-layered gas-diffusional cathode was fabricated using Toray paper as current collector by compression of hydrophobic layer of teflonized Vulcan XC72 carbon black with catalytic layer of CNT-based buckypaper, where BOx was deposited. The glucose bioanode was built on buckypaper, modified with methylene green, acting as a mediator for NAD^+/NADH-dependent GDH. The enzyme was mixed with the chitosan/CNTs composite and applied on the anode surface. In PBS containing 100 mM glucose and 1 mM NAD^+ this sEFC had OCV of 0.56 V and could provide power up to 13.1 mW (1.07 mW cm^{-2}) for current pulse of 4 mA cm^{-2} with reasonable stability evaluated at 0.4 mA cm^{-2} for 4200 cycles (3 days of operation).

Efficient combination of bioelectrocatalytic and supercapacitive components was performed by Kizling et al. [70–72]. In these works, nanocellulose/polypyrrole composite was used to immobilize FDH on the anodic side, whereas biocathode was based on naphthylated CNTs [70–72] or on cellulose/polypyrrole composite [72] modified with laccase, directly adsorbed on the CNTs or polymer surface or entrapped into the electropolymerized polypyrrole layer in the presence of mediator (ABTS). A maximum power output of 2.1 mW cm^{-2} under a load of 1 kΩ and OCV of 0.59 V were achieved for the sEFC with mediated cathode in oxygenated pH 5.3 McIlvaine buffer with 100 mM fructose.

Knoche et al. reported a glucose/O_2 hybrid biodevice, consisting of oxygen-reducing biocathode, where BOx was immobilized with antracene-functionalized CNTs, and tetrabutylammonium bromide-modified Nafion on carbon felt, coupled with glucose-oxidizing bioanode, consisting of GDH mixed with CNTs and ferrocene-containing redox polymer [73]. OCV value of 0.6 V and peak power density of 3 mW cm^{-2}, obtained at O_2–sparged 0.2 M phosphate/citrate buffer (pH 6.5) with 200 mM glucose, indicate perspectives of future employment of redox polymers in sEFC design.

A novel approach for the development of the sEFCs, defined as "Nernstian biosupercapacitor," was introduced by Prof. W. Schuhmann and coworkers [66]. In this work, both enzymatic anode and cathode were based on the same osmium-containing redox polymer (OsRP), biomodified with GDH and BOx, respectively. The potential of OsRP, acting simultaneously as a mediator and charge-storing component, changes in accordance with Nernst equation, and the charge is accumulated as a difference in the activity ratio of reduced and oxidized forms of osmium. The OCV value of the biodevice in oxygen-saturated PBS containing 20 mM glucose changes from 0 V in a fully discharged state to maximum 0.45 V because of self-charging enzymatic activity in the presence of fuel and oxidant. The average power density, when the sEFC was discharged under a constant load of

10 kΩ, was 11.3 μW cm^{-2} (3.9 mW g^{-1}), which was enough to power an e-ink display. Relatively high specific power density coupled with notable operational and storage stability in the absence of fuel and oxidant makes this concept a viable alternative to porous sEFC with extended surface area.

Similar design intended to operate as an autonomous pulse generator has been independently developed by Xiao et al. [74], where poly(3,4-ethylenedioxythiophene) was electrodeposited with OsRP on dealloyed nanoporous gold surface with co-immobilization of GDH or BOx. OCV value achieved in a fully charged state was about 0.46 V, for example, similar to the concept described above, whereas maximum pulse power density was 0.61 mW cm^{-2}, 470 times higher compared to the peak characteristics from a traditional EFC configuration. In the recent example of sEFCs flexible nanoporous gold electrodes modified with different OsRPs for anodic and cathodic sides were used in a quasi solid-state biosupercapacitor [75] GOx-based bioanode and BOx-based biocathode were separated with a poly(vinyl alcohol) hydrogel film acting as electrolyte, which was pre-soaked in PBS containing 100 mM glucose. The assembled biodevice exhibited OCV of *ca.* 0.3 V and a half-life-time of about 11 h, tested by applying 620 discharge pulses of 100 μA cm^{-2}. Power output achieved in continuous mode reached up to 1.0 μW cm^{-2}, which is comparable with that for sEFCs of similar design, operating in solution [74, 75]. This is a significant step toward the development of wearable EFC, which does not require liquid media for operation.

The idea of Nernstian biosupercapacitor was further elaborated in a work by Alsaoub et al. [76], where OsRP with relatively high redox potential (0.55 V vs. SHE) has been employed in the bioanode, whereas "low-potential" OsRP with potential of 0.41 V vs. SHE was used on the cathodic site. Self-charging processes, when fuel and oxidant were added to the electrolyte, redistribute the electrode potentials from negative OCV value in a discharged state to 0.4–0.5 V in fully charged conditions. This device can be a good illustration of potential development of sEFCs, incapable to operate in continuous mode, but efficiently functioning in a pulse regime.

The performance of sEFCs in conditions, when concentration of fuel or/and oxidant is too low to extract sufficient continuous power output, was evaluated by Bobrowski et al. [77] ITO nanoparticles were spray-coated on the ITO surface and subsequently chemically modified for immobilization of BOx and CDH. The assembled sEFC was tested in air saturated buffer solution, mimicking glucose concentration (50 μM) in human lacrimal fluid followed by the prospect of possible application of sEFCs in human tears to overcome low performance of glucose/oxygen EFC previously suggested in several works [78, 79]. Peak stable power density generated in a pulse mode was 30 μW cm^{-2}; however, low OCV (0.27 V for the first self-charge/discharge cycle and 0.15–0.17 V after the third cycle) and poor stability make this concept still far from practical application.

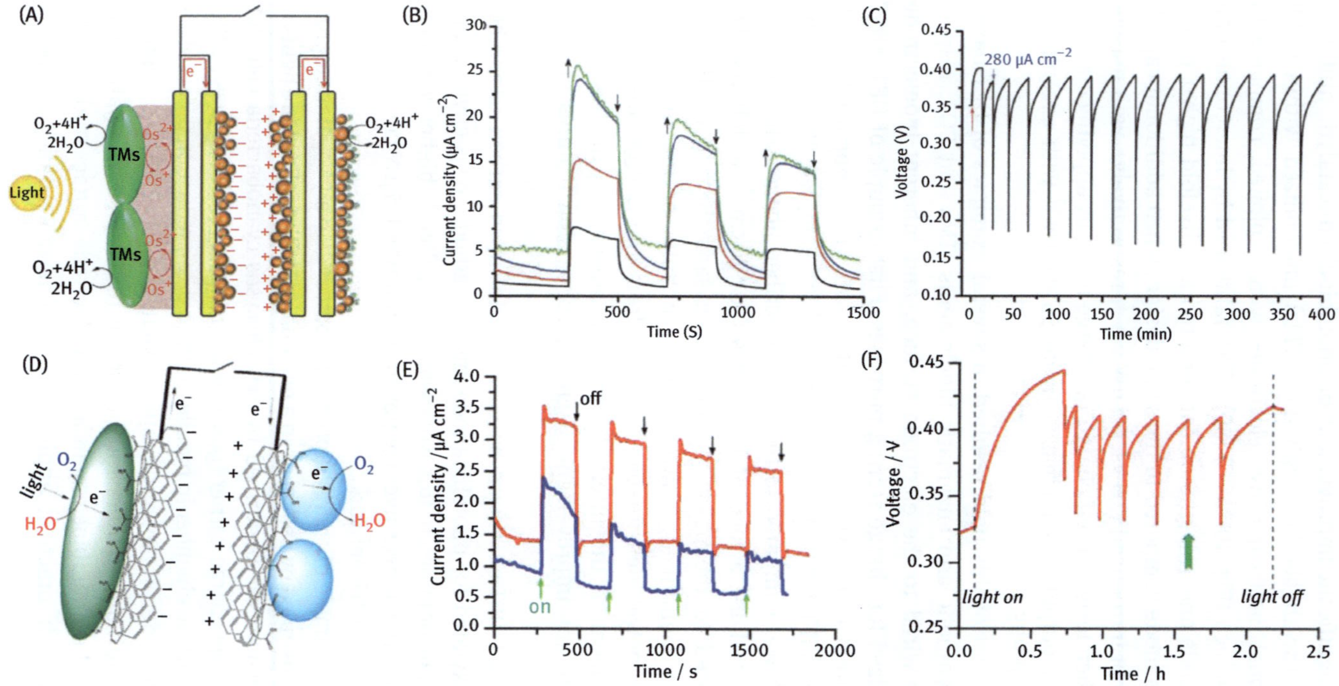

Fig. 10.2: Schematic representation and performance of supercapacitive biosolar cells with mediated (A–C) and direct electron transfer based (D–F) photobioanodes. (A, D): yellow rectangles – gold support, orange spheres – gold nanoparticles, brown rectangle – osmium redox polymer, green globules in (A) and blue globules in (D) – bilirubin oxidase, green ellipses – thylakoid membranes. (B) Amperograms at 0.55 V vs. SHE for conventional (black curve) and supercapacitive photobioanodes, containing additional capacitive element of 0.2, 0.5, and 1.8 mF (red, blue, and green curves, respectively). (C) Self-charge/discharge curve of complete sBSC; discharge was carried out by applying a pulse current of 280 μA cm^{-2} for 3 s. (E) Amperograms at 0.60 V vs. SHE for DET-photobioanodes based on carboxylated (blue curve) and amidated (red curve) CNTs. (F) Self-charge/discharge curve of assembled DET–sBSC; discharge was carried out by applying a pulse current of 140 μA cm^{-2} for 3 s or 420 μA cm^{-2} for 1 s (green arrow). (A–C) Adapted from Pankratova et al. [80] from John Wiley and Sons. (D–F) Adapted from Pankratov et al. [81] with permission from the American Chemical Society.

The concept of sEFCs was further expanded to supercapacitive photobioelectrodes and biosolar cells (sBSCs) aimed to convert light energy into electric power coupled with the possibility of storing it in the form of electric charge. Pankratova et al. demonstrated significant improvement of photobioelectrocatalytic ability of bioanodes based on the thylakoid membranes (TMs) immobilized within OsRP matrix, when additional inert capacitive electrodes were exploited, compared to the conventional approach (Fig. 10.2A–C) [80]. sBSC fabricated by interconnection of BOx-based biocathode and TMs-photobioanode produces fivefold higher power density in continuous mode compared to the same design, when additional capacitive electrodes were not introduced (2.5 vs. 0.5 μW cm^{-2}, respectively), with reasonable OCV of 0.4 V. When tested in self-charge/discharge mode by applying constant current pulses, power output was increased up to 56 μW cm^{-2} with exceptional stability for this kind of BES.

Further development of this field resulted in several examples of DET-based sBSC, where photosynthetic and enzymatic bioelements (TMs and BOx, respectively) were directly wired to the electrode surface, and inherent bioelectrode capacitance was utilized for charge storing purposes. First example of DET–sBSC was built on transparent ITO glass slides retrofitted with ITO nanoparticles [82]. Maximum OCV, achieved by self-charging process of the complete BES in ambient light was twice lower compared to the sBSC described above, whereas peak power extracted by applying a load of 1 MΩ, was 0.6 μW cm^{-2}, which was 120 times higher compared to the power output of the same BES in continuous mode. In should be mentioned, however, that, in spite of the anticipated enhancement of the power density by switching from the continuous to the pulse mode, employment of the internal electrode capacitance in this design does not demonstrate any improvement of energy conversion features of separate electrodes.

Another example of DET-sBSC was based on multiwalled carbon nanotubes (amidated and carboxylated for photobioanode and biocathode, respectively) (Fig. 10.2 D–F) [81]. OCV value obtained in this work was 0.45 V, slightly higher compared to the mediated approach, whereas average power output was 155 μW cm^{-2} because of high capacitance of the CNT network.

Recently, another example of energy storage and power delivery separation processes in BSC was demonstrated by Saar et al. in a two-chamber flow cell, where light-to-electric energy conversion was performed by *Synechocystis* cyanobacteria cells followed by reduction of the charge carrier (potassium ferricyanide), which was further oxidized in power extraction chamber in parallel to the oxygen reduction process [83].

In spite of the early development of supercapacitive photo-BES, the perspectives of their application are obvious and important to disclose new patterns of interactions between the photobioelement and an electrode surface and to make photo-BESs close to their practical application as renewable biological power sources.

10.3 Supercapacitive microbial fuel cells

10.3.1 Principle of a supercapacitive microbial fuel cell

MFCs relies on the redox reactions occurring on both anode and cathode electrodes. In a representative MFC case with air-breathing cathode in a membraneless configuration, anode and cathode are mainly made of similar materials based on carbonaceous constituents that often are connected to a metallic current collector. Therefore, initially, the difference in potential within anode and cathode sharing the same electrolyte is negligible and practically equal to zero [84].

Once the electrolyte containing organics and bacteria is inserted within the membraneless MFC, bacteria colonize the anode bringing the electrode potential towards negative values. The presence of oxygen at the cathode instead pushes the electrode potential towards positive values. Two separate zones are therefore created within the cell: (i) anaerobic zone in which bacteria consume oxygen and (ii) aerobic zone in which oxygen is present. Anode and cathode of the MFC are then self-polarized because of the redox reactions and the environments created within the system. Specifically, anode and cathode surfaces become negatively and positively polarized, respectively. They, therefore, behave like the negative and positive electrodes of an internal supercapacitor [84]. The negative electrode attracts positive ions from the electrolyte whereas negative ions are accumulated on the positive electrode (Fig. 10.3). The energy that is electrostatically

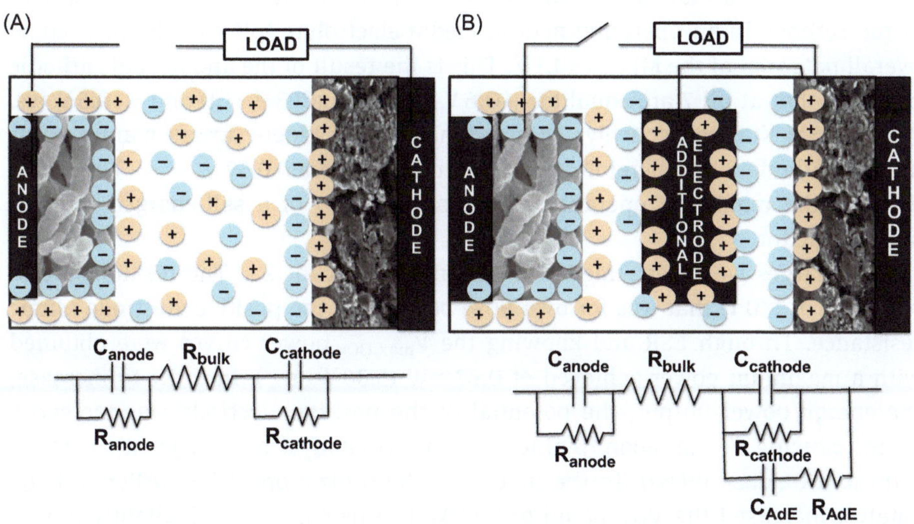

Fig. 10.3: Scheme of a supercapacitive microbial fuel cell (SC-MFC) with anode and cathode as the negative and positive electrodes of an internal supercapacitor (A). Scheme of SC-MFC with a supercapacitive additional electrode (AdE) short-circuited with the cathode (B). Adapted from Santoro et al. [84] with permission of Elsevier.

stored can be delivered through galvanostatic discharges whereby ions are released back to the electrolyte. The electrodes after discharges restore their initial potential when left in rest conditions and therefore the electrodes are self-recharged and another discharge can occur [85, 86].

10.3.2 Supercapacitive microbial fuel cell operations

Capacitive features of the anode electrode and the overall MFC were initially investigated by the group of Wageningen University and Wetsus in 2012 [87–90]. Simultaneously, supercapacitive features of C-Type cytochromes using conductive nanostructured networks of living bacteria was presented by Prof. Lovley group [91]. Scientists jointly from the University of New Mexico and the University of Bologna expanded the concept of supercapacitive electrodes to the entire MFC system in 2016 [84]. In that case, the anode and cathode electrodes of the MFC were used as the negative and positive electrode of an internal supercapacitor (Fig. 10.3) [84]. This concept of supercapacitive-microbial fuel cell (SC–MFC) was tested in a simple configuration based on a single bottle membraneless MFC with carbon brush as anode and activated carbon based air-breathing cathode. The SC–MFC was fed with activated sludge and 0.1 M K–PB (ratio 1:1) and 3 gL^{-1} sodium acetate as readily available organic substrate.

Considering acetate as organic substrate at the anode and oxygen as oxidant at the cathode in an operating neutral media electrolyte (pH = 7), the theoretical overall potential of the MFC is ≈ 1.1 V. This is the result of the anodic and cathodic potential that at pH 7 are equal to ≈ +0.6 V (vs. Ag/AgCl 3M KCl) and ≈ −0.5 V (vs. Ag/AgCl 3M KCl) respectively. Thermodynamic and kinetic losses mainly occur at the cathode and therefore the OCV recorded was 0.59 V [84]. The latter value is the maximum operating potential ($V_{max,OC}$) of a self-charged internal supercapacitor.

Discharges were performed at a current of 3 mA for 2 s. The results showed high ESR of 130 Ω that was for over 90% because of the positive electrode ohmic resistance. Through ESR and knowing the $V_{max,OC}$, power curves were obtained with a maximum power achieved of 0.67 mW (5.36 W m^{-3}). In order to increase, the energy/power output, the potential of the positive electrode was increased by introducing Fe–aminoantipyrine (Fe–AAPyr) catalyst and enzymatic catalyst (bilirubin oxidase [BOx]). In the first case, the utilization of Fe–AAPyr cathode catalyst increased the $V_{max,OC}$ up to 0.65 V. In this case the ESR slightly diminished to 125 Ω and combined with the higher $V_{max,OC}$, Fe–AAPyr-based cathode SC–MFC had a maximum power of 0.9 mW (7.2 W m^{-3}). Further improvement was obtained using BOx catalyst with $V_{max,OC}$ of 0.79 V, ESR of 105 and P_{max} of 1.47 mW (11.7 W m^{-3}) (Fig. 10.4A). Increasing the positive electrode seems to be a

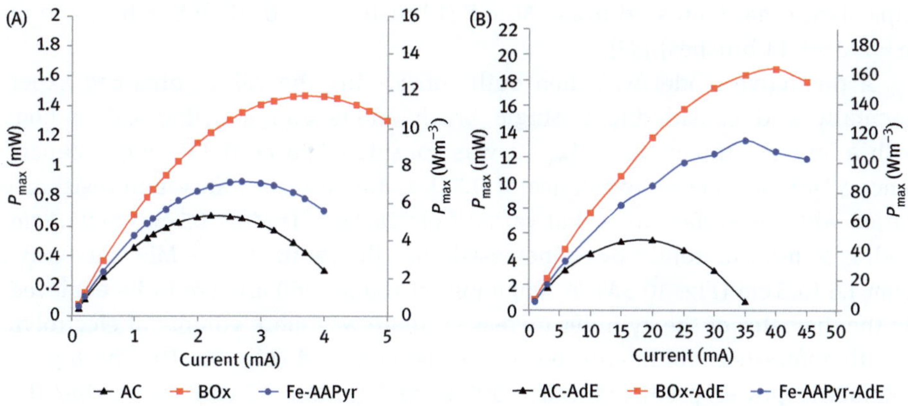

Fig.10.4: Maximum power curves (P_{max}) in absolute and volumetric terms for SC–MFC (A) and SC–MFC–AdE (B). Adapted from Santoro et al. [84] with permission of Elsevier.

suitable way to pursue enhancing power output but unfortunately the ohmic resistance given by the cathode was still very high counting over 90% and being over 100 Ω [84].

In order to optimize the energy/power output even more, an additional electrode (AdE) based on a small carbon brush coated with activated carbon was short-circuited with the positive electrode and immerged into the electrolyte. ESR dropped dramatically to 16–22 Ω that was 6–8 times lower than the previously recorded ESRs. SC-MFC-AdE had a P_{max} of 6 mW (49 Wm^{-3}), 14 mW (112 Wm^{-3}), and 19 mW (152 Wm^{-3}) for AC, Fe–AAPyr, BOx cathode, respectively (Fig. 10.4 B). This corresponded to boost up power of more than one order of magnitude because of the addition of the AdE [84].

In another experiment, operating in similar conditions and electrolyte, it was shown that the increase in the positive and negative electrode area brought an increase in the power output [92]. Different than before, the air-breathing cathode was built with conductive teflonized carbon black with Fe–AAPyr as catalyst. Moreover, the cathode area exposed to the solution increased from 2.25 cm^2 to 2.54 cm^2. The combination of these two variations led to half the positive electrode ohmic resistance during galvanostatic discharges that was calculated to be 57 ± 2.6 Ω. This value further decreased to 36 ± 1.4 Ω and 28 ± 4.2 Ω when the cathode area increased to 3.67 cm^2 and 5.09 cm^2. Also the anode area was increased inserting a second and a third brush within the electrolyte but the negative electrode ohmic resistance remained in any case below 1 Ω. In these operating conditions and with this design, the ohmic resistance of the positive electrode (cathode) was significantly limiting the output. The increase in cathode area also led to a slight increase in the cathode capacitance (*Cc*) that increased from 51 ± 1.9 mF to 73 ± 1.3 mF. More significant was the increase in anodic capacitance that increased from ≈50 mF (1 brush) to ≈ 120 mF (2 brushes) and up to ≈ 190 mF (3 brushes) [92].

A predictive model was then built considering the values obtained experimentally and considering a single brush anode and a cylindrical cathode with a larger cathode area. $V_{max,OC}$ was considered to be 0.8 V, anode capacitance 53 mF and anode resistance of 0.5 Ω. Cathode capacitance and resistance varied with the surface area that varied from 28.3 cm^2 (radius 1.5 cm) to 94.2 cm^2 (radius 5 cm). The total power increased with the radius of the MFC increasing from 1.5 to 5 cm (Fig. 10.5A). A maximum of roughly 80 mW could be achieved. As the diameter of the cylinder increased, there was more volume of electrolyte and therefore less volumetric power can be achieved (Fig. 10.5B). The highest volumetric power predicted was 1,300 W m^{-3} (radius 1.5 cm) but higher discharge current can be achieved with larger volumes for practical applications might be of more interest [92].

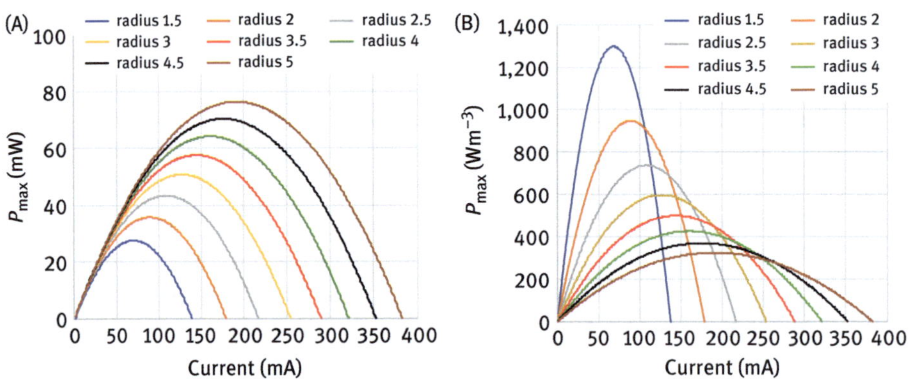

Fig. 10.5: Predictive maximum power in terms of absolute power (A) and in volumetric density (B) considering the variation of radius of the cylindrical SC-MFC. Adapted from Houghton et al. [92], published by Elsevier, CC BY 3.0 (https://creativecommons. org/licenses/by/3.0/).

10.3.3 Supercapacitive microbial desalination cells

As described earlier, MDC is similar to MFC with the additional function of reducing the salinity content within the central chamber that is separated from the anode and cathode chamber through polymeric membranes [36]. Anode and cathode of MDC have the same characteristics of the one described above for the MFC case. Anode and cathode are self-charged because of the redox reaction occurring on both electrodes, therefore the electrodes can be considered as the electrode of an internal supercapacitor as shown in Fig. 10.6A [93].

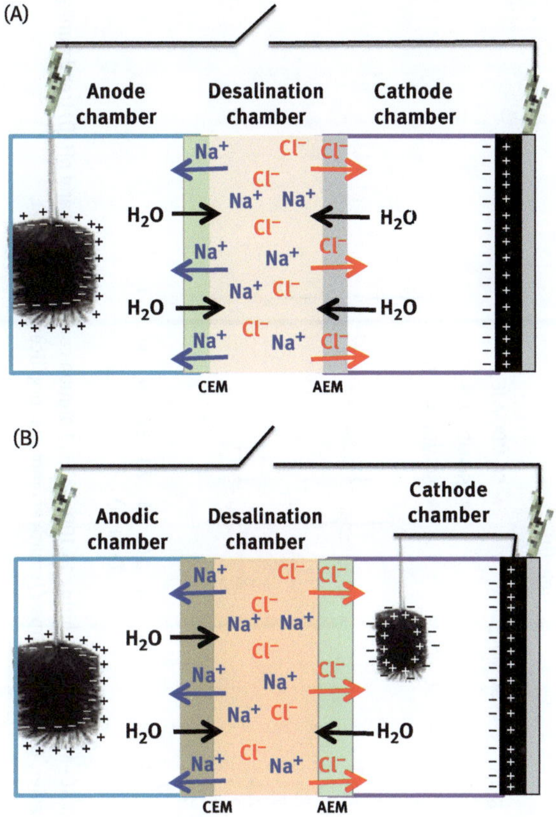

Fig. 10.6: Schematic of the supercapacitive MDC (A) and supercapacitive MDC with additional electrode (B). Adapted from Santoro et al. [93], published by Elsevier, CC BY 3.0 (https://creative-commons. org/licenses/by/3.0/).

Similar electrodes were used with a carbon brush as anode and a cathode containing AC/PTFE/CB decorated with Fe–AAPyr in air-breathing configuration. Three different solutions were used: (i) anode chamber (volume 35 mL) containing 50 mM K–PB and activated sludge (1:1 ratio) with 3 gL^{-1} NaAc; (ii) seawater from the Pacific Ocean within the desalination chamber (volume 11 mL); (iii) 23 mM K-PB in the cathode chamber (volume 35 mL) [93].

Galvanostatic discharges were performed showing an ESR of 110 Ω with the contribution of the cathode to be determined in 100 Ω. The same strategy adopted before for SC–MFC was here adopted and an AdE was short-circuited with the cathode electrode and inserted in the cathode chamber. Cathode ohmic resistance then decreased to roughly 40 Ω. Maximum power achieved for SC–MDC was 1.14 mW (14.07 Wm^{-3}) that increased up to 2.11 mW (26.05 Wm^{-3}) with the additional electrode [93].

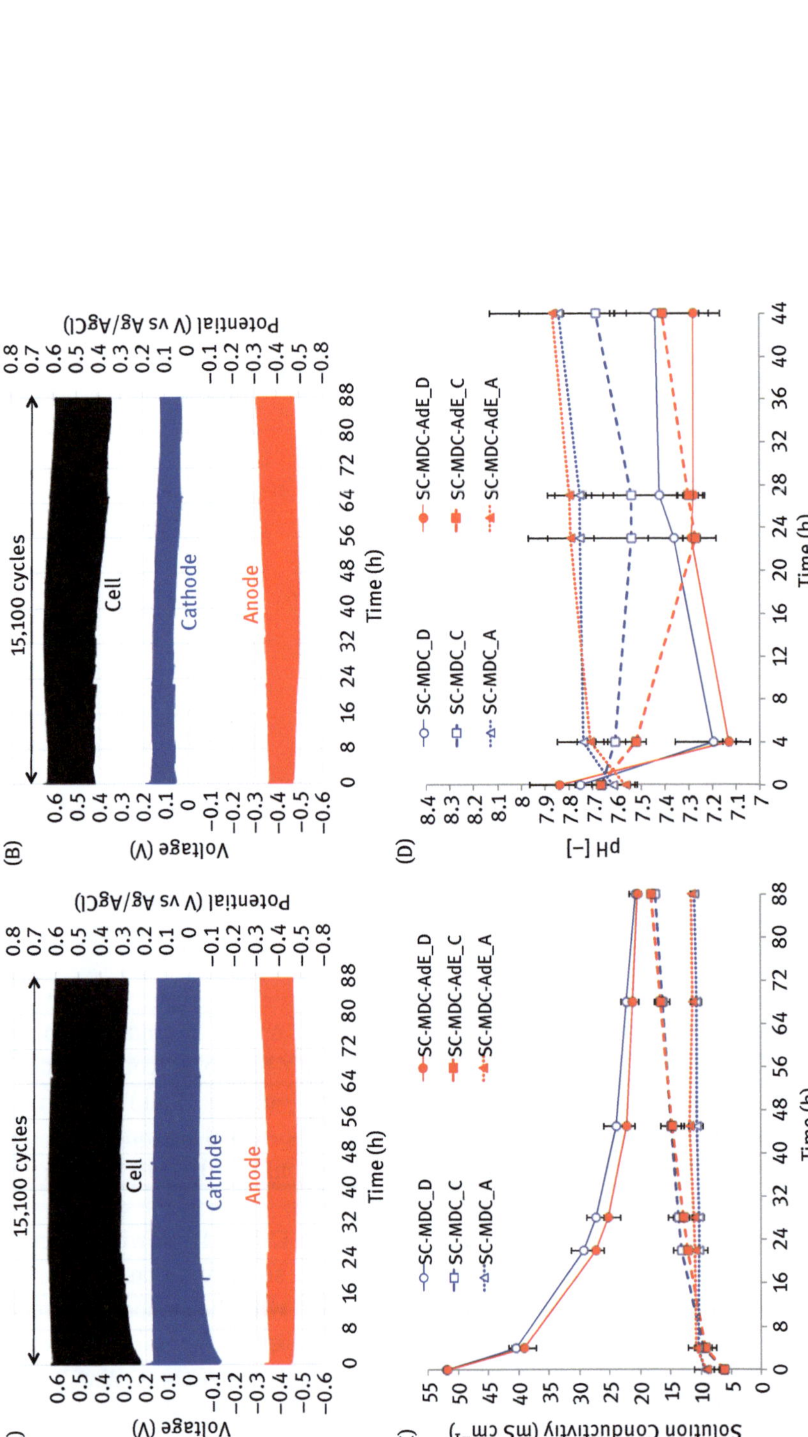

Fig. 10.7: Overall cell and single electrode profiles of SC–MFC (A) and SC–MDC–AdE (B) during 15,100 cycles. Solution conductivity (C) and pH (D) trend during 88 h operations. Adapted from Santoro et al. [93], published by Elsevier, CC BY 3.0 (https://creativecommons. org/licenses/by/3.0/).

Durability tests were conducted with discharges at current of 2 mA for 1 s and rest time of 20 s [93]. A total of 88 h discharges and self-recharges were conducted with a total of 15,100 cycles. ESR of SC–MDC was higher that the one from SC–MDC–AdE (Fig. 10.7A and B). As the anodic profile was similar, the cathodic losses decreased significantly with the additional electrode short-circuited with the cathode (Fig. 10.7A and B). Other parameters such as solution conductivity and pH were monitored during the long-term operations (Fig. 10.7C and D). Interestingly, the solution conductivity of the desalination chamber decreased from roughly 52 mS cm^{-1} to 23.85 ± 2.04 mS cm^{-1} (SC–MDC) and to 22.12 ± 1.39 mS cm^{-1} (SC–MDC–AdE) after 45 h operations. At the end of the experiment, the solution conductivity dropped even more to 20.4 ± 1.5 mS cm^{-1} (Fig. 10.7C). Simultaneously, the solution conductivity of the anodic chamber and the cathodic chamber increased during the experiments. The values reached after 88 h 11.1 ± 0.5 mS cm^{-1} and 17.7 ± 0.4 mS cm^{-1} respectively (Fig. 10.7C).

The variation of solution conductivity among the chambers supports the movement of ions from the desalination chamber to the anode and cathode chambers. Different than classic MDC in which ORR occur at the cathode and the pH tends to increase toward alkaline values, in SC–MDC and SC–MDC–AdE, the pH remained stable varying between 7.2 and 8.1 and was independent from the chambers over 88 h. This indicates that the electrochemical phenomena occurring on the electrodes are mainly of electrostatic nature [93].

10.4 Conclusions

The integration of supercapacitors with EFC and MFC opens new frontiers and horizons within the so called Water-Energy nexus. In fact, supercapacitors cannot only be used externally but they can be monolithically integrated within the electrodes of the fuel cells an internal supercapacitor. The redox reaction occurring at the anode and cathode electrode allows to self-polarize the electrodes of the supercapacitors that can deliver high pulses of current/power during discharges. The capability of producing current/power in pulses that are one or more orders of magnitude compared to traditional enzymatic or MFC makes this supercapacitive system interesting and suitable for applications. Moreover, the system is simpler and more compact compared to the one in which the supercapacitor is externally connected to the fuel cell. As both enzymatic and MFC operate with aqueous media, the implementation of the biosupercapacitors is restricted within the water window stability. The low overall voltage that EFCs and MFCs operate also restricts the operation. Further research should be done on enhancing the supercapacitive features of the electrodes keeping low the overall cost of enzymatic and microbial biosupercapacitors.

References

[1] Rinaldi, A., Mecheri, B., Garavaglia, V., Licoccia, S., Di Nardo, P., & Traversa, E. Engineering materials and biology to boost performance of microbial fuel cells: a critical review. Energy Environ Sci. 2008, 1, 417–429.
[2] Santoro, C., Arbizzani, C., Erable, B., & Ieropoulos, I. Microbial fuel cells: from fundamentals to applications. A review. J Power Sources. 2017, 356, 225–244.
[3] Atanassov, P., Apblett, C., Banta, S., et al. Enzymatic biofuel cells. Electrochem Soc Interface. 2007, 16, 28–31.
[4] Kinoshita, K. Carbon: electrochemical and Physicochemical Properties. New York, NY. John Wiley Sons, 1988.
[5] Kinoshita, K. Electrochemical Oxygen Technology. New York, NY. John Wiley Sons, 1988.
[6] Mazurenko, I., de Poulpiquet, A., & Lojou, E. Recent developments in high surface area bioelectrodes for enzymatic fuel cells. Curr Opin Electrochem. 2017, 5, 74–84.
[7] Macazo, F. C., & Minteer, S. D. Enzyme cascades in biofuel cells. Curr Opin Electrochem. 2017, 5, 114–120.
[8] Sekretaryova, A. N., Beni, V., Eriksson, M., Karyakin, A. A., Turner, A. P., & Vagin, M. Y. Cholesterol self-powered biosensor. Anal Chem. 2014, 86, 9540–9547.
[9] Ji, C., Hou, J., Wang, K., Hau Ng, Y., & Chen, V. Single-enzyme biofuel cells. Angew Chem. 2017, 56, 9762–9766.
[10] Krikstolaityte, V., Oztekin, Y., Kuliesius, J., et al. Biofuel cell based on anode and cathode modified by glucose oxidase. Electroanalysis. 2013, 25, 2677–2683.
[11] http://enzyme-database.org/stats.php.
[12] Cadet, M., Gounel, S., Stinès-Chaumeil, C., et al. An enzymatic glucose/O2 biofuel cell operating in human blood. Biosens Bioelectron. 2016, 83, 60–67.
[13] Bollella, P., Fusco, G., Stevar, D., et al. A glucose/oxygen enzymatic fuel cell based on gold nanoparticles modified graphene screen-printed electrode. Proof-of-concept in human saliva Sens Actuators, B. 2018, 256, 921–930.
[14] Pankratov, D., Ohlsson, L., Gudmundsson, P., et al. Ex vivo electric power generation in human blood using an enzymatic fuel cell in a vein replica. RSC Adv. 2016, 6, 70215–70220.
[15] Cinquin, P., Gongran, C., Giroud, F., et al. A glucose biofuel cell implanted in rats. PLOS ONE. 2010, 5, e10476.
[16] El Ichi-Ribault, S., Alcaraz, J-P., Boucher, F., et al. Remote wireless control of an enzymatic biofuel cell implanted in a rabbit for 2 months. Electrochim Acta. 2018, 269, 360–366.
[17] Zebda, A., Cosnier, S., Alcaraz, J-P., et al. Single glucose biofuel cells implanted in rats power electronic devices. Sci Rep. 2013, 3, 1516.
[18] Yahiro, A. T., Lee, S. M., & Kimble, D. O. Bioelectrochemistry: I. Enzyme utilizing bio-fuel cell studies. Biochim Biophys Acta. 1964, 88, 375–383.
[19] Zhao, C., Gai, P., Song, R., Chen, Y., Zhang, J., & Zhu, -J.-J. Nanostructured material-based biofuel cells: recent advances and future prospects. Chem Soc Rev. 2017, 46, 1545–1564.
[20] Zebda, A., Gondran, C., Le Goff, A., Holzinger, M., Cinquin, P., & Cosnier, S. Mediatorless high-power glucose biofuel cells based on compressed carbon nanotube-enzyme electrodes. Nature Commun. 2011, 2, 370.
[21] González-Arribas, E., Pankratov, D., Gounel, S., Mano, N., Blum, Z., & Shleev, S. Transparent and capacitive bioanode based on specifically engineered glucose oxidase. Electroanalysis. 2016, 28, 1290–1297.
[22] Bollella, P., Ludwig, R., & Gorton, L. Cellobiose dehydrogenase: insights on the nanostructuration of electrodes for improved development of biosensors and biofuel cells. Appl Mater Today. 2017, 9, 319–332.

[23] Milton, RD., Hichey, DP., Abdellaoui, S., et al. Rational design of quinones for high power density biofuel cells. Chem Sci. 2015, 6, 4867–4875.
[24] Ó Conghaile, P., Falk, M., MacAodha, D., et al. Fully enzymatic membraneless glucose/oxygen fuel cell that provides 0.275 mA cm^{-2} in 5 mM glucose, operates in human physiological solutions, and powers transmission of sensing data. Anal Chem. 2016, 88, 2156–2163.
[25] Cosnier, S., Gross, A. J., Le Goff, A., & Holzinger, M. Recent advances on enzymatic glucose/oxygen and hydrogen/oxygen biofuel cells: achievements and limitations. J Power Sources. 2016, 325, 252–263.
[26] Umasankar, Y., Adhikari, B.-R., & Chen, A. Effective immobilization of alcohol dehydrogenase on carbon nanoscaffolds for ethanol biofuel cell. Bioelectrochem. 2017, 118, 83–90.
[27] Mano, N., & de Poulpiquet, A. O2 reduction in enzymatic biofuel cells. Chem Rev. 2018, 118, 2392–2468.
[28] Gomez, C., Shipovskov, S., & Ferapontova, E. E. Peroxidase biocathodes for a biofuel cell development. J Renew Sustain Energy. 2010, 2, 013103.
[29] Pandey, P., Shinde, V. N., Deopurkar, R. L., Kale, S. P., Patil, S. A., & Pant, D. Recent advances in the use of different substrates in microbial fuel cells toward wastewater treatment and simultaneous energy recovery. Appl Energy. (2016), 168, 706–723.
[30] Pant, D., Van Bogaert, G., Diels, L., & Vanbroekhoven, K. A review of the substrates used in microbial fuel cells (MFCs) for sustainable energy production. Bioresour Technol. 2010, 101, 1533–1543.
[31] Kumar, A., Hsu, L.H.H., Kavanagh, P., et al. The ins and outs of microorganism–electrode electron transfer reactions. Nat Rev Chem. 2017, 1, 24.
[32] Ucar, D., Zhang, Y., & Angelidaki, I. An overview of electron acceptors in microbial fuel cells. Front Microbiol. 2017, 8, 643.
[33] Santoro, C., Gokhale, R., Mecheri, B., et al. Design of iron (II) pthalocyanine (FePc) derived oxygen reduction electrocatalysts for high power density microbial fuel cells. ChemSusChem. 2017, 10, 3243–3251.
[34] Santoro, C., Rojas-Carbonell, S., Awais, R., et al. Influence of platinum group metal-free catalyst synthesis on microbial fuel cell performance. J Power Sources. 2018, 375, 11–20.
[35] Santoro, C., Rezaei Talarposhti, M., Kodali, M., et al. Microbial desalination cells with efficient platinun-group-metal-free cathode catalysts. ChemElectroChem. 2017, 4, 3322–3330.
[36] Cao, X., Huang, X., Liang, P., et al. A new method for water desalination using microbial desalination cells. Environ Sci Technol. 2009, 43, 7148–7152.
[37] Lazzari, M., Soavi, F., & Mastragostino, M. Dynamic pulse power and energy of ionic-liquid-based supercapacitor for HEV application. J Electrochem Soc. 2009, 156, A661–6616.
[38] Arbizzani, C., Mastragostino, M., & Soavi, F. New trends in electrochemical supercapacitors. J Power Sources. 2001, 100, 164–170.
[39] González, A., Goikolea, E., Barrena, J. A., & Mysyk, R. Review on supercapacitors: technologies and materials. Renew Sustain Energ Rev. 2016, 58, 1189–1206.
[40] Zhang, L. L., & Zhao, X. S. Carbon-based materials as supercapacitor electrodes. Chem Soc Rev. 2009, 38, 2520–2531
[41] Béguin, F., Presser, V., Balducci, A., & Frackowiak, E. Carbons and electrolytes for advanced supercapacitors. Adv Mater. 2014, 26, 2219–2251
[42] Conway, B. E. Electrochemical Supercapacitors: scientific Fundamentals and Technological Applications. New York, NY, Kluwer Academic/Plenum Publishers, New York, 1999.(1999).
[43] Soavi, F., Bettini, LG., Piseri, P., et al. Miniaturized supercapacitors: key materials and structures towards autonomous and sustainable devices and systems. J Power Sources. 2016, 326, 717–725.

[44] Santoro, C., Soavi, F., Arbizzani, C., et al. Co-generation of hydrogen and power/current pulses from supercapacitive MFCs using novel HER iron-based catalysts. Electrochim Acta. 2016, 220, 672–682.
[45] Papaharalabos, G., Greenman, J., Melhuish, C., et al. Increased power output from micro porous layer (MPL) cathode microbial fuel cells (MFC). Int J Hydrogen Energy. 2013, 38, 11552–11558.
[46] Ieropoulos, I., Greenman, J., & Melhuish, C. Microbial fuel cells based on carbon veil electrodes: stack configuration and scalability. Int J Energy Res. 2008, 32, 1228–1240.
[47] Wang, H., Park, J. D., & Ren, Z. J. Practical energy harvesting for microbial fuel cells: a review. Environ Sci Technol. 2015, 49, 3267–3277.
[48] Grattieri, M., Hasan, K., & Minteer, S. Bioelectrochemical systems as a multipurpose biosensing tool: present perspective and future outlook. ChemElectroChem. 2017, 4, 834–842.
[49] Grattieri, M., & Minteer, S. D. Self-powered biosensors. ACS Sensors. 2018, 3, 44–53.
[50] Zhou, M., & Dong, S. Bioelectrochemical interface engineering: toward the fabrication of electrochemical biosensors, biofuel cells, and self-powered logic biosensors. Acc Chem Res. 2011, 44, 1232–1243.
[51] Di Lorenzo, M., Curtis, T. P., Head, I. M., & Scott, K. A single-chamber microbial fuel cell as a biosensor for wastewaters. Water Res. 2009, 43, 3145–3154.
[52] Tender, LM., Gray, SA., Groveman, E., et al. The first demonstration of a microbial fuel cell as a viable power supply: powering a meteorological buoy. J Power Sources. 2008, 179, 571–575.
[53] Meriah Arias-Thode, Y., Hsu, L., Anderson, G., et al. Demonstration of the SeptiStrand benthic microbial fuel cell powering a magnetometer for ship detection. J Power Sources. 2017, 356, 419–429.
[54] Shantaram, A., Beyenal, H., Veluchamy, R. R. A., & Lewandowski, Z. Wireless sensors powered by microbial fuel cells. Environ Sci Technol. 2005, 39, 5037–5042.
[55] Wilkinson, S. "Gastrobots"—benefits and challenges of microbial fuel cells in foodpowered robot applications. Auton Robots. 2000, 9, 99–111.
[56] Ieropoulos, I. A., Melhuish, C., & Greenman, J. Artificial metabolism: towards true energetic autonomy in artificial life. European conference in artificial life. Lect Notes Artif Intell. 2003, 2801, 792–799.
[57] Ieropoulos, I., Melhuish, C., Greenman, J., & Horsfield, I. EcoBot-II: an artificial agent with a natural metabolism. Int J Adv Robotic Systems. 2005, 2, 31
[58] Ieropoulos, I., Greenman, J., Melhuish, C., & Horsfield, I. EcoBot-III-a robot with guts. ALIFE. 2010, 733–740.
[59] Philamore, H., Rossiter, J., Stinchcombe, A., & Ieropoulos, I.. Row-bot: an energetically autonomous artificial water boatman. IEEE/RSJ International Conference on Intelligent Robots and Systems (IROS) 2015.
[60] Cosnier, S., Holzinger, M., Le Golf, A., & Agnes, C. Biocompatible electrochemical supercapacitor. FR Patent Application. 2014, PCT/FR2014/050482 2014135787.
[61] Shleev, S., Pankratov, D., & Blum, Z. Charge-storing fuel cell. SE Patent Application. 2014, PCT/EP2014/057291 2014167063.
[62] Pankratov, D., Falkman, P., Blum, Z., & Shleev, S. A hybrid electric power device for simultaneous generation and storage of electric energy. Energ Environ Sci. 2014, 7, 989–993.
[63] Pankratov, D., Blum, Z., Suyatin, D. B., Popov, V. O., & Shleev, S. Self-charging electrochemical biocapacitor. Chem Electro Chem. 2014, 1, 343–346.
[64] Parunova, YM., Bushnev, SO., Gonzales-Arribas, E., et al. Potentially implantable biocathode with the function of charge accumulation based on nanocomposite of polyaniline/carbon nanotubes. Russ J Electrochem. 2016, 52, 1166–1171.
[65] Agnes, C., Holzinger, M., Le Goff, A., et al. Supercapacitor/biofuel cell hybrids based on wired enzymes on carbon nanotube matrices: autonomous reloading after high power pulses in neutral buffered glucose solutions. Energ Environmen Sci. 2014, 7, 1884–1888.

[66] Pankratov, D., Conzuelo, F., Pinyou, P., Alsaoub, S., Schuhmann, W., & Shleev, S. A Nernstian biosupercapacitor. Angew Chem. 2016, 55, 15434–15438.
[67] Majdecka, D., & Bilewicz, R. Nanostructuring carbon supports for optimal electrode performance in biofuel cells and hybrid fuel cells. J Solid State Electrochem. 2016, 20, 949–955.
[68] Pankratov, D., Blum, Z., & Shleev, S. Hybrid electric power biodevices. Chem Electro Chem. 2014, 1, 1798–1807.
[69] Narvaez Villarrubia, CW., Soavi, F., Santoro, C., et al. Self-feeding paper based biofuel cell/self-powered hybrid mu-supercapacitor integrated system. Biosens Bioelectron. 2016, 86, 459–465.
[70] Kizling, M., Draminska, S., Stolarczyk, K., et al. Biosupercapacitors for powering oxygen sensing devices. Bioelectrochem. 2015, 106, 34–40.
[71] Kizling, M., Stolarczyk, K., Sim Sin Kiat, J., et al. Pseudocapacitive polypyrrole-nanocellulose composite for sugar-air enzymatic fuel cells. Electrochem Commun. 2015, 50, 55–59.
[72] Kizling, M., Stolarczyk, K., & Tammela, O. Bioelectrodes based on pseudocapacitive cellulose/polypyrrole composite improve performance of biofuel cell. Bioelectrochem. 2016, 112, 184–190.
[73] Knoche, K. L., Hickey, D. P., Milton, R. D., Curcoe, C. L., & Minteer, S. D. Hybrid glucose/O_2 biobattery and supercapacitor utilizing a pseudocapacitive dimethylferrocene redox polymer at the bioanode. ACS Energy Lett. 2016, 1, 380–385.
[74] Xiao, X., O Conghaile, P., Leech, D., Ludwig, R., & Magner, E. A symmetric supercapacitor/biofuel cell hybrid device based on enzyme-modified nanoporous gold: an autonomous pulse generator. Biosens Bioelectron. 2017, 90, 96–102.
[75] Xiao, X., & Magner, E. A quasi-solid-state and self-powered biosupercapacitor based on flexible nanoporous gold electrodes. Chem Commun. 2018, 54, 5823–5826
[76] Alsaoub, S., Ruff, A., Conzuelo, F., et al. An intrinsic self-charging biosupercapacitor comprised of a high-potential bioanode and a low-potential biocathode. ChemPlusChem. 2017, 82, 576–583.
[77] Bobrowski, T., Gonzales-Arribas, E., Ludwig, R., Toscano, M. D., Shleev, S., & Schumann, W. Rechargeable, flexible and mediator-free biosupercapacitor based on transparent ITO nanoparticle modified electrodes acting in mu M glucose containing buffers. Biosens Bioelectron. 2018, 101, 84–89.
[78] Blum, Z., Pankratov, D., & Shleev, S. Powering electronic contact lenses: current achievements, challenges, and perspectives. Expert Rev Ophthalmol. 2014, 9, 269–273.
[79] Pankratov, D., Gonzalez-Arribas, E., Blum, Z., & Shleev, S. Tear based bioelectronics. Electroanalysis. 2016, 28, 1250–1266.
[80] Pankratova, G., Pankratov, D., Hasan, K., et al. Supercapacitive photo-bioanodes and biosolar cells: a novel approach for solar energy harnessing. Adv Energy Mater. 2017, 7, 1602285
[81] Pankratov, D., Pankratova, G., Dyachkova, P., et al. Supercapacitive biosolar cell driven by direct on electron transfer between photosynthetic membranes and CNT networks with enhanced performance. ACS Energy Lett. 2017, 2, 2635–2639.
[82] Gonzalez-Arribas, E., Aleksejeva, O., Bobrowski, T., et al. Solar biosupercapacitor. Electrochem Commun. 2017, 74, 9–13.
[83] Saar, KL., Bombelli, P., Lea-Smith, DJ., et al. Enhancing power density of biophotovoltaics by decoupling storage and power delivery. Nature Energy. 2018, 3, 75–81.
[84] Santoro, C., Soavi, F., Serov, A., Arbizzani, C., & Atanassov, P. Self-powered supercapacitive microbial fuel cell: the ultimate way of boosting and harvesting power. Biosens Bioelectron. 2016, 78, 229–235.
[85] Santoro, C., Flores-Cadengo, C., Soavi, F., et al. Ceramic microbial fuel cells stack: power generation in standard and supercapacitive mode. Sci Rep. 2018, 8, 3281.
[86] Santoro, C., Kodali, M., Kabir, S., Soavi, F., Serov, A., & Atanassov, P. Three-dimensional graphene nanosheets as cathode catalysts in standard and supercapacitive microbial fuel cell. J Power Sources. 2017, 356, 371–380.

[87] Deeke, A., Sleutels, T. H. J. A., Hamelers, H. V. M., & Buisman, C. J. N. Capacitive bioanodes enable renewable energy storage in microbial fuel cells. Environ Sci Technol. 2012, 46(6), 3554–3560.
[88] Borsje, C., Liu, D., Sleutels, T. H. J. A., Buisman, C. J. N., & Ter Heijne, A. Performance of single carbon granules as perspective for larger scale capacitive bioanodes. J Power Sources. 2016, 325, 690–696.
[89] Deeke, A., Sleutels, T. H. J. A., Ter Heijne, A., Hamelers, H. V. M., & Buisman, C. J. N. Influence of the thickness of the capacitive layer on the performance of bioanodes in microbial fuel cells. J Power Sources. 2013, 243, 611–616.
[90] Deeke, A., Sleutels, T. H. J. A., Donkers, T. F. W., Hamelers, H. V. M., Buisman, C. J. N., & Ter Heijne, A. Fluidized capacitive bioanode as a novel reactor concept for the microbial fuel cell. Environ Sci Technol. 2015, 49, 1929–1935.
[91] Malvankar, N. S., Mester, T., Tuominen, M. T., & Lovley, D. R. Supercapacitors based on c-type cytochromes using conductive nanostructured networks of living bacteria. ChemPhysChem. 2012, 13, 463–468.
[92] Houghton, J., Santoro, C., Soavi, F., et al. Supercapacitive microbial fuel cell: characterization and analysis for improved charge storage/delivery performance. Bioresour Technol. 2016, 218, 552–560.
[93] Santoro, C., Benito Abad, F., Serov, A., et al. Supercapacitive microbial desalination cells: new class of power generating devices for reduction of salinity content. Appl Energy. 2017, 208, 25–36.

Sergey Shleev, Magnus Falk, Stefan Cirovic, Zoltan Blum

11 Wearable bioelectronic devices

11.1 Introduction

Bioelectronics is a rapidly progressing multidisciplinary scientific area, having emerged from the combination of biology, electronics, bio-, and nanotechnologies [1]. Bioelectronics aims to integrate biomaterials (proteins, enzymes, organelles, and living cells) and electric or electronic elements (electrodes, chips, field-effect transistors, piezoelectric crystals, etc.) into functional devices [2]. Such devices may function as discrete biosensors, biocomputing devices (biodiode, biotransistor, and bioelectronics circuitry), biological electric power sources (biofuel cell, biobattery, and biosupercapacitor), as well as complete self-contained (i.e., self-powered and wireless) bioelectronic devices (Fig. 11.1).

One of the intended applications of bioelectronics, which attracts a lot of attention nowadays, is implantable self-contained biodevices [3]. Indeed, bioelectronic devices are presently studied not only in vitro but also in vivo (Fig. 11.1). Unfortunately, the main impediment with regard to bioelectronic devices is the limited operational stability. It is clear that the full potential of implantable bioelectronics will be realized only when they become robust and reliable enough for real practical applications. On the one hand, there is a lot of research going on to improve the operational lifetime of bioelectronic devices. On the other hand, a limited number of studies address biocompatibility and safety issues [4]. Thus, it seems that practically usable implanted bioelectronics is quite challenging to achieve at least in the immediate future.

At the beginning of the 21st century, several scientific papers appeared, disclosing flexible, attachable, adhesive electronics. One of the most fascinating papers introduced new classes of electronic systems that achieved thicknesses, effective elastic moduli, bending stiffness, and areal mass densities matched to the human epidermis (Fig. 11.2) [11]. Following that report, papers concerning biosensors and biofuel cells operating ex vivo (Fig. 11.1), that is, in tears, sweat, urine, and saliva, were published; however, not all of these publications were directly attributed to wearable biodevices.

Nevertheless, fabrication and characterization of bioelectronics operating in human physiological fluids pave the way toward bodyNET – an intriguing scientific and technological direction, drawing on a network of sensors, screens, and smart devices woven into our clothing, worn on our skin, and implanted in our bodies [12].

Sergey Shleev, Magnus Falk, Stefan Cirovic, Zoltan Blum, Department of Biomedical Science, Faculty of Health and Society, Malmö University, Malmö, Sweden

https://doi.org/10.1515/9783110570526-011

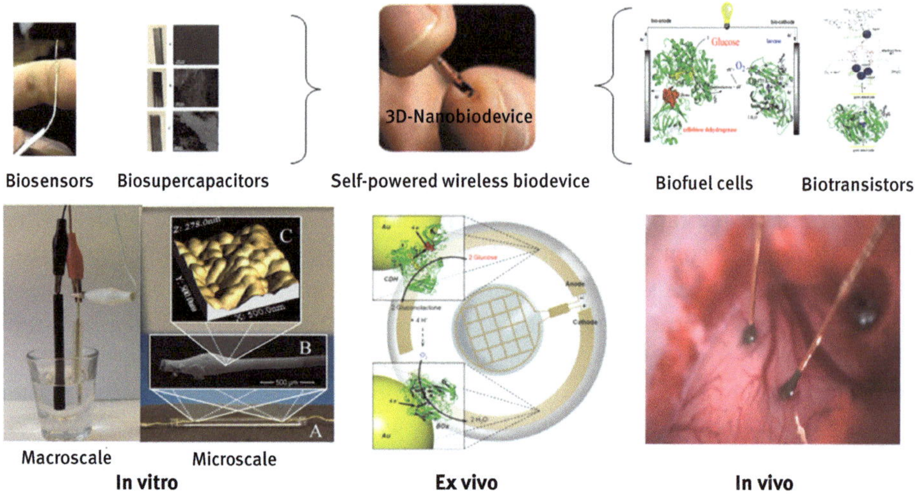

Fig. 11.1: Schematic illustration of a combination of separate (bio)electronic elements into a functional bioelectronic device, which could operate in vitro, ex vivo, and in vivo. Biosensor – a flexible biosensor from Andoralov et al. [5]; biosupercapacitor – a self-charging electrochemical biocapacitor from Pankratov et al. [6]; biofuel cell – a direct electron transfer–based enzymatic fuel cell from Coman et al. [7]; biotransistor – a laccase-gold transistor from Shleev and Ruzgas [8]; ex vivo – a direct electron transfer–based enzymatic fuel cell on contact lens from Falk et al. [9]; and in vivo – a direct electron transfer–based microscale enzymatic fuel cell in rat brain from Andoralov et al. [10].

11.2 Wearable bioelectronic devices

The development of bioelectronics operating in tears, sweat, saliva, and urine has seen a massive research interest in recent years, owing to potential use in diagnostics and as power sources for personal electronics, including wearables. A brief description of bioelectronic devices, which operate (or could potentially operate) in excreted human physiological fluids, is presented, with a special focus on wearable bioelectronic devices, operating autonomously (without the active participation of the user) and preferably continuously.

11.2.1 Bioelectronics in tears

Human tears (lachrymal fluid/liquid) is a complex physiological fluid (98% water), which contains both low and high molecular weight compounds (2% solids), viz. water, electrolytes, small organic molecules, proteins, and enzymes, secreted by lachrymal glands [13]. Many of these compounds have diagnostic potential, as summarized in the review concerning contact lens sensors [14], and some of them are considered as potential biofuels, for example, lactate, glucose, and ascorbate, as reviewed recently by our group [15].

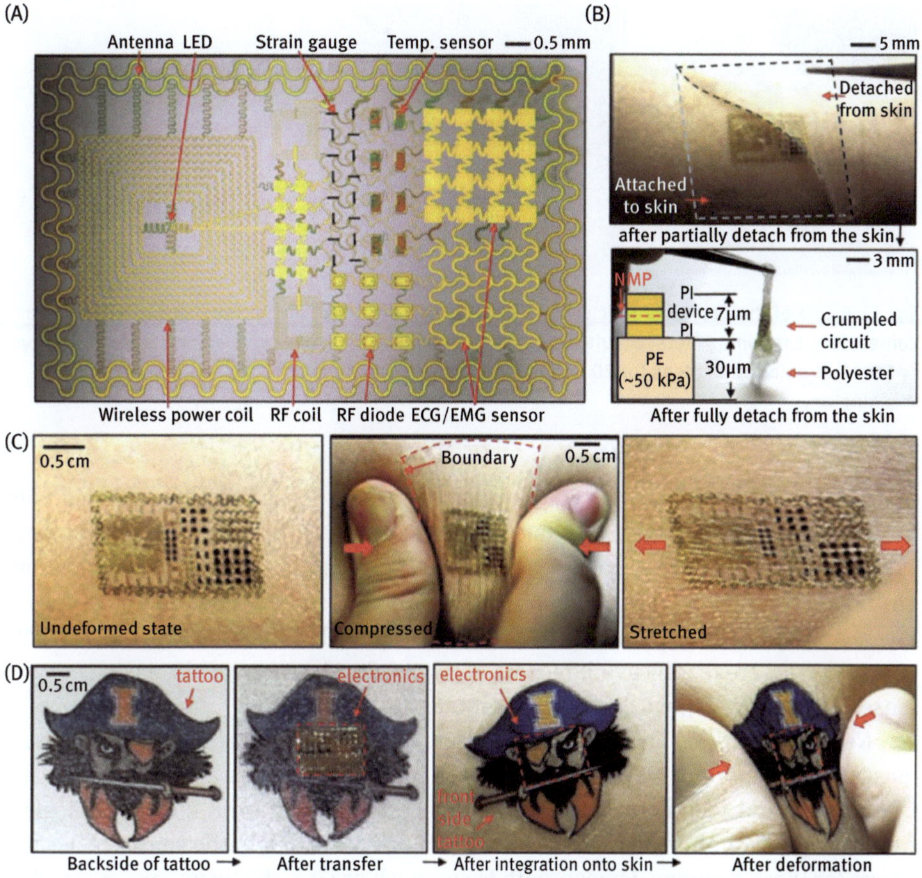

Fig. 11.2: Epidermal electronics. (A) An image of a demonstration platform for multifunctional electronics with physical properties matched to the epidermis. (B) Epidermal electronic system partially (top) and fully (bottom) peeled away from the skin. (Inset) A representative cross-sectional illustration of the structure, with the neutral mechanical plane defined by a red dashed line. (C) Multifunctional epidermal electronic system on skin: undeformed (left), compressed (middle), and stretched (right). (D) A commercial temporary transfer tattoo provides an alternative to polyester/polyvinyl alcohol for the substrate; in this case, the system includes an adhesive to improve bonding to the skin. Images are of the backside of a tattoo (far left), electronics integrated onto this surface (middle left), and attached to skin with electronics facing down in undeformed (middle right) and compressed (far right) states. Reproduced from Kim [11] with permission from the American Association for the Advancement of Science.

Among different ex vivo operating bioelectronics, tear-based bioelectronic devices were fabricated and tested first in 2009. As usual in the bioelectrochemical field in general, and bioelectronics in particular, the first reports were concerned with glucose sensing [16] in human tears. In 2008–2009, the first prototypes of electronic contact lenses were presented (Fig.11.3) by a group of researchers headed by Babak A. Parviz, a

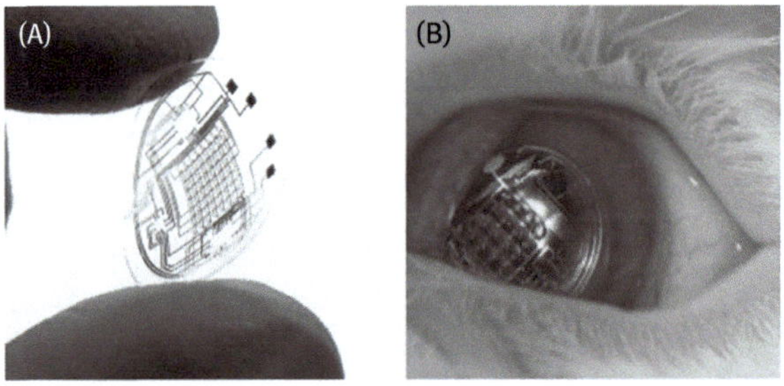

Fig. 11.3: One of the first electronic contact lenses. (a) A picture of one of the first prototypes of a contact lens containing metal circuit structures. (b) An illustration of ex vivo animal trials (rabbit eye) of the electronic contact lens for 20 min at a time with no adverse effects. Credit: University of Washington.

pioneer of this research field [17]. Later, such contact lenses were augmented with biosensors for glucose [16, 18], lactate [19], and ethanol [20].

Almost all these biodevices were composed in a similar manner, that is, an enzymatic layer, which is specific toward a particular substrate, a product diffusion layer, and a working electrode (Fig. 11.4). Despite the fact that smart electronic contact lenses are readily available (Fig. 11.5), the development of biosensing contact lenses is ongoing [21], without usable prototypes.

Nevertheless, the development of useful self-contained smart bionic contact lenses is an important direction, in spite of certain challenges at the current stage of scientific and technological development, especially when it comes to bioelectronics implementation. One of the main problems is related to the source of electric power for these complex devices, with issues such as size, thickness, and mass, that is, the overall form factor, along with many other parameters including biocompatibility and safety.

In the case of bionic contact lenses, not only biosensors but also other bioelectronic devices, that is, biological power sources, can be very useful to provide the electric energy needed, as described later. The power can be generated by converting in situ available chemical energy from different biofuels present in human tears (*vide supra*) into electric energy, drawing on the mainstay of fuel cell technology. The idea of utilizing biofuels in tears to generate electric power is necessitated because of the complex restrictions offered by a contact lens, leaving conventional, even state-of-the-art, power management technology at an impasse. As early as in 2010 a patent application titled "Flexible Biofuel Cell, Device and Method" was filed, with the priority date 24 March 2010 [22], describing the possible usage of biological fuel cells to power bionic contact lenses. The first paper detailing a biofuel cell as a power source for electronic contact lenses appeared a couple of years later, in 2012 [9]. The authors made a nanostructured (to improve the biodevice characteristics) glucose/oxygen enzymatic fuel cell (Fig. 11.6), based on direct electron transfer to

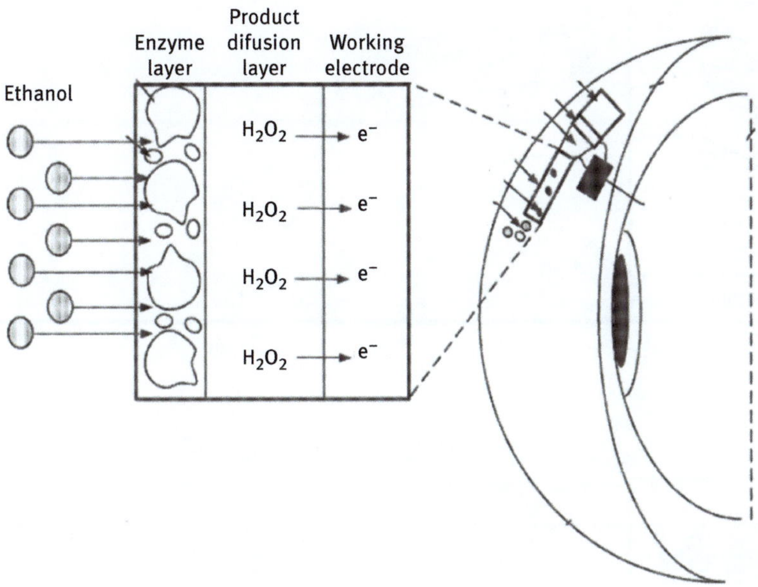

Fig. 11.4: Schematic representation of contact lens–based biosensors. According to Liu and Parviz [20] with some modifications and changes.

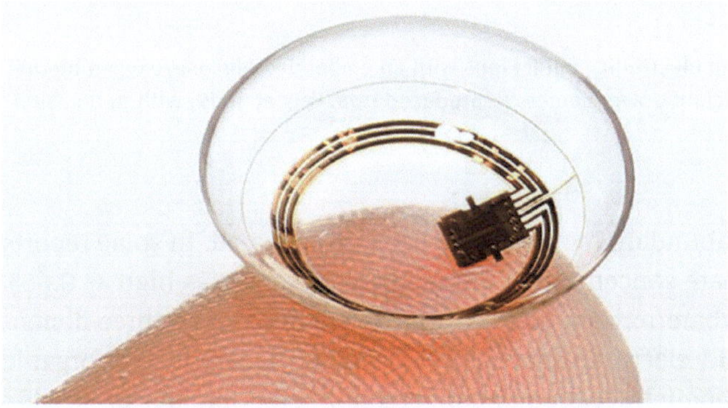

Fig. 11.5: A photograph of a smart contact lens for ocular pressure monitoring, from Sensimed AG (Lausanne, Switzerland).

avoid mediators, thus simplifying the biodevice construction. Although tear glucose has no function as a biofuel for cornea cells and indeed can be completely depleted in vivo without any apparent disadvantages for human eyes [13], the low glucose concentration in tears (ca. 50 μM) and the limited amount of this physiological fluid (ca. 1 μL) [15] make glucose an unlikely on-lens energy source, if conventional biofuel cells are to be employed. To circumvent the problem of low-power glucose-

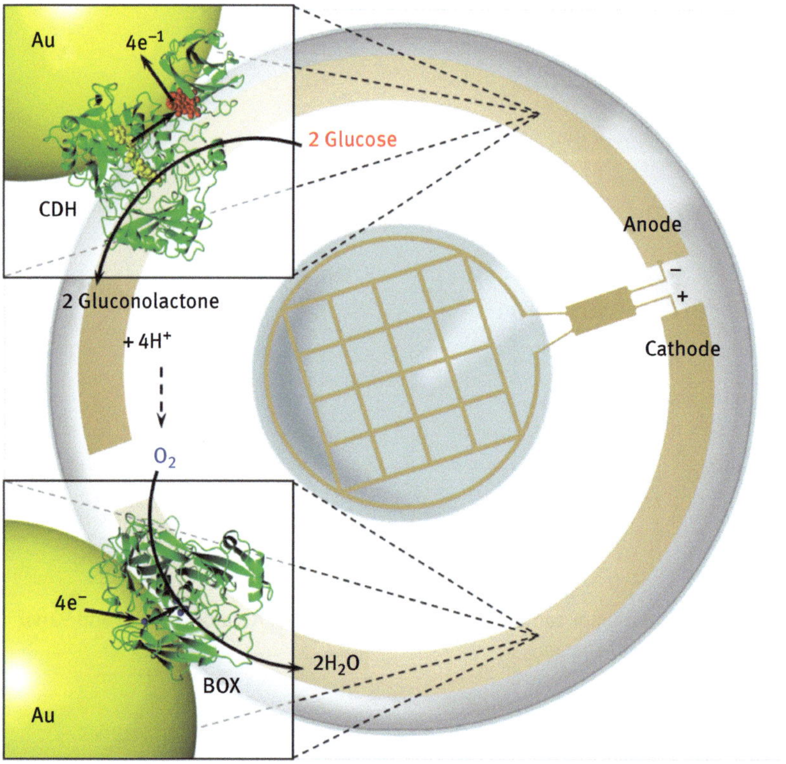

Fig. 11.6: Principle scheme of an electronic contact lens with an embedded glucose/oxygen biofuel cell acting as a green self-sufficient power source. Reproduced from Falk et al. [9] with permission from Elsevier.

based biofuel cells, more abundant biofuels have been investigated. In some reports it is stated that the ascorbate concentration in human tears can be as high as 0.665 mM [23]. Indeed, an ascorbate/oxygen fuel cell was assembled, using three-dimensional nanostructured gold electrodes covered with abiotic (a conductive organic complex, tetrathiafulvalene-tetracyanoquinodimethane) and biological (redox enzyme and bilirubin oxidase) materials functioning as efficient anodic and cathodic catalysts, respectively [24].

Nevertheless, when operated in human tears, the biodevice exhibited quite poor characteristics, viz. an open circuit voltage of 0.54 V (the theoretical OCV is about 0.74 V, as evident from the equilibrium redox potentials of oxygen/water (0.82 V) and dehydroascorbate/ascorbate (0.08 V) redox couples (pH 7)) and a maximal power density of 3.1 μW cm^{-2} at 0.25 V. Recent investigations have shown that the ascorbate concentration in tear fluid is only 17 ± 6 μM [25] and the average realistic electric power obtained from ascorbate as a biofuel in continuous mode is less than 0.06 μW, that is, even lower than that from glucose. Thus, also taking into account the physiological importance of ascorbate for the human eye, that is, the antioxidant capacity, and since it cannot be completely depleted in vivo without disadvantages for human eyes, it is an unlikely on-lens energy source. If conventional biological fuel cells are to be employed, based on the low half-reaction potential of lactate oxidation (ca. −0.19 V) and the high concentration in tears (up to 5 mM) [15], lactate seems to be the only option as an on-lens energy source. Hence, a report concerning a lactate/oxygen biofuel cell has been published [26]. Contact lens biofuel cell testing was performed in a synthetic tear solution at 35 °C, but in spite of the relatively high maximal theoretical voltage of a lactate/oxygen fuel cell, that is, about 1 V, the open-circuit voltage of the biodevice was only around 0.4 V, in all likelihood due to biocathode depolarization and opportunistic oxidation of a high redox potential biofuel, viz. ascorbate, on the anode. However, because of the high surface area of the electrodes, maximum current and power densities were calculated to be ca. 61 μA cm^{-2} and 8 μW cm^{-2}, respectively. It seems that current research should be focused on improvement of open circuit and especially operating potentials of biocathodes and bioanodes in human tears, rather than apparent increase of current densities in artificial (synthetic) electrolytes in unlimited volumes. Nevertheless, the only solution to further increase power is the usage of modern, recently disclosed biodevices, which are able to operate in both continuous and pulse modes, as discussed later.

In the beginning of 2013 an entirely new kind of electric power devices were disclosed by Shleev's and Cosnier's groups, in which chemical energy is directly converted into electric energy, which is capacitively stored within a singular contrivance [27, 28]. Such hybrid devices [29], including biodevices [30], are built based on dual-function electrodes, viz. electrodes manifesting simultaneous electrocatalytic and charge-storage features [31]. Recently, several different self-charging biosupercapacitors have been assembled and characterized [32], including the first transparent and flexible biodevices, which are able to operate at low glucose concentrations (Fig. 11.7) [33].

By running a charge-storing biofuel cell in pulse mode, and appropriately timing the cycle or pulse length, low- and high-abundance biofuels can be efficiently converted and the demands of consistent, temporally stable levels of power with regard to the sensing, encoding, and transmission events can be reproducibly met. Thus, *hybrid biodevices*, including tear-based biodevices, seem to the best choice for wearable bioelectronics.

11.2.2 Bioelectronics in sweat

Sweat is exuded by eccrine and porcine glands on the skin and is mainly composed of water (99%) and electrolytes, with a pH ranging between 4 and 7 [34]. In addition, sweat is abundant in not only low-molecular weight compounds, such as urea, glucose, cortisol, ethanol, and lactate, but also proteins and peptides, typically present at very low concentrations (μM to mM for low-molecular weight compounds

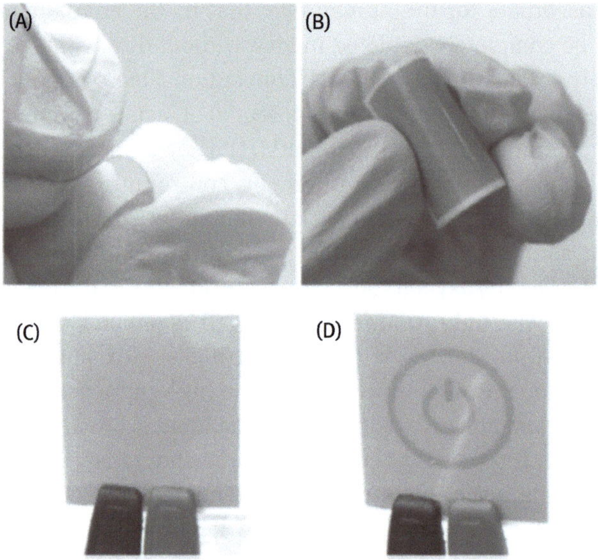

Fig. 11.7: Photographic images of (A) a flexible and transparent nanostructured electrode, which is bent without any damage, (B) a flexible low voltage display which is powered by the flexible biosupercapacitor, (C) and (D) connected display in OFF and ON states, respectively. Reproduced from Bobrowski et al. [33] with permission from Elsevier.

and pM for peptides and proteins) [35]. These compounds can either be assessed using biosensors and/or used as power sources, employing biofuel cells and biosupercapacitors. Such devices can be fabricated on substrates that are soft and deformable, such as fabrics, paper, and soft polymers, to provide comfort while worn and still be resilient to mechanical strains that arise while moving. As sweat can be sampled noninvasively, generated, for example, due to heat, exercise, or via chemical stimulation (e.g., by applying pilocarpine), this makes sweat analysis an attractive option for monitoring of the health as well as fitness status, but since the composition of sweat can be expected to vary depending on the cause of sweating, intrinsic variations need to be considered in the analysis [36]. Sweat as a diagnostic tool is also problematic due to the difficulty to produce enough sweat for analysis, sample evaporation, and contamination, as well as the fact that the, relation between sweat analyte levels and health status is still poorly understood [35, 37]. Changing health conditions can, for example, alter sweat composition by either permuting the concentration of common components, or by introducing new components. High sweat levels of chloride and urea have been linked to cystic fibrosis and kidney failure, respectively [38, 39]. In addition, correlation between both sweat ethanol as well as sweat glucose with respective blood concentrations has been demonstrated [40, 41]. The development of flexible bioelectronics offers a route of using sweat as a diagnostic tool or for continuous monitoring of main bioanalytes. However, because of low bioanalyte concentrations, highly sensitive chemical sensors are required, which is something biosensors offer, and they can be placed in close proximity to where the sweat is created. Even though direct correlation with blood values may not be possible, establishing a personal baseline through continuous sweat monitoring would be a highly valuable diagnostic tool, where deviations from the baseline could indicate changing health and/or fitness status. Effective sampling and transport of sweat to the sensor is crucial for successful analysis, and for continuous monitoring of the general health status a low-volume requirement is important. Typical sweat volume ranges from about 1 up to 20 nL min^{-1} per gland, with a gland density of 100–550 glands cm^{-2} [35]. Considering sweat uptake from a 1 cm^2 patch, typically ca. 0.1 up to 11 μL min^{-1} sweat will be released, placing a very high demand on the device design to be able to facilitate successful sampling for continuous monitoring. Integrating microfluidics in wearable bioelectronics is a possible solution to ensure that real-time measurements are conducted as close as possible and also to minimize evaporation [37]. The data collected from sensors could then be managed with a flexible electronic board, controlling sensor operation, and wireless data transmission, for example, to a smart phone.

Because of the generally low volumes of perspiration produced under typical conditions, most bioelectronics devices reported so far operating in human sweat have been tested during physical exercise to ensure sufficient supply of sweat for these early device designs. A key difficulty in the design of flexible electronics is the power supply, where traditional bulky batteries are not suitable for a wearable skinlike device. A possible solution could be to integrate a biofuel cell, which can be designed as a flexible device converting the chemical energy in sweat to electrical energy, thus powering the biosensor or other intended wearable application. Wang and coworkers have done extensive work in this area, leading the development of wearable, flexible bioelectronics. The group was the first to develop tattoo bioelectronics, where the first device design was a lactate biofuel cell fabricated by screen-printing on temporary tattoo paper, where the bioanode was modified with lactate oxidase and tetrathiafulvalene as mediator, whereas the cathode was composed of platinum black for reduction of oxygen [42]. The biofuel cell generated a power density of up to 70 μW cm^{-2} in sweat generated during exercise. Instead of using the tattoo anode to design a biofuel cell, the group also used it as a lactate sensor by instead combining it with printed silver ink and carbon ink reference and counter electrodes, monitoring the change in lactate concentration during physical exercise (Fig. 11.8) [43].

An year later, the group headed by Shleev, fabricated and characterized the first direct electron transfer glucose/oxygen enzymatic fuel cell, which operated in induced human sweat [44]. Because of the mediatorless approach, as well as usage of glucose as fuel instead of lactate (concentrations differ by almost three orders of magnitude), less power was obtained, viz. only ca. 0.3 W cm^{-2}. Later, Wang and coworkers also exploited the tattoo platform toward developing an iontophoretic-amperometric system for glucose detection, using glucose oxidase instead of lactate oxidase [45]. However, the

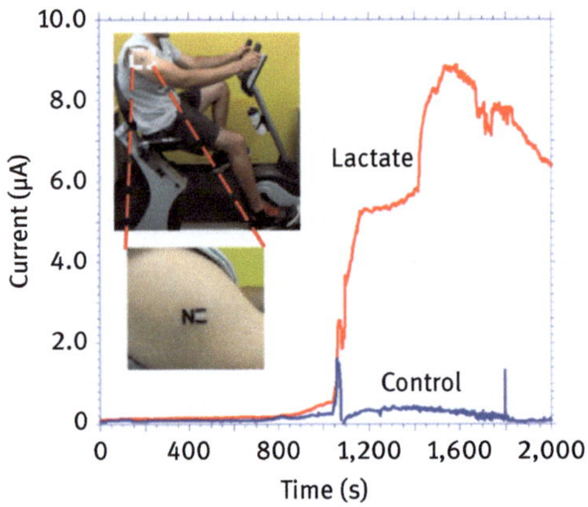

Fig. 11.8: Examples of tattoo bioelectronics, showing a lactate sensor operating in sweat generated upon physical exercise. Reproduced from Jia et al. [43] with permission from the American Chemical Society.

system employed reverse iontophoresis, where a current is applied to electro-osmotically extract interstitial fluid instead of using sweat. This worked on the same principle as the "GlucoWatch," an electrochemical wearable continuous glucose monitor that was commercialized, but ultimately discontinued in part due to user discomfort by the reverse iontophoresis protocol and need for calibration with blood glucose [46]. Instead of relying on extraction of interstitial fluid, sweating can be stimulated by iontophoresis, where a drug that enhances the sweat rate is driven beneath the skin by an applied current, stimulating perspiration. Wang et al. used this method of induced sweating to demonstrate a wearable tattoo–based alcohol-biosensing system for noninvasive alcohol monitoring [47]. Alcohol was detected amperometrically using alcohol dehydrogenase combined with Prussian blue to allow for electrochemical detection of the enzyme-produced hydrogen peroxide. Instead of relying on biosensors based on amperometric detection, Prasad and coworkers developed portable, impedance-based detection of glucose and cortisol in sweat [48-50]. By employing monoclonal antibodies toward glucose oxidase, the enzyme was immobilized on flexible zinc oxide, where the operation of the enzyme in the presence of glucose caused a change in conductivity and relative permittivity of the electrical double layer, proportional to the glucose concentration, measured by electrochemical impedance spectroscopy [48]. Similarly, cortisol antibodies were used for detection of cortisol, where surface charges were altered upon binding. Instead of zinc oxide, the group also demonstrated a cortisol sensor based on molybdenum disulfide nanosheets combined with a flexible polyamide, allowing direct contact of the polymer with sweat on the skin for wearable detection, with a limit of detection of 1 ng mL^{-1} (Fig. 11.9).

Kim and coworkers made significant progress toward epidermal glucose sensing by incorporating a soft-material–based amperometric glucose biosensor with a drug

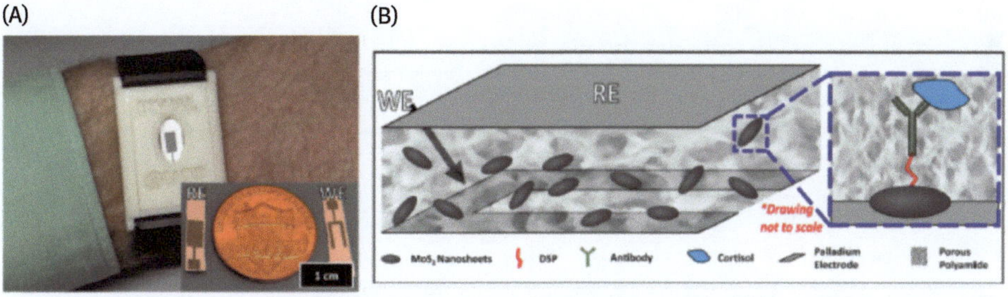

Fig. 11.9: Wearable device prototype for cortisol using electrochemical impedance to monitor the biosensor response. Reproduced from Kinnamon et al. [49] with permission from Springer Nature.

delivery module, capable of regulating blood glucose values [51, 52]. By introducing multiple porous sweat-uptake and water-proof layers on an ultrathin and stretchable patch-type device, sensing was achieved in volumes down to 1 µL [52]. To extract as much information as possible from sweat analysis, considering the complex correlation among all biophysical and biochemical information, simultaneous multiplexed detection combined with onsite data processing and communication has the potential to deepen our understanding and unveil previously unidentified correlations. Two examples of fully integrated devices with biosensors combined with a flexible electronics platform for processing and data transmission were designed by researchers at Berkley, using a flexible printed circuit board onto a flexible polyethylene terephthalate–based substrate [53, 54]. Recently, they demonstrated a wearable miniaturized iontophoresis interface, where a glucose oxidase biosensor was employed to continuously monitor the glucose concentration in stimulated sweat (Fig. 11.10) [54].

The wearable, mechanically flexible device also included sensors for sodium and chloride and could be programmed to induce sweating with various secretion profiles for real-time analysis, temporarily elevating the local rate of perspiration without any perceived discomfort for the test subject in the clinical evaluation of the device. Earlier, the same group reported a fully integrated wearable array–based carbon nanotube/chitosan biosensor for the detection of glucose and lactate (using glucose oxidase and lactate oxidase), multiplexed with sensors measuring electrolytes and skin temperature [53]. Perspiration was not stimulated, but the device operation was demonstrated during physical exercise.

The aforementioned integrated devices were powered by a rechargeable lithium ion polymer battery. Designing a flexible power supply for wearables is a key difficulty. A self-powered wearable biosensing platform was recently reported by Han et al. (Fig. 11.11) [55]. The device incorporated biosensors for glucose, uric acid, lactate, and urea, using the enzymes glucose oxidase, uricase, lactate oxidase, and urease, respectively, and could monitor the bioanalyte concentration in sweat generated upon physical exercise. The enzymes

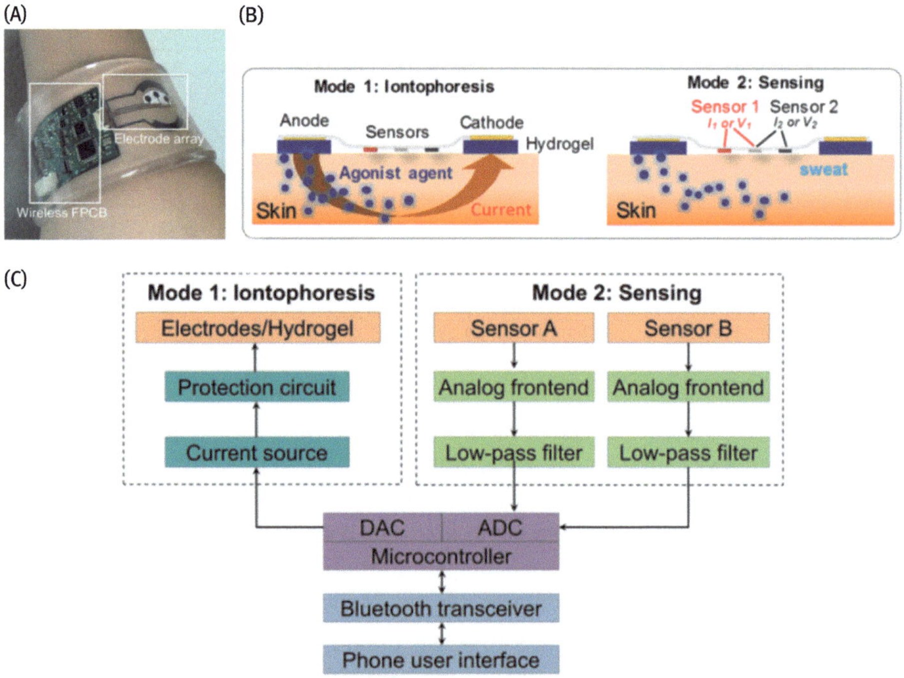

Fig. 11.10: A photograph of wearable autonomous sensing platform with iontophoresis electrodes to stimulate sweating (A) as well as mode of operation (B) and block diagram of the integrated platform operation (C). Reproduced from Emaminejad et al. [54] with permission from National Academy of Sciences of USA.

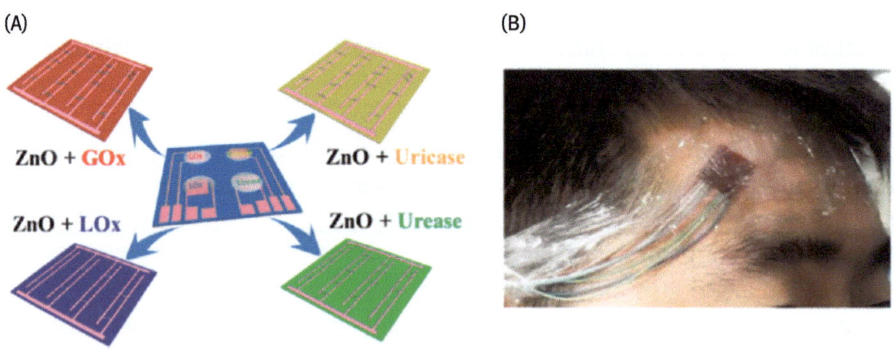

Fig. 11.11: An example of piezo-biosensing skin, where ZnO nanowires were modified with appropriate enzymes to allow for self-powered biosensing. Reproduced from Han et al. [55] with permission from the American Chemical Society.

were combined with zinc oxide nanoarrays to realize a piezo-biosensing unit, where the piezoelectric signal driven by body movement was dependent on the analyte concentration present in the perspiration. Under applied compressive force, the piezo-biosensing unit actively output a piezoelectric voltage. As the enzymatic reactions change the surface carrier density of the zinc oxide nanowire, the piezoelectric output upon deformation was changed, proportional to the analyte concentration.

Besides tattoo-based bioelectronics, the research group of Wang has done extensive work with other flexible bioelectronics devices. His team built a highly flexible and stretchable carbon nanotube–based biosensor and biofuel cell, integrating an amperometric glucose sensor and glucose biofuel cell, capable of repeated stretching by 300% and 180° torsional twisting without any negative impact on device performance (Fig. 11.12) [56].

Using screen printable stretchable inks, combining the electrical and mechanical properties of carbon nanotubes with the elastomeric properties of polyurethane is highly attractive as a material for wearable bioelectronics. A flexible lactate biosensor was also included with an electrocardiogram sensor for a combined hybrid sensing system [57]. The group also deployed bioelectronics in textiles. A lactate biofuel cell was integrated with a DC/DC converter wearable printed circuit board embedded in a textile headband to demonstrate the flexibility of the design [58]. Recently, the group also designed a carbon nanotube-silver nanoink–based glucose and lactate biofuel cell on a textile substrate, generating a maximum power density of 160 and 250 μW cm^{-2}, respectively [59].

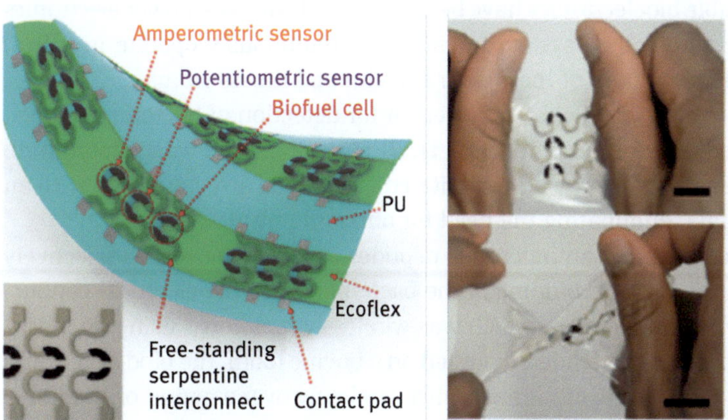

Fig. 11.12: Demonstrating the flexibility of wearable bioelectronics, manufactured using printed ink of carbon nanotubes combined with elastomeric polyurethane binder. Reproduced from Bandodkar et al. [56] with permission from the American Chemical Society.

11.2.3 Bioelectronics in saliva

Saliva has received increasing attention as a noninvasive diagnostic tool for investigating different medical conditions, as well as a tool for monitoring fitness status, for example, in sports science, being highly attractive due to the ease of sampling and abundant availability [60-64]. Saliva is excreted by several different glands of varying size, controlled by the autonomic nervous system, where adults typically secrete 0.5–1 L daily [64]. Besides excretion from the glands, saliva also include gingival fluid, serum components, bacteria, and other cells. Saliva is mainly composed of water (94–99%), but also contain many peptides and proteins, lipids, carbohydrates, and salt and includes the majority of blood disease markers such as antibodies, interleukins, and neoplasma markers, where the constituents not directly excreted also enter saliva through the blood via passive diffusion, active transport, or extracellular ultrafiltration [61, 64] While some studies show good correlation between saliva and blood concentrations of different analytes, others have shown no correlation, leading to a certain ambiguity for using saliva, for example, for glucose monitoring [65-68]. Saliva is predominantly collected by spitting or a passive drool method, with a period of fasting prior to collection. However, no uniform collection criteria exists, despite the significant influence the collection method may have on the composition. Analysis of saliva is already used in laboratory investigations for, for example, infectious disease diagnosis, autoimmune diseases, and hormonal analysis (using, e.g., chromatography or microarrays), and a variety of biosensors as well as biofuel cells have also been designed that operate in this physiological fluid [44, 62, 68, 69]. However, these have not been wearable devices, but operating in samples extracted from the mouth. Recently, a few examples of wearable biosensor devices enabling continuous monitoring have been reported. Using soft materials and nanofabrication technologies, wearable bioelectronics have been combined with low-power electronics to design miniaturized wireless system intended to continuously operate in saliva, providing real-time information of the wearer's health and fitness status.

In 2012 Mannoor et al. developed a biosensor for detection of bacteria in saliva, which was mounted on tooth enamel (Fig. 11.13) [70]. Graphene was printed onto water-soluble silk and contacted by interdigitated electrodes, also patterned with an inductive coil antenna, eliminating the need for onboard power.

The biosensor consisted of antimicrobial peptides on graphene, which selectively bind to bacteria in saliva. Upon binding of the bacterial target, the electrical conductivity of the graphene film changes, which was wirelessly monitored using an inductively coupled radio frequency reader. Mounted on a bovine tooth, the biodevice could detect 100 bacteria cells in 1 μL of human saliva wirelessly over a period of 10 min.

Instead of attaching the sensor system directly on a tooth, the research group of Wang integrated biosensors into mouth guards [71, 72]. In the first device reported by the group, lactate oxidase was immobilized on the working electrode, and the three-electrode system was screen printed onto the mouth guard [71]. Upon addition of

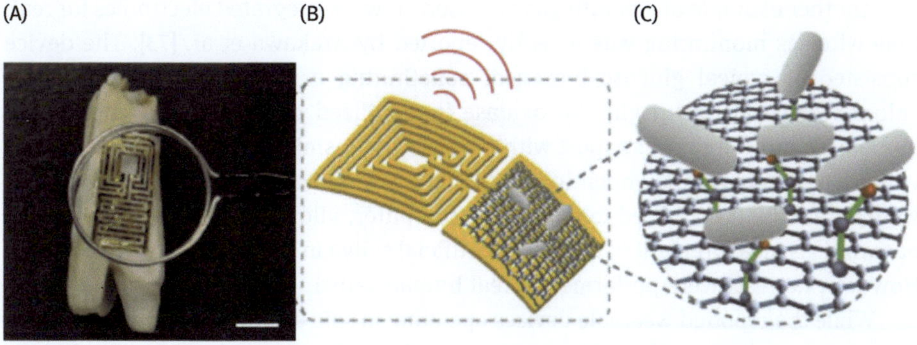

Fig. 11.13: A photograph of the sensor mounted on a tooth (A) along with a schematic drawing showing the nanosensing architecture (B) and sensing element (C). Reproduced with permission from Mannoor et al. [70] with permission from Macmillan Publishers Ltd.

lactate, the enzyme produces hydrogen peroxide from the conversion of lactate to pyruvic acid, which is then sensed by the printed Prussian blue transducer layer. The sensor was investigated in human saliva and the response was linear up to 1 mM with a detection limit of 50 μM lactate, giving a stable response for 2 h. In a similar manner, the group also developed a mouth guard biosensor toward uric acid, utilizing uricase to produce hydrogen peroxide in the presence of uric acid (Fig. 11.14) [72]. In addition to the biosensor, a miniaturized potentiostat, a microcontroller, and a bluetooth low-energy transceiver were incorporated in the mouth guard. This allowed for real-time wireless monitoring of salivary uric acid, and the device was tested in human saliva collected by passive drool. Similar to the lactate biosensor, the urate biosensor gave a linear response up to 1 mM and it was stable for over 2 h.

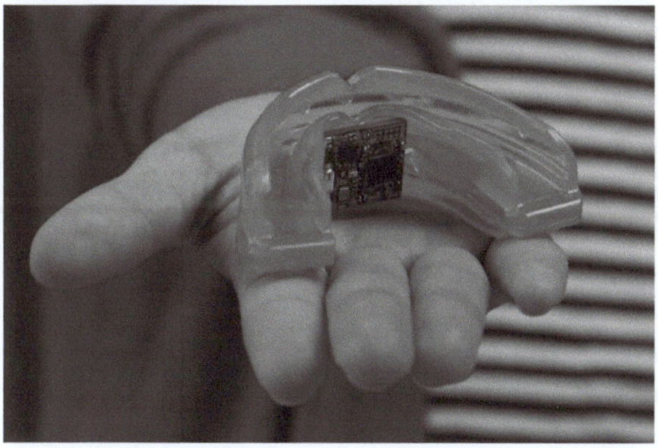

Fig. 11.14: A photograph of an electronic mouth guard. Credit: Jacobs School of Engineering, UC San Diego.

Another example of a mouth guard biosensor with integrated electronics for real-time wireless monitoring was recently reported by Arakawa et al. [73]. The device consisted of typical glucose biosensor constituting a platinum and silver/silver chloride electrode, with glucose oxidase immobilized on top, on a custom-fitted monolithic mouth-guard support with a wireless transmitter. The sensor was tested with a phantom jaw in an open-loop artificial saliva injection system and the results were wirelessly transmitted to a personal computer, allowing for real-time continuous wireless measurement of glucose in artificial saliva in the range of 0.05–1.00 mM. However, no tests were performed in real human saliva.

While the reported wearable devices operating in saliva show promising results, no measurements have so far been performed using actually worn devices by human subjects. To allow for continuous in-mouth use, potential toxicity needs to be considered and all materials used in the device design must be fully biocompatible. In addition, the biosensor needs to be highly specific, since active chemicals from food could significantly interfere with the signal, as well as detrimentally affect the stability of the biosensor. The device also needs to be resistant to the mechanical stress resulting from masticating.

11.2.4 Bioelectronics in urine

The first bioelectronic devices for excreted human physiological fluids were intendend to operate in urine, rather than saliva, sweat, or tears. In all likelihood, this is because of the large volume of the fluid, which can be easily obtained from an individual, as well as wide application of this physiological fluid in clinical analysis already in the second part of the last century [74]. Urine is excreted by an adult human in the average amount of 1.4 L per day, consisting of water (about 91–96%), inorganic salts, and organic compounds, including proteins, hormones, and a wide range of metabolites [75]. Already in the 1970s, oxygen-sensitive enzymatic electrodes for the detection of uric acid in urine were fabricated and characterized [76]. Later, hydrogen peroxide–sensitive enzymatic electrodes for the detection of glucose in urine were designed and tested [77]. Since urine glucose assay is subject to severe interferences due to the low concentration of the bioanalyte, ca. 0.2 mM, specific membranes were used in biodevice constructions, for example, a cellulose nitrate–glucose oxidase modified membrane [78]. Despite of the obvious possibility to design wearable urine biosensors operating in diapers, to the best of our knowledge, such bioelectronic devices have not been realized yet.

Surprisingly, contrary to biosensors, an example of a wearable biological fuel cell operating in urine is reported in the literature. Specifically, a self-sufficient system, powered by a wearable energy generator based on a microbial fuel cell operating in urine, was designed (Fig. 11.15). Microbial fuel cells made from compliant materials were developed in the frame of a pair of socks, which was fed by urine via a manual gating

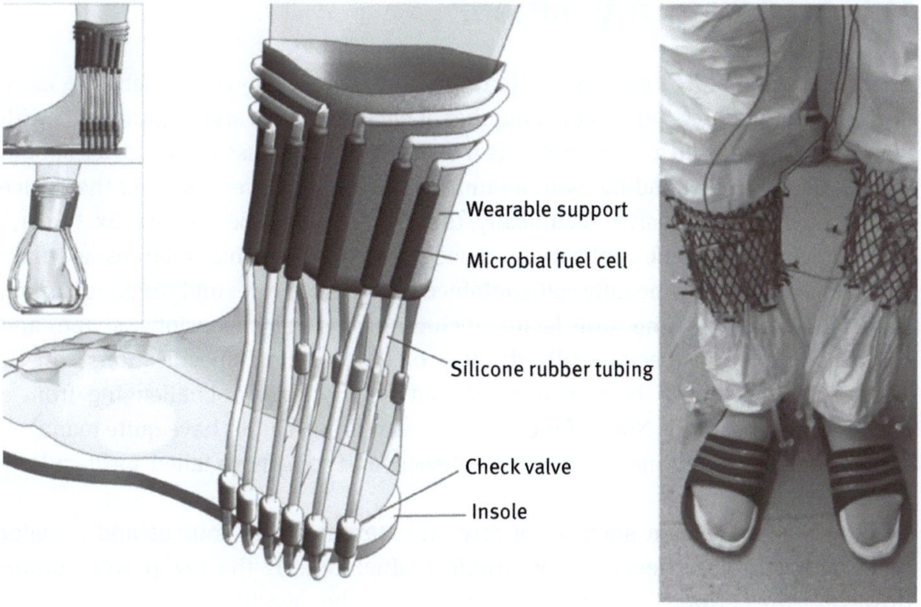

Fig. 11.15: Wearable microbial fuel cell: (left) schematic drawing and (right) image of the developed wearable generator. From UPI (https://www.upi.com/Science_News/2015/12/11/Scientists-unveil-urine-powered-wearable-energy-generator/7221449853868) with some modifications and additions; Credit: WE Bristol/Ioannis Ieropoulos.

pump [79]. The simple and single-loop cardiovascular fish circulatory system was used as the inspiration for the design of the manual pump. The maximum achievable power was about 110 μW, which was generated when a load of 30 kΩ was connected to 12 biofuel cells in series. A wireless programmable communication module, engineered to operate within the range of the generated electricity, was employed, which opens a new avenue for research in the utilization of waste products from humans for powering portable as well as wearable electronics.

In fact, many papers have been published by Ieropoulos and colleagues [80–83], and also by other authors [84, 85], describing different microbial fuel cells operating in urine, whereas there are only a few reports regarding enzymatic fuel cell performance in urine [86]. This is because of low concentrations of usual biofuels, such as glucose, lactate, and ascorbate, in this physiological fluid. Indeed, galvanodynamic performance measurements in urine showed a significant loss of the maximum power density of an enzymatic fuel cell compared to the biodevice in 5 mM glucose containing buffer, that is, down to 12% from the initial value of about 100 μW cm^{-2} [86].

Taken together, the earlier examples illustrate the multifaceted approaches used by the scientific community, focusing on noninvasive methods, taking stock of the feasibility and the availability of the various physiological fluids.

11.3 Summary and outlook

As far as “wearability” goes, given that the background bioevent/bioanalyte monitoring is adequately attended to, a number of *end user* issues and requirements with regard to convenience and cosmetics can be identified. The use of the device should be technically undemanding, with a minimum of user intervention, and the device itself should be unobtrusive. Technically, the device needs to be easy to affix, “wear,” and remove, and in some applications, devices need to be flexible and transparent. In addition, devices must be fully self-contained in terms of power and communication. Regarding cosmetics, the form factor, including physical dimensions, shape, and color, should be discreetly realized. Most of the cosmetic specifications can be fulfilled if devices can be miniaturized, but downscaling is challenging from a technical point of view. None of the designs exemplified earlier have quite managed to collect all the requirements in a single device, and some have failed spectacularly regarding cosmetics.

Power management should not rely on external power sources and a major issue with currently developed wearable biofuel cells is the low-power output. However, the power output can momentarily be significantly enhanced by designing so-called hybrid devices, integrating biofuel cells with supercapacitors, where the energy amassed is continuously stored until needed [31, 32]. Achieving a sensor reading every few minutes is sufficient for most applications to obtain useful information from a wearable device. Depending on the monitoring scenario, the bioanalyte to be monitored and the biofuel needed for the fuel cell to charge the integrated supercapacitor could be the same compound, for example, glucose. If the hybrid device is well characterized and well behaved, the pulse frequency will be related to the bioanalyte concentration. Unless the reading is trivial, for example, a number or a color change, presented directly on the device, communication, that is, delivery of complex data, should draw on one of the wireless protocols currently available. On device data processing is at present unrealistic, and the heavy lifting, for example, applying AI or signal deconvolution, should be done externally, using a smartphone or a relaying device.

Provided that the basic, background biochemistry can be unraveled, that is, by individually or collectively authenticated bioanalyte concentration correspondences between blood and various physiological fluids, the importance of wearable bioelectronic devices cannot be overestimated. In fact, by pooling data from a large number of devices, the correlation models developed to catalogue blood/physiological fluid interrelationships will be improved constantly, increasing the reliability of the measurements. As for the technical challenges, regarding, for example, device size, power management, data communication, biocompatibility, and operational stability, tremendous improvements the last couple of decades promise an imminent breakthrough.

References

[1] Willner, I., & Katz, E., Integration of layered redox proteins and conductive supports for bioelectronic applications. Angew Chem Int Ed. 2000, 39, 1181–1218.

[2] Katz, E., Bioelectronics. Electroanalysis. 2006, 18, 1855–1857.

[3] Falk, M., Alcalde, M., Bartlett, P. N., De Lacey, A. L., Gorton, L., Gutierrez-Sanchez, C., Haddad, R., Kilburn, J., Leech, D., Ludwig, R., Magner, E., Mate, D. M., Conghaile, P. O., Ortiz, R., Pita, M., Poeller, S., Ruzgas, T., Salaj-Kosla, U., Schuhmann, W., Sebelius, F., Shao, M., Stoica, L., Sygmund, C., Tilly, J., Toscano, M. D., Vivekananthan, J., Wright, E., & Shleev, S. Self-powered wireless carbohydrate/oxygen sensitive biodevice based on radio signal transmission. PLoS One. 2014, 9, e109104/1-e/9, 9 pp.

[4] Zebda, A., Alcaraz, J.-P., Vadgama, P., Shleev, S., Minteer, S. D., Boucher, F., Cinquin, P., & Martin, D. K. Challenges for successful implantation of biofuel cells. Bioelectrochemistry. 2018, 124, 57–72.

[5] Andoralov, V., Shleev, S., Arnebrant, T., & Ruzgas, T. Flexible micro(bio)sensors for quantitative analysis of bioanalytes in a nanovolume of human lachrymal liquid. Anal and Bioanal Chem. 2013, 405, 3871–3879.

[6] Pankratov, D., Blum, Z., Suyatin, D. B., Popov, V. O., & Shleev, S. Self-charging electrochemical biocapacitor. ChemElectroChem. 2014, 1, 343–346.

[7] Coman, V., Vaz-Dominguez, C., Ludwig, R., Harreither, W., Haltrich, D., De Lacey, A. L., Ruzgas, T., Gorton, L., & Shleev, S. A membrane-, mediator-, cofactor-less glucose/oxygen biofuel cell. Phys Chem Chem Phys. 2008, 10, 6093–6096.

[8] Shleev, S., & Ruzgas, T. Transistor-like behavior of a fungal laccase. Angew Chem Int Ed. 2008, 47, 7270–7274.

[9] Falk, M., Andoralov, V., Blum, Z., Sotres, J., Suyatin, D. B., Ruzgas, T., Arnebrant, T., & Shleev, S. Biofuel cell as a power source for electronic contact lenses. Biosens Bioelectron. 2012, 37, 38–45.

[10] Andoralov, V., Falk, M., Suyatin Dmitry, B., Granmo, M., Sotres, J., Ludwig, R., Popov Vladimir, O., Schouenborg, J., Blum, Z., & Shleev, S. Biofuel cell based on microscale nanostructured electrodes with inductive coupling to rat brain neurons. Sci Rep. 2013, 3, 3270.

[11] Kim, D. H. Epidermal electronics. Science. 2011, 333, 1703.

[12] Chu, B., Burnett, W., Chung, J. W., & Bao, Z. Bring on the bodyNET. Nature. 2017, 549, 328–330.

[13] Berman, E. R. Biochemistry of the eye. 1991, New York, Plenum Press.

[14] Farandos, N. M., Yetisen, A. K., Monteiro, M. J., Lowe, C. R., & Yun, S. H. Contact lens sensors in ocular diagnostics. Adv Healthcare Mater. 2015, 4, 792–810.

[15] Pankratov, D., Gonzalez-Arribas, E., Blum, Z., & Shleev, S. Tear based bioelectronics. Electroanalysis. 2016, 28, 1250–1266.

[16] Shum, A. J., Cowan, M., Laehdesmaeki, I., Lingley, A., Otis, B., & Parviz, B. A. Functional modular contact lens. Proc SPIE. 2009, 7397, 73970K/1-K/8.

[17] Parviz, B. A. For your eye only. Ieee Spectrum. 2009, 46, 36–41.

[18] Yao, H., Shum, A. J., Cowan, M., Lahdesmaki, I., & Parviz, B. A. A contact lens with embedded sensor for monitoring tear glucose level. Biosens Bioelectron. 2011, 26, 3290–3296.

[19] Thomas, N., Laehdesmaeki, I., & Parviz, B. A. A contact lens with an integrated lactate sensor. Sens Actuat B. 2012, 162, 128–134.

[20] Liu, Z., & Parviz, B. Sensor to detect ethanol concentration in blood. 2015, (Google Inc., USA), WO Patent Application, 30 pp.

[21] Park, J., Kim, J., Kim, S.-Y., Cheong, W. H., Jang, J., Park, Y.-G., Na, K., Kim, Y.-T., Heo, J. H., Lee, C. Y., Lee, J. H., Bien, F., & Park, J.-U. Soft, smart contact lenses with integrations of wireless circuits, glucose sensors, and displays. Sci Adv. 2018, 4, eaap9841. DOI: 10.1126/sciadv.aap9841.

[22] Ruzgas, T., Shleev, S., & Arnebrant, T. Flexible biofuel cell, device and method. 2011, WO Patent Application 31pp.
[23] Gogia, R., Richer, S. P., & Rose, R. C. Tear fluid content of electrochemically active components including water soluble antioxidants. Curr Eye Res. 1998, 17, 257–263.
[24] Falk, M., Andoralov, V., Silow, M., Toscano, M. D., & Shleev, S. Miniature biofuel cell as a potential power source for glucose-sensing contact lenses. Anal Chem. 2013, 85, 6342–6348.
[25] Choy, C. K., Cho, P., Chung, W. Y., & Benzie, I. F. Water-soluble antioxidants in human tears: effect of the collection method. Invest Ophthalmol Vis Sci. 2001, 42, 3130–3134.
[26] Reid, R. C., Minteer, S. D., & Gale, B. K. Contact lens biofuel cell tested in a synthetic tear solution. Biosens Bioelectron. 2015, 68, 142–148.
[27] Shleev, S., Pankratov, D., & Blum, Z. Charge-storing fuel cell. 2014, WO Patent Application, 59 pp.
[28] Cosnier, S., Holzinger, M., Le, G. A., & Agnes, C. Electrochemical supercapacitor. 2014, WO Patent Application, 25 pp.
[29] Pankratov, D., Falkman, P., Blum, Z., & Shleev, S. A hybrid electric power device for simultaneous generation and storage of electric energy. Energy Environ Sci. 2014, 7, 989–993.
[30] Agnes, C., Holzinger, M., Le Goff, A., Reuillard, B., Elouarzaki, K., Tingry, S., & Cosnier, S. Supercapacitor/biofuel cell hybrids based on wired enzymes on carbon nanotube matrices: autonomous reloading after high power pulses in neutral buffered glucose solutions. Energy Environ Sci. 2014, 7, 1884–1888.
[31] Pankratov, D., Blum, Z., & Shleev, S. Hybrid electric power biodevices. ChemElectroChem. 2014, 1, 1798–1807.
[32] Shleev, S., Gonzalez-Arribas, E., & Falk, M. Biosupercapacitors. Curr Opin Electrochem. 2017, 5, 226–233.
[33] Bobrowski, T., Gonzalez Arribas, E., Ludwig, R., Toscano, M. D., Shleev, S., & Schuhmann, W. Rechargeable, flexible and mediator-free biosupercapacitor based on transparent ITO nanoparticle modified electrodes acting in μM glucose containing buffers. Biosens Bioelectron. 2018, 101, 84–89.
[34] Sato, K., Kang, W. H., Saga, K., & Sato, K. T. Biology of sweat glands and their disorders. I. Normal sweat gland function. J Am Acad Dermatol. 1989, 20, 537–563.
[35] Sonner, Z., Wilder, E., Heikenfeld, J., Kasting, G., Beyette, F., Swaile, D., Sherman, F., Joyce, J., Hagen, J., Kelley-Loughnane, N., & Naik, R. The microfluidics of the eccrine sweat gland, including biomarker partitioning, transport, and biosensing implications. Biomicrofluidics. 2015, 9, 031301.
[36] Bariya, M., Nyein, H. Y. Y., & Javey, A. Wearable sweat sensors. Nature Electron. 2018, 1, 160–171.
[37] Heikenfeld, J. Non-invasive analyte access and sensing through eccrine sweat: challenges and outlook circa. Electroanalysis. 2016, 2016, 28, 1242–1249.
[38] O'Sullivan, B. P., & Freedman, S. D. Cystic fibrosis. Lancet. 2009, 373, 1891–1904.
[39] Keller, R., & Sands, J. Localization of urea transporters UT-A1 and UT-B in human eccrine clear cell (LB719). FASEB J. 2014, 28, LB719.
[40] Buono, J. M. Sweat ethanol concentrations are highly correlated with co-existing blood values in humans. Experimental Physiol. 1999, 84, 401–404.
[41] Moyer, J., Wilson, D., Finkelshtein, I., Wong, B., & Potts, R. Correlation between sweat glucose and blood glucose in subjects with diabetes. Diabetes Technol Therapeutics. 2012, 14, 398–402.
[42] Jia, W., Valdés-Ramírez, G., Bandodkar Amay, J., Windmiller Joshua, R., & Wang, J. Epidermal biofuel cells: energy harvesting from human perspiration. Angew Chem Int Ed. 2013, 52, 7233–7236.

[43] Jia, W., Bandodkar, A. J., Valdés-Ramírez, G., Windmiller, J. R., Yang, Z., Ramírez, J., Chan, G., & Wang, J. Electrochemical tattoo biosensors for real-time noninvasive lactate monitoring in human perspiration. Anal Chem. 2013, 85, 6553–6560.
[44] Falk, M., Pankratov, D., Lindh, L., Arnebrant, T., & Shleev, S. Miniature direct electron transfer based enzymatic fuel cell operating in human sweat and saliva. Fuel Cells. 2014, 14, 1050–1056.
[45] Bandodkar, A. J., Jia, W., Yardımcı, C., Wang, X., Ramirez, J., & Wang, J. Tattoo-based noninvasive glucose monitoring: a proof-of-concept study. Anal Chem. 2015, 87, 394–398.
[46] Tierney, M. J., Tamada, J. A., Potts, R. O., Jovanovic, L., & Garg, S. Clinical evaluation of the GlucoWatch® biographer: a continual, non-invasive glucose monitor for patients with diabetes. Biosens Bioelectron. 2001, 16, 621–629.
[47] Kim, J., Jeerapan, I., Imani, S., Cho, T. N., Bandodkar, A., Cinti, S., Mercier, P. P., & Wang, J. Noninvasive alcohol monitoring using a wearable tattoo-based iontophoretic-biosensing system. ACS Sensors. 2016, 1, 1011–1019.
[48] Munje, R. D., Muthukumar, S., & Prasad, S. Lancet-free and label-free diagnostics of glucose in sweat using zinc oxide based flexible bioelectronics. Sensors Actuat B. 2017, 238, 482–490.
[49] Kinnamon, D., Ghanta, R., Lin, K.-C., Muthukumar, S., & Prasad, S. Portable biosensor for monitoring cortisol in low-volume perspired human sweat. Sci Rep. 2017, 7, 13312.
[50] Sankhala, D., Muthukumar, S., & Prasad, S. A four-channel electrical impedance spectroscopy module for cortisol biosensing in sweat-based wearable applications. Slas Technol. 2018, 23, 529–539.
[51] Lee, H., Choi, T. K., Lee, Y. B., Cho, H. R., Ghaffari, R., Wang, L., Choi, H. J., Chung, T. D., Lu, N., Hyeon, T., Choi, S. H., & Kim, D.-H. A graphene-based electrochemical device with thermoresponsive microneedles for diabetes monitoring and therapy. Nature Nanotechnol. 2016, 11, 566.
[52] Lee, H., Song, C., Hong, Y. S., Kim, M. S., Cho, H. R., Kang, T., Shin, K., Choi, S. H., Hyeon, T., & Kim, D.-H. Wearable/disposable sweat-based glucose monitoring device with multistage transdermal drug delivery module. Sci Adv. 2017, 3, e1601314. DOI: 10.1126/sciadv.1601314.
[53] Gao, W., Emaminejad, S., Nyein, H. Y. Y., Challa, S., Chen, K., Peck, A., Fahad, H. M., Ota, H., Shiraki, H., Kiriya, D., Lien, D.-H., Brooks, G. A., Davis, R. W., & Javey, A. Fully integrated wearable sensor arrays for multiplexed in situ perspiration analysis. Nature. 2016, 529, 509.
[54] Emaminejad, S., Gao, W., Wu, E., Davies, Z. A., Yin Yin Nyein, H., Challa, S., Sp, R., Hm, F., Chen, K., Shahpar, Z., Talebi, S., Milla, C., Javey, A., & Davis, R. W. Autonomous sweat extraction and analysis applied to cystic fibrosis and glucose monitoring using a fully integrated wearable platform. Proc Natl Acad Sci USA. 2017.
[55] Han, W., He, H., Zhang, L., Dong, C., Zeng, H., Dai, Y., Xing, L., Zhang, Y., & Xue, X. A self-powered wearable noninvasive electronic-skin for perspiration analysis based on piezo-biosensing unit matrix of enzyme/ZnO nanoarrays. ACS Appl Mater Interface. 2017, 9, 29526–29537.
[56] Bandodkar, A. J., Jeerapan, I., You, J.-M., Nuñez-Flores, R., & Wang, J. Highly stretchable fully-printed CNT-based electrochemical sensors and biofuel cells: combining intrinsic and design-induced stretchability. Nano Lett. 2016, 16, 721–727.
[57] Imani, S., Bandodkar, A. J., Mohan, A. M. V., Kumar, R., Yu, S., Wang, J., & Mercier, P. P. A wearable chemical–electrophysiological hybrid biosensing system for real-time health and fitness monitoring. Nat Commun. 2016, 7, 11650.
[58] Jia, W., Wang, X., Imani, S., Bandodkar, A. J., Ramirez, J., Mercier, P. P., & Wang, J. Wearable textile biofuel cells for powering electronics. J Mater Chem A. 2014, 2, 18184–18189.
[59] Jeerapan, I., Sempionatto, J. R., Pavinatto, A., & You J-M, W. J. Stretchable biofuel cells as wearable textile-based self-powered sensors. J Mater Chem A. 2016, 4, 18342–18353.

[60] Lima, D. P., Diniz, D. G., Moimaz, S. A. S., Sumida, D. H., & Okamoto, A. C. Saliva: reflection of the body. Int J Infectious Diseases. 2010, 14, e184–18e8.
[61] Javaid, M. A., Ahmed, A. S., Durand, R., & Tran, S. D. Saliva as a diagnostic tool for oral and systemic diseases. J Oral Biol Craniofacial Res. 2016, 6, 67–76.
[62] Mishra, S., Saadat, D., Kwon, O., Lee, Y., Choi, W.-S., Kim, J.-H., & Yeo, W.-H. Recent advances in salivary cancer diagnostics enabled by biosensors and bioelectronics. Biosens Bioelectron. 2016, 81, 181–197.
[63] Viswanath, B., Choi, C. S., Lee, K., & Kim, S. Recent trends in the development of diagnostic tools for diabetes mellitus using patient saliva. TrAC Trends Anal Chem. 2017, 89, 60–67.
[64] Chojnowska, S., Baran, T., Wilińska, I., Sienicka, P., Cabaj-Wiater, I., & Knaś, M. Human saliva as a diagnostic material. Adv Med Sci. 2018, 63, 185–191.
[65] Aguirre, A., Testa-Weintraub, L. A., Banderas, J. A., Haraszthy, G. G., Reddy, M. S., & Levine, M. J. Sialochemistry: a diagnostic tool? Critical Reviews in Oral. Biol Med. 1993, 4, 343–350.
[66] Tékus, É., Kaj, M., Szabó, E., Szénási, N., Kerepesi, I., Figler, M., Gábriel, R., & Wilhelm, M. Comparison of blood and saliva lactate level after maximum intensity exercise. Acta Biol Hungarica. 2012, 63, 89–98.
[67] Sashikumar, R., & Kannan, R. Salivary glucose levels and oral candidal carriage in type II diabetics. oral surgery, oral medicine, oral pathology. Oral Radiol Endodontol. 2010, 109, 706–711.
[68] Malon, R. S. P., Sadir, S., & Balakrishnan, M. #xf3, rcoles EP. Saliva-based biosensors: noninvasive monitoring tool for clinical diagnostics. BioMed Res Int. 2014, 20, 962903.
[69] Mink, J. E., Qaisi, R. M., Logan, B. E., & Hussain, M. M. Energy harvesting from organic liquids in micro-sized microbial fuel cells. Npg Asia Mater. 2014, 6, e89.
[70] Mannoor, M. S., Tao, H., Clayton, J. D., Sengupta, A., Kaplan, D. L., Naik, R. R., Verma, N., Omenetto, F. G., & McAlpine, M. C. Graphene-based wireless bacteria detection on tooth enamel. Nat Commun. 2012, 3, 763.
[71] Kim, J., Valdes-Ramirez, G., Bandodkar, A. J., Jia, W., Martinez, A. G., Ramirez, J., Mercier, P., & Wang, J. Non-invasive mouthguard biosensor for continuous salivary monitoring of metabolites. Analyst. 2014, 139, 1632–1636.
[72] Kim, J., Imani, S., de Araujo, W. R., Warchall, J., Valdés-Ramírez, G., Paixão, T. R. L. C., Mercier, P. P., & Wang, J. Wearable salivary uric acid mouthguard biosensor with integrated wireless electronics. Biosens Bioelectron. 2015, 74, 1061–1068.
[73] Arakawa, T., Kuroki, Y., Nitta, H., Chouhan, P., Toma, K., Sawada, S.-I., Takeuchi, S., Sekita, T., Akiyoshi, K., Minakuchi, S., & Mitsubayashi, K. Mouthguard biosensor with telemetry system for monitoring of saliva glucose: a novel cavitas sensor. Biosens Bioelectron. 2016, 84, 106–111.
[74] Free, A. H. Urinalysis in clinical laboratory practice. 1975, University of California, USA: CRC Press, 284 pp.
[75] Brunzel, N. A. Fundamentals of urine and body fluid analysis. 2012, Amsterdam, The Nethelands: Elsevier, 464pp.
[76] Nanjo, M., & Guilbault, G. G. Enzyme electrode sensing oxygen for uric acid in serum and urine. Anal Chem. 1974, 46, 1769–1772.
[77] Scheller, F. W., Schubert, F., Neumann, B., Pfeiffer, D., Hintsche, R., Dransfeld, I., Wollenberger, U., Renneberg, R., Warsinke, A. et al. Second generation biosensors. Biosens Bioelectron. 1991, 6, 245–253.
[78] Hanke, G., Nentwig, J., & Scheller, F. The use of polyurethane membranes in the enzymatic-amperometric determination of glucose in urine. Zeitschrift fur medizinische Laboratoriumsdiagnostik. 1988, 29, 390–394.

[79] Taghavi, M., Stinchcombe, A., Greenman, J., Mattoli, V., Beccai, L., Mazzolai, B., Melhuish, C., & Ieropoulos, I. A. Self sufficient wireless transmitter powered by foot-pumped urine operating wearable MFC. Bioinspir Biomim. 2015, 11, 016001.
[80] Ieropoulos, I., Greenman, J., & Melhuish, C. Urine utilisation by microbial fuel cells; energy fuel for the future. Phys Chem Chem Phys. 2012, 14, 94–98.
[81] Ieropoulos, I. A., Greenman, J., & Melhuish, C. Miniature microbial fuel cells and stacks for urine utilisation. Int J Hydrogen Energy. 2013, 38, 492–496.
[82] You, J., Greenman, J., Melhuish, C., & Ieropoulos, I. Electricity generation and struvite recovery from human urine using microbial fuel cells. J Chem Technol Biotechnol. 2016, 91, 647–654.
[83] Walter, X. A., Stinchcombe, A., Greenman, J., & Ieropoulos, I. Urine transduction to usable energy: a modular MFC approach for smartphone and remote system charging. Appl Energy. 2017, 192, 575–581.
[84] Shreeram, D. D., Hassett, D. J., & Schaefer, D. W. Urine-powered microbial fuel cell using a hyperpiliated pilT mutant of Pseudomonas aeruginosa. J Industrial Microbiol Biotechnol. 2016, 43, 103–107.
[85] Karthik, K., Vivek, D., Rahul, R., Swati, K., & Karishma, R. Production of electricity using microbial fuel cell utilizing untreated human urine. Int J Adv Sci Res Manage. 2016, 1, 224–228.
[86] Goebel, G., Beltran, M. L., Mundhenk, J., Heinlein, T., Schneider, J., & Lisdat, F. Operation of a carbon nanotube-based glucose/oxygen biofuel cell in human body liquids-Performance factors and characteristics. Electrochim Acta. 2016, 218, 278–284.

Sergey Shleev, Olga Aleksejeva, Magnus Falk, Zoltan Blum

12 Biodegradable electric power devices

12.1 Introduction

Transient electronics is an emerging technology with a huge potential for different applications [1–3]. The key attribute of transient electronics is an ability to dissipate in an orderly manner after a period of stable operation [4]. Potential applications of biodegradable semiconductor devices include environmental friendly sensors, hardware-secure memory modules, and temporary biomedical implants [5]. In the design of transient electronics, requirements include biodegradability, biocompatibility, and that hardware components are environmentally benign in the broadest possible sense [6]. Although the transient concept has only been actively developed in a few years, this unique semiconductor technology is believed to find more opportunities in the fast development of advanced electronics. Perhaps stating the obvious, semiconductor devices need to be electrically powered, and thus, the electric power supplies used for transient electronics should be biocompatible, biodegradable, and also environmental friendly. Prior to the description of the state of the art of biodegradable electrochemical cells, the objective of this chapter, a brief account of biodegradable electric power supplies from a general viewpoint, including a possible classification protocol, is called for.

Biodegradable power supplies are essential components of transient electronics (Fig. 12.1). In spite of the short time dedicated to the development of these, several demonstrated strategies, including degradable radio frequency power transfer modules [7], silicon-based photovoltaics [4], harvesters of mechanical energy [8], and electrochemical cells (vide infra), have been implemented.

Electrochemical cells are devices that supply electric energy using electrochemical reactions occurring at two different electrodes, one positive and one negative electrode. Simplistically, electrochemical cells as electric power supplies can be divided into three major groups, viz. fuel cells, batteries, and electrochemical capacitors. Fuel cells and batteries are devices that convert chemical energy directly into electric energy, that is, they generate electric power, whereas capacitors and modern accumulators (i.e., secondary batteries) are rechargeable devices, and, perhaps yet again stating the obvious, neither electrochemical capacitors nor accumulators produce electric power by themselves and hence need to be externally charged. The major difference between fuel cells and batteries is that the former are open systems and will work as long as the fuel and oxidant supply is maintained, whereas the latter are closed systems and need to be replenished,

Sergey Shleev, Olga Aleksejeva, Magnus Falk, Zoltan Blum, Department of Biomedical Science, Faculty of Health and Society, Malmö University, Malmö, Sweden

https://doi.org/10.1515/9783110570526-012

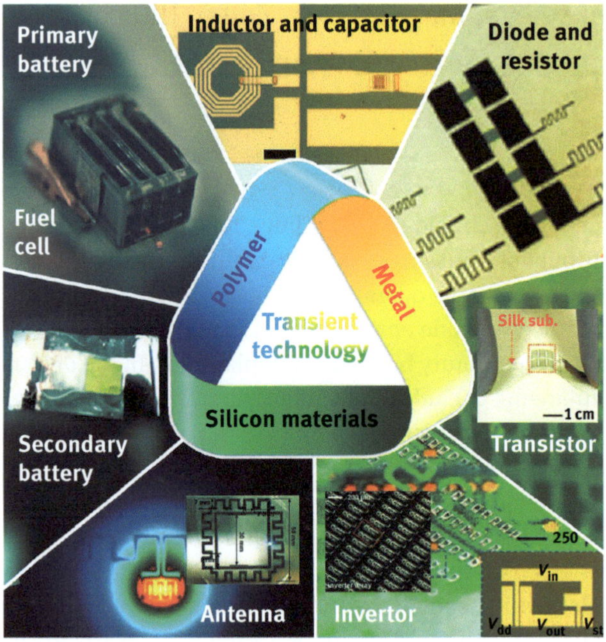

Fig. 12.1: An overview of transient electronics: materials and devices (according to Fu et al. [6]. with some additions and changes).

when fuels/oxidants are consumed. Hybrid devices do exist as well, for example, commercially available Zn/Air batteries, in which one electrode represents a primary battery anode, whereas the cathode is an air-breathing electrode of a fuel cell (Fig. 12.2). Biodegradable, or at least transient, prototypes of fuel cells, batteries, and electrochemical capacitors have all been reported in the literature, as exemplified later.

12.2 Biodegradable electrochemical cells

Following the classification given earlier of electrochemical cells as electric power supplies, the development of biodegradable fuel cells, batteries, and capacitors is discussed in chronological order. However, prior to that, a brief description of transient materials versus truly biodegradable materials is provided.

In broad terms, biodegradable materials can be disintegrated by biological means, while in a strict sense a biodegradable material is (can be) consumed by bacteria or fungi. Some dielectric materials, such as silicon dioxide and silicon

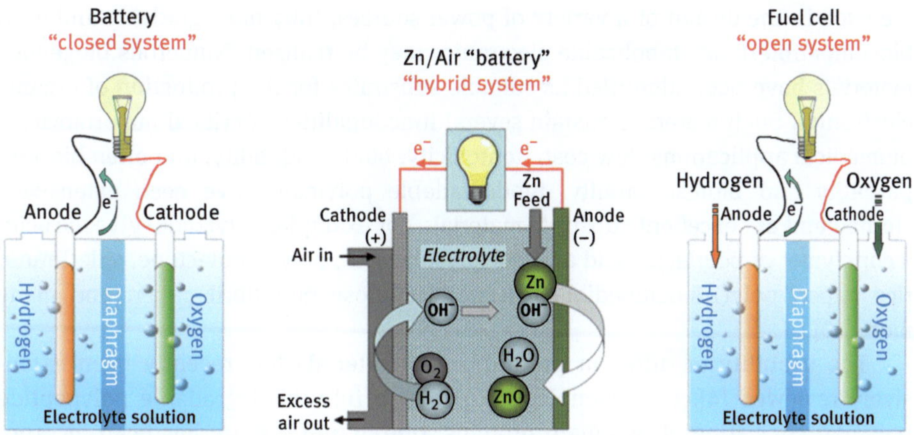

Fig. 12.2: A schematic representation of the function of hypothetical hydrogen/oxygen battery and fuel cell, as well as real Zn/Air battery.

nitride, have been found to be biodegradable [2], whereas metallic films of, for example, magnesium, zinc, iron, and molybdenum can all be degraded in complex biochemical and aqueous *inorganic* processes [2]. Thus, metals are not biodegradable in the strict sense, at least not all of them, but rather transient. Moreover, traditional materials used in power sources, such as heavy metals, strong electrolytes, precious metal catalysts, and synthetic polymers, are typically nondegradable, nonbiocompatible, and often directly or indirectly toxic. Nonetheless, in spite of the complex features of biodegradation, research into the development of biodegradable materials for electronics has attracted a great deal of attention in recent years [9–11]. Early attempts were focused on organic materials including natural or synthetic biodegradable polymers, and partially degradable devices have been realized. For example, organic bioelectronics have been used to develop disposable (biodegradable) electronics [12, 13] for food packaging [4] or applications based on single-use appliances [14]. So far, the demonstrated transient devices are mostly associated with nonbiogenic degradation in aqueous solution, and thus, the devices are not necessarily fully biodegradable [6]. If the efforts are met with success, introducing biodegradable power sources to consumer electronics or environmental monitors is expected to greatly alleviate landfill and environmental issues caused by electronic waste [15]. In addition, a fully biodegradable power source has the advantage of eliminating potential retention of materials in in vivo applications.

In the development of biodegradable power sources, using biodegradable organic materials in the device design is particularly promising. Various organic materials, including different compounds of natural origin, with applications in electronics are biodegradable, safe, and nontoxic [1, 10, 11, 13]. When such materials are used in the design of a variety of power sources, fully biodegradable and even biocompatible/biometabolizable electronics may be realized. Numerous biogenous materials have been identified as suitable substrates for the production of organic electronics. Such materials sustain several functionalities of critical importance for biomedical applications: low cost, nontoxicity, biodegradability, and often biocompatibility and bioresorbability. Biodegradable polymers have been extensively studied and are excellent substrate materials, for example, polylactic–glycolic acid, a copolymer of polylactic acid and polyglycolic acid, polycaprolactone, silk fibroin, rice paper, poly(1,8-octanediol-*co*-citrate), cellulose nanofibril paper, and many more [10].

The usability of different protein-based materials has recently been extensively reviewed [4]. For example, silk, being a fully biodegradable polypeptide polymer, consisting of two main proteins, fibroin and sericin, has been incorporated in electronics [4]. Gelatin is another example of a fully biocompatible and biodegradable protein-based material used in the design of, for example, edible energy storage devices (vide infra). Aside from protein-based polymers, polysaccharides, such as chitosan and dextran, can also be used as biocompatible substrate materials [16], and chitosan in particular has been extensively investigated as a novel material for applications in energy storage and conversion devices [17]. The natural material that has attracted the most attention in recent years is paper based on plant-derived cellulose, with the added benefits of being inexpensive, lightweight, flexible, and environmental friendly. Recent advancements have seen the development of numerous biogenic paper materials, such as graphene paper, carbon microfiber paper, carbon nanotube paper, and composite paper, incorporating conductive polymers, forming a conductive paper material [14, 18, 19].

Incorporation of conductive polymers, such as polypyrrole or polyaniline, into the active materials of different energy storage and conversion devices has now found widespread use [16, 20] However, conducting polymers are typically not biodegradable, but different strategies have been developed to improve their ability to biodegrade [9]. By designing composite materials, where conductive polymers are combined with some natural degradable polymer, such as polylactic acid, chitosan, or paper, the amount of nonbiodegradable conductive polymer can be minimized, where the issue is to still maintain a high conductivity of the material. In addition to conductive polymers, a variety of different carbon materials, such as graphenic fillers and carbon nanotubes, have been used to improve and tune the electrical conductivity and mechanical properties of biodegradable/biocompatible polymers [21]. Carbon materials have also found widespread applications in different energy storage and conversion devices on their own [22–24]. However, biodegradation of carbon nanotubes and graphene is only feasible under certain conditions, catalyzed by some peroxidases [25–28].

The synthesis of exhaustively degradable electrical materials has still not been realized, and current energy storage and conversion systems are typically not fully biodegradable. Significant research efforts have recently been directed toward the development of intrinsically biodegradable materials for electrochemical cells, such as batteries, supercapacitors, and fuel cells.

12.2.1 Biodegradable fuel cells

Ordinarily, fuel cells require catalysts based on transition metals, such as platinum, rhodium, ruthenium, palladium, gold, and silver, or nickel and chromium, as well as alloy-derived materials, such as Raney nickel, and employ nonbiodegradable separators. In contrast, a biofuel cell is a special type of fuel cell, which instead of the typical metal-based catalysts utilizes biodegradable biological catalysts, such as enzymes, organelles, or even whole living cells, to convert chemical energy into electric energy. Biofuel cells have been identified as appropriate wearable and implantable energy sources [29, 30].

Taking into account that catalysts, fuels, and products are biodegradable, the inherent environmental impact of biofuel cells compared to fuel cells is significantly reduced, making biological fuel cells particularly interesting when trying to design exhaustively biodegradable devices. The downsides of using biofuel cells as energy harvesting devices are the rather low power output, at best about a mW cm^{-2}, and the limited operational stability [31]. While biofuel cells can be considered green in terms of their contribution to renewable energy, most devices still employ polymeric membranes, potentially toxic redox species, and/or nonbiodegradable carbon materials as the core parts, rendering the devices as a whole nonbiodegradable and will thus contribute toward anthropogenic waste. Recently, however, nonconventional biodegradable materials have been successfully applied in a variety of different fuel cell designs.

Fuel cells based on paper, an eco-friendly and cost-effective material, have recently attracted considerable research interest. Reported devices, while potentially disposable, are only to varying extents biodegradable (for example, still employing nonbiodegradable packaging material or the conductivity of the electrode material rely on a nonbiodegradable material). Three different examples of recently reported paper-based fuel cells are shown in Fig. 12.3.

Sabaté and coworkers reported on a PMMA encased "single-use fuel cell," with a hydrogen generator based on a magnesium–iron alloy and an alkaline fuel cell, using a paper strip blotted with liquid electrolyte as separator; the fuel cell consumed the alloy-generated hydrogen to produce electrical power (Fig. 12.3A) [32]. The device produced a maximum power of ca. 103 mW cm^{-2}, enough to sustain a variety of portable applications. The same group has also reported on several different paper-based enzymatic fuel cells, utilizing redox polymers and carbon nanotubes to

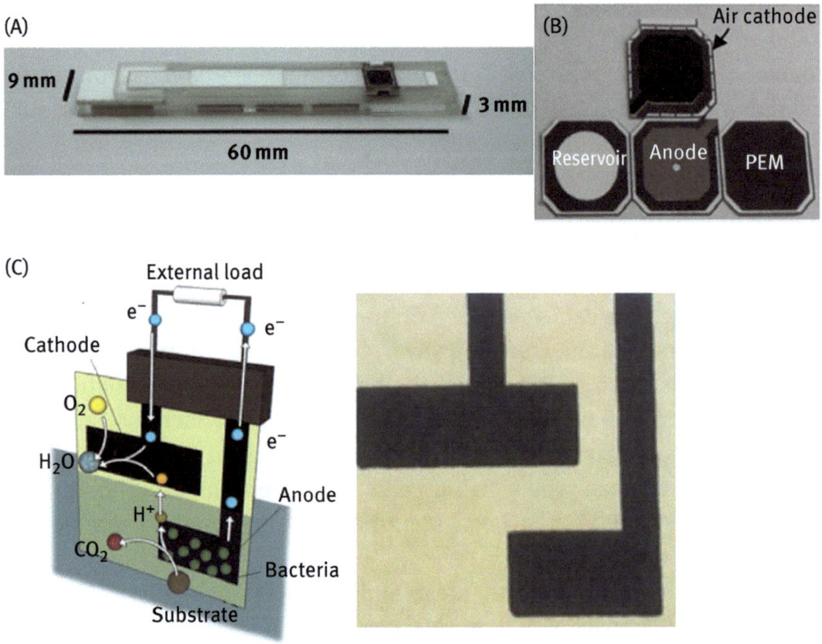

Fig. 12.3: Examples of paper-based fuel cells. (A) A single-use disposable paper hydrogen fuel cell, encased in PMMA. (B) A foldable screen-printed paper-based microbial fuel cell. (C) Principle of operation along with a photo of a printed microbial on filter paper. Reprinted with permission from Esquivel et al. [32], González-Guerrero et al. [33], and Chouler et al. [39], respectively.

electrically connect the enzymes to the electrode surface [35]. The research group of Choi has done extensive work on different paper-based biofuel cells, both microbial- and enzyme-based biodevices, with the goal of designing disposable power sources [33, 36–39]. In a recent study, the group reported on a printed foldable paper-based microbial fuel cell, integrating the anode, reservoir, cation-exchange membrane, and air cathode into one device (Fig. 12.3B). The paper was modified with a graphite-polymer composite (using PTFE) and graphite ink with activated carbon (AC) to create the anode, combined with a hydrophobic wax-based membrane and an air cathode. The device delivered a maximum power output of about 140 $\mu W\ cm^{-2}$. Another recent example of a single-component paper-based microbial fuel cell, fabricated by screen printing a conductive ink onto a single sheet of paper, was recently reported by Chouler et al. (Fig. 12.3C) [39]. The conductive ink, that is, a suspension of carbon nanotubes, graphite powder, and cellulose, was modified with chitosan to enhance biofilm formation. This allowed a low production cost for the device; however, the power output was well below $\mu W\ cm^{-2}$.

While the design of the aforementioned devices was focused on making disposable devices, the biodegradability was not investigated per se. Ieropoulos and coworkers recently designed a microbial fuel cell based on biodegradable materials and also investigated the degradation of the device (Fig. 12.4) [40] Natural rubber was used as a substitute to a conventional membrane, combined with, for example, paper, gelatin, polylactic acid, and lanolin, which are all biodegradable. The materials were verified to be biodegradable by investigating the degradation when placed in the ground, as shown in Fig 12.4A–C. Carbon was used as the conductive element for both electrodes, using carbon veil for the anode and conductive paint with graphite powder for the cathode. The microbial fuel cell was assembled into a stack (Fig. 12.4D) and reached a maximum power density of about 4 $\mu W\ cm^{-2}$.

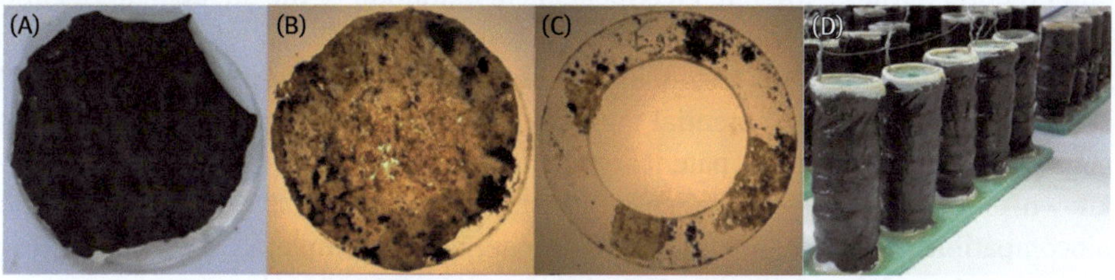

Fig. 12.4: Biodegradable microbial fuel cell. Biodegradation of egg-based cathode on natural rubber (A) before composting, (B) after 2 months, and (C) after four months. (D) Assembled microbial fuel cell stack. Reprinted with permission from Winfield et al. [40].

When designing fuel cells using sustainable materials, and thus excluding platinum catalysts and potentially toxic redox species, the device performance will be compromised, at least to some extent. In addition, the power output of reported disposable biofuel cells is several orders of magnitude lower when compared with a conventional fuel cell, typically delivering power far below mW cm^{-2}, at an output voltage too low to drive a useful application and needing several devices in series to reach sufficient voltage. While potentially green and environmental friendly, all of the described devices contained at least some materials that will not biodegrade, such as certain types of conductive polymers, or conductive materials made up of carbon nanotubes or graphite, redox mediators, including redox polymers, and/or nonbiodegradable polymers, such as PMMA or PTFE. Significant research efforts are currently being devoted toward developing disposable energy sources, where the use of biological catalysts to design fuel cells, combined with new biodegradable conductive materials, presents an exciting opportunity to make green energy harvesting devices, for example, for wearable applications, where the energy source after use then simply could be thrown away. However, the performance of the devices reported so far is inadequate for any practically useful application, and significant improvements are called for.

12.2.2 Biodegradable batteries

As mentioned earlier, batteries are closed systems, which do not produce electric energy, when fuels and/or oxidants are consumed. Thus, the electrochemical corrosion of materials within physiological medium found its logical application in biodegradable batteries, utilizing the energy released from metallic corrosion reactions. However, parasitic corrosion, which may discharge a battery prior to the intended use, as well as chemistries of all the components should be taken into consideration in design of a biodegradable energy source with a stable shelf life [41–43].

Taking into consideration a US patent application with the priority date 15 July 2011, it is reasonable to assume that the very first fully biodegradable battery was realized by the Tyco Healthcare Group LP, USA. According to the applicants, the title biodegradable battery includes an anode and a cathode, both having an inner surface and an outer surface, wherein electrochemical oxidation of the anode material and electrochemical reduction of the cathodic material results in the formation of two reaction products that are essentially nontoxic and wherein the cathode material carries a larger standard potential than the anode material. However, it should be emphasized that despite this patent application some partly degradable devices were known much earlier. For instance, a gastric-fluid-utilizing microbattery based on biocompatible materials including a Zn anode, which completely dissolves during the generation of electricity (Fig. 12.5), is known since 2008 [44].

Fig. 12.5: A photo of gastric-fluid-utilizing microbattery prototype.

In 2013 Kim et al. presented a sodium-ion electrochemical cell using anode/cathode materials composed of AC and λ-manganese oxide (λ-MnO_2), respectively, on a conductive polymer film. The device was capable to realize an energy density of 0.3 Wh kg^{-1}, with a specific capacity of 9.7 mAh g^{-1} upon galvanostatic discharge [45]. Subsequently, the same research group demonstrated a sodium ion cell with improved performance composed of natural melanin anodes and λ-MnO_2 cathodes, which exhibited an initial potential of 1.03 V with a maximum specific capacity of 16.1 mAh g^{-1} [45]. In both studies, the setups had a footprint in the 20–30 mm^2 range and the mechanism of biodegradation was not clearly explained. In 2014 Yin et al. reported water-activated primary batteries based on biodegradable metal foils [5]. These non-rechargeable devices have low output voltages (0.4 V for a single cell) and large dimensions ($3 \times 2 \times 1.3\ cm^3$) for arrays of cells. Also in 2014, Tsang et al. presented a series of microelectromechanical systems (MEMS)-enabled biodegradable batteries composed of biodegradable, polymer-coated Mg anodes, and Fe cathodes in a 0.1 M $MgCl_2$ electrolyte. A microelectromechanical system-enabled poly(glycerol-sebacate)-coated battery showed a capacity and power delivery capability of up to 0.7 mAh and 26 µW, respectively [43]. The mechanism of biodegradation of the device was not clearly illustrated though.

In 2015, Fu et al. described a rechargeable, flexible device with an output voltage of 2.8 V [46], but the electrode materials, current collectors, and substrate dissolved in water within a few seconds/minutes (Fig. 12.6). In the same year, Tsang et al. introduced a microfabricated polycaprolactone-encapsulated Mg/Fe biodegradable battery with a footprint of $0.2\ cm^2$ and a total cell volume of less than $0.02\ cm^3$, that is, the smallest device shown so far. The miniaturized battery was capable to produce 30 µW of electric power during 100 h upon galvanostatic discharge in phosphate buffer saline (gravimetric energy density 694 Wh kg^{-1}), which satisfies the lower limit of IMD power requirements; theoretically, that would be enough to power a low-power neurostimulator for ca. 4 days [47]. The Mg-based anode was able to nearly completely dissolve by day 20, though the long-lasting degradation of the Fe cathode was beyond the time scale of the study [42].

In 2016, Edupuganti et al. replaced the pure Mg anode by a Mg alloy, AZ31, that is, while keeping the Fe cathode. Introduction of the AZ31 alloy increased the lifetime of the device, and improved the battery's capacity (6.41 mAh cm^{-2}) and power output (67 µW). However, the biodegradability profile was not discussed in this study [48]. Jia et al. introduced a partially biodegradable biobattery using a silk-fibroin polypyrrole film cathode coupled with bioresorbable AZ31 Mg alloy, in phosphate buffer saline as an electrolyte. The Mg-air battery exhibited a specific capacity up to 3.9 mAh cm^{-2}, providing a specific energy density of 4.70 mWh cm^{-2}. Upon biodegradation of the silk-fibroin polypyrrole cathode in protease solution, 18 % of the original mass remained [49].

In 2017 Jia et al. demonstrated a biodegradable thin-film magnesium alloy battery with silk fibroin-choline nitrate polymer electrolyte, encapsulated in silk, offering a specific capacity of 0.06 mAh cm^{-2}. The battery lifetime can be encoded by varying the amount of silk protection. The enzymatic degradation of the device takes place over 45 days in the buffered protease solution (Fig. 12.7). However, given the dimensions of the device, that is, $3.6 \times 2.7 \times 0.017\ cm^3$, device miniaturization was clearly not a consideration [50].

To conclude, the design of a biodegradable power source requires several criteria, such as size, biocompatibility, mechanical stability, flexibility, programmable biodegradation rate, and power output, to be met. Different research groups dedicated significant efforts toward finding solutions on how to match all of the

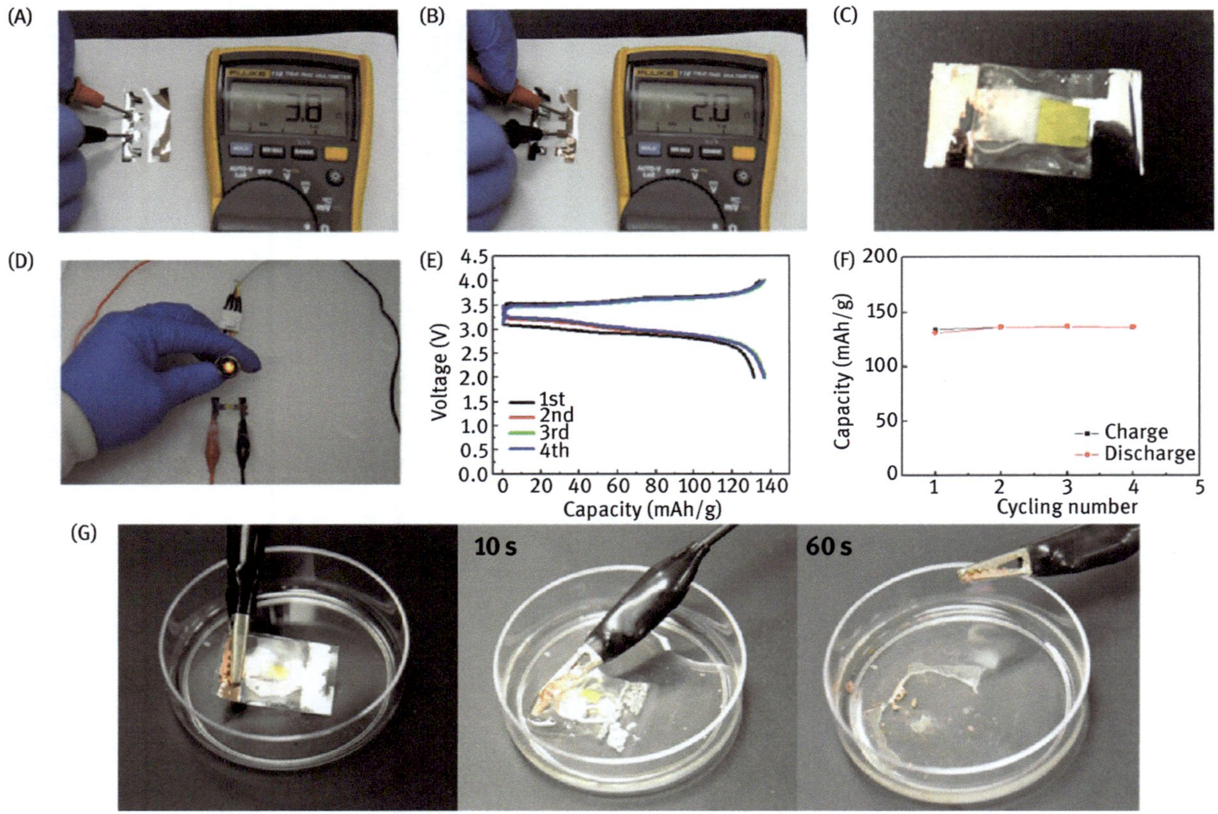

Fig. 12.6: V_2O_5 transient battery performance. (A and B) Resistance measurements of Al and Cu current collector, respectively. (C) Digital image of a single transient battery. (D) Light-emitting diode (LED) powered by a V_2O_5 transient battery. (E) Charge–discharge curves. (F) Cycling performance of the V_2O_5 transient battery. (G) Dissolution of the transient battery in water at room temperature. Reprinted with permission from Fu et al. [46].

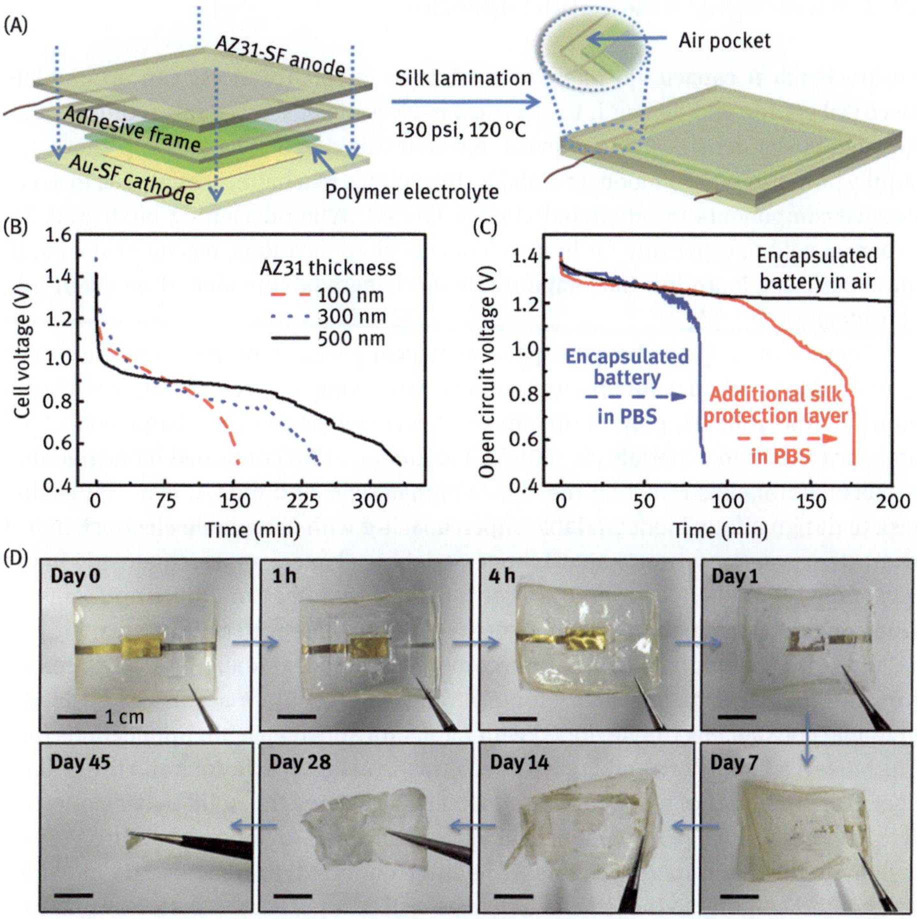

Fig. 12.7: Device structure, discharge performance, and biodegradation profile of an encapsulated Mg thin-film battery. (A) Schematic of an encapsulated battery with a silk pocket. (B) Effect of AZ31 thin-film thickness on the discharge performance of an encapsulated battery at a current density of 10 μA cm^{-2}. (C) In situ OCV changes of an encapsulated battery (AZ31 thickness 500 nm) and with an additional silk protection layer exposed to air and 1mL of PBS. (D) Photographs demonstrating the biodegradation profile of an encapsulated battery (device size of 3.6 × 2.7 × 0.017 cm^3) in buffered protease solution at 37 °C. Reprinted with permission from Jia et al. [50].

aforementioned requirements in a single device. The successful findings are shown, but practical applications relying on such devices are few and far between.

12.2.3 Biodegradable supercapacitors

Electrochemical capacitors, often referred to as supercapacitors, can accumulate electrical energy and deliver it to electrical devices at substantial rates, thus providing high power densities. Such devices are well suited to deliver back-up energy, to supply power in pulsed mode, to stabilize the voltage of a battery, as well as to act as passive components in integrated circuits [51–54]. When designing biodegradable supercapacitors, especially for in vivo biomedical applications, mechanical compliance, softness, footprint, biocompatibility, and degradability should be taken into consideration [10, 55].

Conventional "green" supercapacitors typically consist of the active electrodes (metal oxides, carbon materials, and conjugated polymers), electrode support/carrier substrate (polyimides, polycarbonates or paper), gel electrolyte, charge collectors, and encapsulation materials [56–60]. The requirement for controlled biodegradability sets natural constraints on the choice of materials, and thus, it is a challenging task to design a fully biodegradable supercapacitor with acceptable electrochemical characteristics. Reports on natural biodegradable polymers appeared already in the end of the 1990s [61]. However, it was only in 2006 that Bhat and Kumar presented a study on high polymer blends of polymethyl methacrylate with cellulose acetate and cellulose acetate phthalate, where enzyme based biodegradability was examined [62]. In 2008 the same research group introduced a cellulose acetate-based biodegradable polymer electrolyte doped with $LiClO_4$ for supercapacitor applications [63]. The biodegradability of a solid polymer electrolyte in buffer solutions and the electrochemical properties of a polypyrrole supercapacitor using this solid polymer electrolyte have been tested [63]. Later on, a biodegradable NaCl–agarose gel electrolyte was tested in flexible supercapacitors. The interconnected agarose matrix provides mechanical stability to the gel electrolyte as well as a porous network for enhanced ion transport and mobility [64]. Electrode materials based on composites of biodegradable polymers and conjugated polymers and graphene are considered for supercapacitors; however, properly matching conductivity and biodegradability is essential [65]. In 2015 Kang et al. introduced thin foils of Mo, Fe, W, or Zn as biodegradable substrates and silicate spin-on-glass materials as insulating and encapsulating layers, with applications in transient passive electronic components, that is, capacitors and inductors [66]. Silk from the silkworm *Bombyx Mori* has been used in biomedicine for centuries [67]. Electrode support and encapsulation materials made of biocompatible and biodegradable silk proteins, such as fibroin and sericin [68, 69], allow encoding of proteolytic degradation rates. In 2016 Wang et al. reported on an edible supercapacitor that utilizes soft, food-based materials [70]. The structure involves a relatively thick electrode film (≈120 µm) and a stack-type structure (Fig. 12.8). In the same year, Kumar et al. reported on melanin-based flexible supercapacitors [71]. In 2017 Chen et al. demonstrated the design and construction of an all-wood structured asymmetric supercapacitor (which implies two

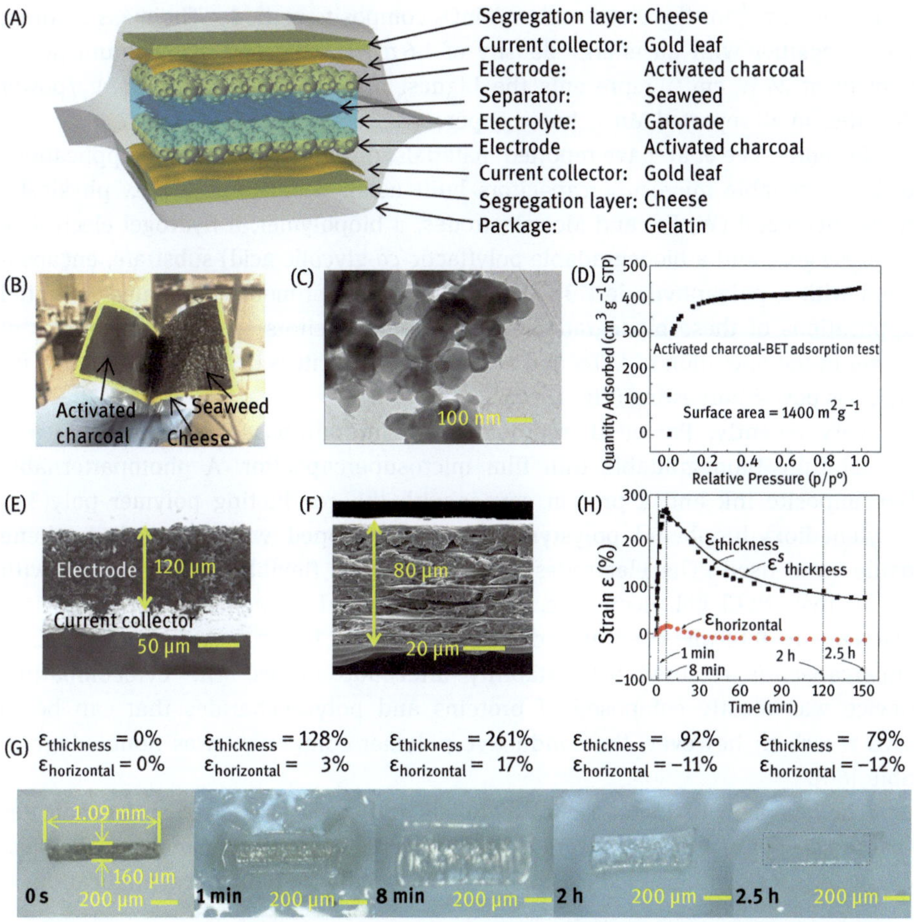

Fig. 12.8: Illustration and materials analysis of the edible supercapacitor. (A) Schematic structure of an edible supercapacitor. (B) An opened supercapacitor showing the activated charcoal electrode, seaweed separator, cheese segregation layer and gelatin package. (C) TEM image showing that the particle size of activated charcoal is about 100 nm. (D) Brunauer–Emmett–Teller test demonstrating that the surface area of the activated charcoal is about 1,400 $m^2\ g^{-1}$. (E) The SEM image of a cross-section of the activated charcoal electrode. (F) The cross-section photograph of the seaweed separator showing the multilayer structure. (G) Dissolution test of gelatin in simulated gastric fluid. The gelatin becomes indistinguishable from water after 2.5 h. (H) Time evolutions of the strains obtained from experiments and simulations. Reprinted with permission from Wang et al. [70].

different electrodes in the construction of the device) based on an activated wood carbon anode, wood membrane separator, and MnO_2/wood carbon cathode [72]. The structural features of the all-wood-structured device – desirable thickness (up to about 1 mm), direct channels with low tortuosity, high electronic, and ionic conductivity – permitted high areal mass loadings, that is, up to 30 mg cm^{-2} for the anode and 75 mg cm^{-2} for the wood carbon/MnO_2 composite cathode. The mass loading, taken together with an energy density of 1.6 mWh cm^{-2} and a maximum power density of 24 W cm^{-2}, represents the highest mass loading, areal energy/power densities in all reported MnO_2-based supercapacitors (Fig. 12.9).

Recently, Lee et al. have reported materials, design strategies, and applications of biodegradable microsupercapacitors built using water-soluble (i.e., physically transient) metal (W, Fe, and Mo) electrodes, a biopolymer, a hydrogel electrolyte (agarose gel), and a biodegradable poly(lactic-*co*-glycolic acid) substrate, encapsulated with a polyanhydride [73]. Demonstration experiments illustrated potential applications of these biodegradable microsupercapacitors as transient sources of power in the operation of LEDs and as charging capacitors in integrated circuits for wireless power harvesting (Fig. 12.10).

Very recently, Pal et al. reported a silk protein-based biocompatible and almost fully biodegradable thin film microsupercapacitor. A photopatternable biocomposite ink and a protein carrier with the conducting polymer poly(3,4-ethylenedioxythiophene) polystyrene sulfonate doped with reduced graphene oxide were used. The electrodes were printed on flexible protein sheets with an agarose–NaCl gel electrolyte. The average areal capacitance of the devices averaged to 9.85 mF cm^{-2} at a current density of 1 A g^{-1}. A micrometer-sized supercapacitor retained 91% stability after 500 cycles. The cytocompatible device was mostly composed of proteins and polysaccharides that can be in vivo resorbed; however, the conductive polymer component was nonbiodegradable [60].

The analysis of supercapacitor materials and devices reported so far indicates that the design of a fully biodegradable supercapacitor with proper electrochemical characteristics continues to be a challenge, especially for biomedical applications.

12.3 Summary and outlook

To conclude, it is fairly obvious that the disruptive potential of biodegradable or transient power sources is yet to be fully realized. Specifically, it appears that the trade-off between comprehensive biodegradability and appropriate power supplied is unfavorably biased. Biodegradability implies that unstable components deliberately are chosen and that the control of the onset, rate, and timing of events is abandoned. Hence, biodegradability more often than not implicates that the intrinsically available energy is uncontrollably dissipated, making the overall performance of transient electronics relying on biodegradable power sources unreliable and unpredictable. It is important to note that lack of control is a general characteristic, and it befalls all kinds of apparently biodegradable electrochemical cells, fuel cells, batteries, and supercapacitors. Moreover, in terms of transient electronics, the active

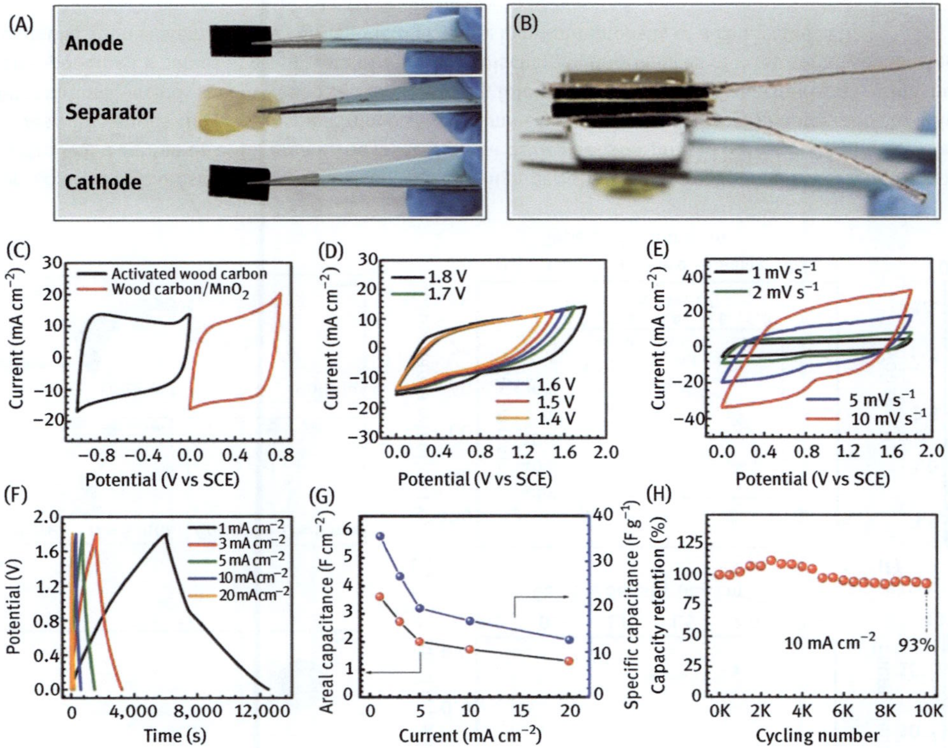

Fig. 12.9: All-wood, low tortuosity, aqueous, biodegradable supercapacitor. (A) Pictures of the activated wood carbon anode, wood separator, and MnO_2@wood carbon cathode. (B) Picture of the all-wood structured all-solid state asymmetric supercapacitor. (C–H) Electrochemical performance of the activated wood carbon//wood separator//MnO_2@wood carbon asymmetric supercapacitor. (C) Typical cyclic voltammograms of the activated wood carbon anode and MnO_2@wood carbon cathode in the −1.0–0.8 V potential range, (D) cyclic voltammograms using different potential ranges, (E) cyclic voltammograms at various scan rates, (F) charge–discharge profiles at different current densities, (G) rate performances, and (H) cycling performance.

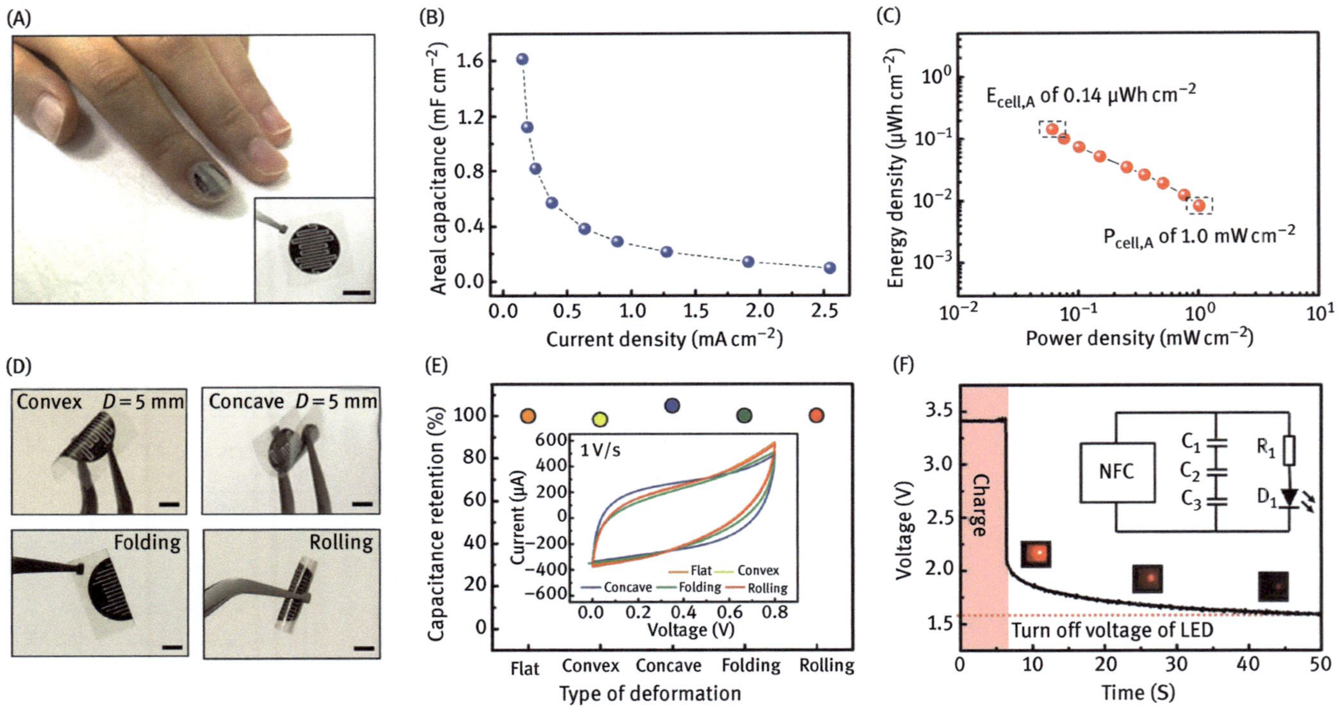

Fig. 12.10: Fully bioresorbable, mechanically flexible high-performance microsupercapacitors. (A) A photograph of a Mo supercapacitor with interdigitated electrodes, mounted on a fingernail. The inset shows a device transfer-printed onto a PLGA film. (B) Areal capacitance calculated from the charge/discharge curves with various current densities from 0.15 to 2.6 mA cm^{-2}. (C) Ragone plot. (D) Photographs of fully biodegradable supercapacitors under various deformations, including convex and concave bending, folding, and rolling. (E) Capacitance retention measured under deformations; the inset shows the corresponding CV curves at a scan rate of 1 V s^{-1}. (F) Discharge performances of three serially connected Mo supercapacitors after charging with a near field communication system and the corresponding change in output intensity of a connected LED.

module, that is, the module that is driven by the power source, also suffers from the same randomness.

To overcome the issues of the associated randomness, which translates into insufficient stability and power generation/storage, as well as to resolve the manufacturability of biodegradable devices will require a concerted effort of (bio)chemistry, engineering, and (bio)materials science, developing new materials, device designs, and manufacturing processes. However, despite these current limitations to transient systems, the pursuit of achieving biodegradable power sources offer the possibility to revolutionize electronics. A biodegradable power supply is an indispensable component to achieve biodegradable electronic systems, which would alleviate waste issues such as landfill build up and environmental hazards caused by the waste. Achieving such systems would have a monumental impact on a vast array of different ubiquitous applications such as wearables, environmental monitoring, active packaging as well as offer bioremediation capabilities, eliminating potential retention of materials for in vivo use, allowing for benign integration into life and environment.

With regard to future developments, the best option may well be to distinguish between the issues that transient electronics aim to correct. The two main areas of concern, that is, environmental friendly consumer electronics and biomedical applications, will require distinctly different methodologies. For consumer electronics the onset problem can be trivially solved since the devise in question is physically accessible; biocompatibility is not critical; and the success of the biodegradation process does not depend on exhaustive disintegration of the construction materials, as long as the ultimate end products are innocuous. Regarding biomedical applications, especially in the case of implanted devises, onset can at best be delayed, and the delay time can only be a more or less rough estimate; since the physiological circumstances are individually settled, the onset delay will vary between individuals. Biocompatibility is essential; degradation needs to be exhaustive, deposits are not tolerated, and soluble components need to be nontoxic. Thus, it is highly unlikely that a common approach will ever succeed; rather the different issues need separate unique solutions.

References

[1] Irimia-Vladu, M., Głowacki, E. D., Voss, G., Bauer, S., & Sariciftci, N. S. Green and biodegradable electronics. Mater Today. 2012, 15, 340–346.

[2] Hwang, S. W., Park, G., Cheng, H., Song, J. K., Kang, S. K., Yin, L., Kim, J. H., Omenetto, F. G., Huang, Y., Lee, K. M., & Rogers, J. A. 25th anniversary article: materials for high-performance biodegradable semiconductor devices. Adv Mater. 2014, 26, 1992–2000.

[3] Feig, V. R., Tran, H., & Bao, Z. Biodegradable polymeric materials in degradable electronic devices. ACS Central Sci. 2018, 4, 337–348.

[4] Hwang, S. W., Tao, H., Kim, D. H., Cheng, H., Song, J. K., Rill, E., Brenckle, M. A., Panilaitis, B., Won, S. M., Kim, Y. S., Song, Y. M., Yu, K. J., Ameen, A., Li, R., Su, Y., Yang, M., Kaplan, D. L., Zakin, M. R., Slepian, M. J., Huang, Y., Omenetto, F. G., & Rogers, J. A. A physically transient form of silicon electronics. Science. 2012, 337, 1640–1644.
[5] Yin, L., Huang, X., Xu, H., Zhang, Y., Lam, J., Cheng, J., & Rogers John, A. Materials, designs, and operational characteristics for fully biodegradable primary batteries. Adv Mater. 2014, 26, 3879–3884.
[6] Fu, K. K., Wang, Z., Dai, J., Carter, M., & Hu, L.Transient electronics: materials and devices. Chem Mater. 2016, 28, 3527–3539.
[7] Hwang, S. W., Huang, X., Seo, J. H., Song, J. K., Kim, S., Hage-Ali, S., Chung, H. J., Tao, H., Omenetto, F. G., Ma, Z., & Rogers, J. A. Materials for bioresorbable radio frequency electronics. Adv Mater. 2013, 25, 3526–3531.
[8] Dagdeviren, C., Hwang, S. W., Su, Y., Kim, S., Cheng, H., Gur, O., Haney, R., Omenetto, F. G., Huang, Y., & Rogers, J. A. Transient, biocompatible electronics and energy harvesters based on ZnO. Smal.l 2013, 9, 3398–3404.
[9] Guo, B., Glavas, L., & Albertsson, A. C. Biodegradable and electrically conducting polymers for biomedical applications. Prog Polym Sci. 2013, 38, 1263–1286.
[10] Irimia-Vladu, M. “Green” electronics: biodegradable and biocompatible materials and devices for sustainable future. Chem Soc Rev. 2014, 43, 588–610.
[11] Li, R., Wang, L., Kong, D., & Yin, L. Recent progress on biodegradable materials and transient electronics. Bioactive Mater. 2017, 3, 322–333.
[12] Bettinger, C. J., & Bao, Z. Organic thin-film transistors fabricated on resorbable biomaterial substrates. Adv Mater. 2010, 22, 651–655.
[13] Irimia-Vladu, M., Troshin, P. A., Reisinger, M., Shmygleva, L., Kanbur, Y., Schwabegger, G., Bodea, M., Schwödiauer, R., Mumyatov, A., Fergus, J. W., Razumov, V. F., Sitter, H., Sariciftci, N. S., & Bauer, S. Biocompatible and biodegradable materials for organic field-effect transistors. Adv Funct Mater. 2010, 20, 4069–4076.
[14] Tobjörk, D., & Österbacka, R. Paper electronics. Adv Mater. 2011, 23, 1935–1961.
[15] Moreno-Merino, L., Jiménez-Hernández, M. E., de la Losa, A., & Huerta-Muñoz, V. Comparative assessment of button cells using a normalized index for potential pollution by heavy metals. Sci Total Environ. 2015, 526, 187–195.
[16] Sadasivuni, K. K., Cabibihan, J., Ponnamma, D., Al-Maadeed, M. A., & Kim, J. eds. Biopolymer composites in electronics. Elsevier, NY, USA, 2016.
[17] Vaghari, H., Jafarizadeh-Malmiri, H., Berenjian, A., & Anarjan, N. Recent advances in application of chitosan in fuel cells. Sustain Chem Processes. 2013, 1, 16.
[18] Liu, H., Qing, H., Li, Z., Han, Y. L., Lin, M., Yang, H., Li, A., Lu, T. J., Li, F., & Xu F., P. A promising material for human-friendly functional wearable electronics. Mater Sci Engineer R. 2017, 112, 1–22.
[19] Du, X., Zhang, Z., Liu, W., & Deng, Y. Nanocellulose-based conductive materials and their emerging applications in energy devices – A review. Nano Energy. 2017, 35, 299–320.
[20] Heinze, J., Frontana-Uribe, B. A., & Ludwigs, S. Electrochemistry of conducting polymers—Persistent models and new concepts. Chem Rev. 2010, 110, 4724–4771.
[21] Sayyar, S., Officer, D. L., & Wallace, G. G. Fabrication of 3D structures from graphene-based biocomposites. J Mater Chem B. 2017, 5, 3462–3482.
[22] Frackowiak, E., & Béguin, F. Carbon materials for the electrochemical storage of energy in capacitors. Carbon. 2001, 39, 937–950.
[23] Aricò, A. S., Bruce, P., Scrosati, B., Tarascon, J. M., & van Schalkwijk, W. Nanostructured materials for advanced energy conversion and storage devices. Nature Mater. 2005, 4, 366.

[24] Pumera, M. Graphene-based nanomaterials for energy storage. Energy Environ Sci. 2011, 4, 668–674.
[25] Bianco, A., Kostarelos, K., & Prato, M. Making carbon nanotubes biocompatible and biodegradable. Chem Commun. 2011, 47, 10182–10188.
[26] Kotchey, G. P., Allen, B. L., Vedala, H., Yanamala, N., Kapralov, A. A., Tyurina, Y. Y., Klein-Seetharaman, J., Kagan, V. E., & Star, A. The enzymatic oxidation of graphene oxide. ACS Nano. 2011, 5, 2098–2108.
[27] Kotchey, G. P., Hasan, S. A., Kapralov, A. A., Ha, S. H., Kim, K., Shvedova, A. A., Kagan, V. E., & Star, A. A natural vanishing act: the enzyme-catalyzed degradation of carbon nanomaterials. Account Chem Res. 2012, 45, 1770–1781.
[28] Kurapati, R., Russier, J., Squillaci, M. A., Treossi, E., Ménard-Moyon, C., Del Rio-Castillo, A. E., Vazquez, E., Samorì, P., Palermo, V., & Bianco, A. Dispersibility-dependent biodegradation of graphene oxide by myeloperoxidase. Small. 2015, 11, 3985–3994.
[29] Falk, M., Narváez Villarrubia, C. W., Babanova, S., Atanassov, P., & Shleev, S. Biofuel cells for biomedical applications: colonizing the animal kingdom. ChemPhysChem. 2013, 14, 2045–2058.
[30] Bandodkar, A. J., & Wang, J. Wearable biofuel cells: a review. Electroanalysis. 2016, 28, 1188–1200.
[31] Davis, F., & Higson, S. P. J. Biofuel cells—Recent advances and applications. Biosens Bioelectron. 2007, 22, 1224–1235.
[32] Esquivel, J. P., Buser, J. R., Lim, C. W., Domínguez, C., Rojas, S., Yager, P., & Sabaté, N. Single-use paper-based hydrogen fuel cells for point-of-care diagnostic applications. J Power Sources. 2017, 342, 442–451.
[33] González-Guerrero, M. J., Del Campo, F. J., Esquivel, J. P., Giroud, F., Minteer, S. D., & Sabaté, N. Paper-based enzymatic microfluidic fuel cell: from a two-stream flow device to a single-stream lateral flow strip. J Power Sources. 2016, 326, 410–416.
[34] Fischer, C., Fraiwan, A., & Choi, S. A 3D paper-based enzymatic fuel cell for self-powered, low-cost glucose monitoring. Biosens Bioelectron. 2016, 79, 193–197.
[35] Mohammadifar, M., Zhang, K., & Choi, S. A saliva-powered paper biobattery for disposable biodevices. In 2017 IEEE 30th International Conference on Micro Electro Mechanical Systems (MEMS). 2017.
[36] Fraiwan, A., Mukherjee, S., Sundermier, S., & Lee H S, C. S. A paper-based microbial fuel cell: instant battery for disposable diagnostic devices. Biosens Bioelectron. 2013, 49, 410–414.
[37] Fraiwan, A., Lee, H., & Choi, S. A multianode paper-based microbial fuel cell: a potential power source for disposable biosensors. IEEE Sensors J. 2014, 14, 3385–3390.
[38] Mohammadifar, M., Zhang, J., Yazgan, I., Sadik, O., & Choi, S. Power-on-paper: origami-inspired fabrication of 3-D microbial fuel cells. Renewable Energy. 2018, 118, 695–700.
[39] Chouler, J., Á, C. I., Rengaraj, S., Scott, J. L., & Di Lorenzo, M. A screen-printed paper microbial fuel cell biosensor for detection of toxic compounds in water. Biosens Bioelectron. 2018, 102, 49–56.
[40] Winfield, J., Chambers, L. D., Rossiter, J., Stinchcombe, A., Walter, X. A., Greenman, J., & Ieropoulos, I. Fade to green: a biodegradable stack of microbial fuel cells. ChemSusChem. 2015, 8, 2705–2712.
[41] Tsang, M., Armutlulu, A., Herrault, F., Sharfer, R. H., Allen, S. A. B., & Allen, M. G. Development of electroplated magnesium microstructures for biodegradable devices and energy sources. J Microelectromech Syst. 2014, 23, 1281–1289.
[42] Tsang, M., Armutlulu, A., Martinez, A. W., Allen, S. A. B., & Allen, M. G. Biodegradable magnesium/iron batteries with polycaprolactone encapsulation: a microfabricated power source for transient implantable devices. Microsyst Nanoeng. 2015, 1, 15024.

[43] Tsang, M., Armutlulu, A., Martinez, A., Herrault, F., Allen, S. A. B., & Allen, M. G. A MEMS-enabled biodegradable battery for powering transient implantable devices. IEEE 27th International Conference on Micro Electro Mechanical Systems (MEMS). 2014, January 26–30, IEEE Xplore, DOI: 10.1109/MEMSYS.2014.6765650.

[44] Jimbo, H., & Miki, N. Gastric-fluid-utilizing micro battery for micro medical devices. Sens Actuators B. 2008, 134, 219–224.

[45] Kim, Y. J., Wu, W., Chun, S. E., Whitacre, J. F., & Bettinger, C. J. Biologically derived melanin electrodes in aqueous sodium-ion energy storage devices. Proc Natl Acad Sci USA. 2013, 110, S/1-S/8, 20912–20917.

[46] Fu, K., Liu, Z., Yao, Y., Wang, Z., Zhao, B., Luo, W., Dai, J., Lacey, S. D., Zhou, L., Shen, F., Kim, M., Swafford, L., Sengupta, L., & Hu, L. Transient rechargeable batteries triggered by cascade reactions. Nano Letters. 2015, 15, 4664–4671.

[47] Wei, X., & Liu, J. Power sources and electrical recharging strategies for implantable medical devices. Frontiers Energy Power Engineer China. 2008, 2, 1–13.

[48] Edupuganti, V., & Solanki, R. Fabrication, characterization, and modeling of a biodegradable battery for transient electronics. J Power Sources. 2016, 336, 447–454.

[49] Jia, X., Wang, C., Zhao, C., Ge, Y., & Wallace, G. G. Toward biodegradable Mg-Air bioelectric batteries composed of silk fibroin-polypyrrole film. Adv Funct Mater. 2016, 26, 1454–1462.

[50] Jia, X., Wang, C., Ranganathan, V., Napier, B., Yu, C., Chao, Y., Forsyth, M., Omenetto, F. G., MacFarlane, D. R., & Wallace, G. G. A biodegradable thin-film magnesium primary battery using silk fibroin-ionic liquid polymer electrolyte. ACS Energy Lett. 2017, 2, 831–836.

[51] Dougal, R. A., Liu, S., & White, R. E. Power and life extension of battery-ultracapacitor hybrids. IEEE Transactions Components Pack Technol. 2002, 25, 120–131.

[52] Mastragostino, M., Arbizzani, C., & Soavi, F. Polymer-based super-capacitors. J Power Sources. 2001, 97–98, 812–815.

[53] Huggins, R. A. Supercapacitors and electrochemical pulse sources. Solid State Ionics. 2000, 134, 179–195.

[54] Zhao, X., Sanchez, B. M., Dobson, P. J., & Grant, P. S. The role of nanomaterials in redox-based supercapacitors for next generation energy storage devices. Nanoscale. 2011, 3, 839–855.

[55] Pech, D., Brunet, M., Durou, H., Huang, P., Mochalin, V., Gogotsi, Y., P-L, T., & Simon, P. Ultrahigh-power micrometre-sized supercapacitors based on onion-like carbon. Nat Nanotechnol. 2010, 5, 651–654.

[56] Dong, L., Xu, C., Li, Y., Huang, Z. H., Kang, F., & Yang Q-H, Z. X. Flexible electrodes and supercapacitors for wearable energy storage: a review by category. J Mater Chem A. 2016, 4, 4659–4685.

[57] Wang, G., Zhang, L., & Zhang, J. A review of electrode materials for electrochemical supercapacitors. Chem Soc Rev. 2012, 41, 797–828.

[58] Zheng, Y., Yang, Y., Chen, S., & Yuan, Q. Smart, stretchable and wearable supercapacitors: prospects and challenges. CrystEngComm. 2016, 18, 4218–4235.

[59] Zequine, C., Ranaweera, C. K., Wang, Z., Dvornic, P. R., Kahol, P. K., Singh, S., Tripathi, P., Srivastava, O. N., Singh, S., Gupta, B. K., Gupta, G., & Gupta, R. K. High-performance flexible supercapacitors obtained via recycled jute: bio-waste to energy storage approach. Sci Rep. 2017, 7, 1–12.

[60] Pal, R., Kundu, S. C., & Yadavalli, V. K. Fabrication of flexible, fully organic, degradable energy storage devices using silk proteins. ACS Appl Mater Interfaces. 2018, 10, 9620–9628.

[61] Chandra, R., & Rustgi, R. Biodegradable polymers. Prog Polym Sci. 1998, 23, 1273–1335.

[62] Bhat, D. K., & Kumar, M. S. Biodegradability of PMMA blends with some cellulose derivatives. J Polym Environ. 2006, 14, 385–392.

[63] Selvakumar, M., & Bhat, D. K. LiClO4 doped cellulose acetate as biodegradable polymer electrolyte for supercapacitors. J Appl Polym Sci. 2008, 110, 594–602.
[64] Moon, W. G., Kim, G. P., Lee, M., Song, H. D., & Yi, J. A biodegradable gel electrolyte for use in high-performance flexible supercapacitors. ACS Appl Mater Interfaces. 2015, 7, 3503–3511.
[65] Htut, K. Z., Kim, M., Lee, E., Lee, G., Baeck, S. H., & Shim, S. E. Biodegradable polymer-modified graphene/polyaniline electrodes for supercapacitors. Synth Met. 2017, 227, 61–70.
[66] Kang, S. K., Hwang, S. W., Yu, S., Seo, J. H., Corbin, E. A., Shin, J., Wie, D. S., Bashir, R., Ma, Z., & Rogers, J. A. Biodegradable thin metal foils and spin-on glass materials for transient electronics. Adv Funct Mater. 2015, 25, 1789–1797.
[67] Altman, G. H., Diaz, F., Jakuba, C., Calabro, T., Horan, R. L., Chen, J., Lu, H., Richmond, J., & Kaplan, D. L. Silk-based biomaterials. Biomaterials. 2003, 24, 401–416.
[68] Pal, R. K., Farghaly, A. A., Wang, C., Collinson, M. M., Kundu, S. C., & Yadavalli, V. K. Conducting polymer-silk biocomposites for flexible and biodegradable electrochemical sensors. Biosens Bioelectron. 2016, 81, 294–302.
[69] Kim, D. H., Viventi, J., Amsden, J. J., Xiao, J., Vigeland, L., Kim, Y. S., Blanco, J. A., Panilaitis, B., Frechette, E. S., Contreras, D., Kaplan, D. L., Omenetto, F. G., Huang, Y., Hwang, K. C., Zakin, M. R., Litt, B., & Rogers, J. A. Dissolvable films of silk fibroin for ultrathin conformal bio-integrated electronics. Nat Mater. 2010, 9, 511–517.
[70] Wang, X., Xu, W., Chatterjee, P., Lv, C., Popovich, J., Song, Z., Dai, L., Kalani, M. Y. S., Haydel, S. E., & Jiang, H. Food-materials-based edible supercapacitors. Adv Mater Technol. 2016, 1, DOI: 10.1002/admt.201600059.
[71] Kumar, P., Di Mauro, E., Zhang, S., Pezzella, A., Soavi, F., Santato, C., & Cicoira, F. Melanin-based flexible supercapacitors. J Mater Chem C. 2016, 4, 9516–9525.
[72] Chen, C., Zhang, Y., Li, Y., Dai, J., Song, J., Yao, Y., Gong, Y., Kierzewski, I., Xie, J., & Hu, L. All-wood, low tortuosity, aqueous, biodegradable supercapacitors with ultra-high capacitance. Energy Environ Sci. 2017, 10, 538–545.
[73] Lee, G., Kang, S. K., Won, S. M., Gutruf, P., Jeong, Y. R., Koo, J., Lee, S. S., Rogers, J. A., & Ha, J. S. Fully biodegradable microsupercapacitor for power storage in transient electronics. Adv Energy Mater. 2017, 7, DOI: 10.1002/aenm.201700157.

Timothé Philippon, Thomas Flinois, Estelle Lebègue,
Nazua L. Costa, Frédéric Barrière, Joanna Rogińska,
Mathieu Etienne

13 Current trends for water treatment with microbial electrodes

13.1 Introduction

Microbial electrodes are of interest to the electrochemists because the electrode catalyst is alive, it self-connects to the electrode, it can oxidize or reduce a large variety of substrates, it is stable in time (thanks to the self-replication), and it is three-dimensional as the microbes organize as thick catalytic biofilms on the electrode surface. To the microbiologists, however, these properties may simply be viewed as a special case of cellular respiration and metabolism. One instinctively relates respiration to the breathing of air; oxygen is indeed the terminal electron acceptor in the respiratory chain of mitochondria as it is in the metabolism of aerobic bacteria. Electroactive bacteria, however, directly use solid extracellular electron donors or acceptors like iron oxides for respiration. With appropriate selective pressure and conditions, electroactive bacteria develop on solid electrodes and connect their metabolism to them. Their catalytic properties can then be studied and exploited by both electrochemists and biotechnologists. One potential application of these catalytic microbial electrodes is waste treatment coupled to the recovery of electricity in the so-called microbial fuel cell (MFC) configuration. Nevertheless, several technological locks have to be lifted before a microbial bioelectrochemical system can be scaled-up and efficiently implemented in industry. This chapter covers the fundamentals of bacteria-electrode electron transfer, and the current approaches under development for the future applications of microbial bioelectrochemical system to wastewater treatment and pollution bioremediation.

13.2 Fundamentals of the microbe-electrode electron transfer

Environmentally ubiquitous microorganisms, from soil, municipal waters, biocorroded surfaces, or heavy metal-contaminated environments, can be used as a living catalyst to produce electrical current, hydrogen, and several added-value

Timothé Philippon, Thomas Flinois, Estelle Lebègue, Nazua L. Costa, Frédéric Barrière, Univ Rennes, CNRS, ISCR (Institut des Sciences Chimiques de Rennes) - UMR 6226, F-35000, Rennes, France
Joanna Rogińska, Mathieu Etienne, Université de Lorraine, CNRS, LCPME, F-54000, Nancy, France

https://doi.org/10.1515/9783110570526-013

compounds. Electroactive bacteria hold a great potential to be used in a plethora of microbial electrochemical technologies with different architectures and characteristics due to their ability to act as catalysts in the conversion of the chemical energy stored in reduced organic compounds and in oxidized inorganic species [1].

Currently, more than 90 species of electroactive bacteria have been isolated from natural environment or in MFC consortia [2]. Most electroactive organisms are considered as dissimilatory metal-reducing bacteria since they are able to reduce insoluble metal oxides and high oxidation state metal ions such as Fe (III), Mn (IV), Cr (VI), U (VI), Tc (VII), Co (III), Mo (VI), Hg (II), and Au (I, III) [3] while performing their anaerobic respiration. The dissimilatory metal-reducing ability was originally discovered in bacteria belonging to Geobacter and Shewanella spp [4, 5]. and still to date *Geobacter sulfurreducens* PCA and *Shewanella oneidensis* MR1 are the most well-characterized dissimilatory metal-reducing bacteria. Because of their ability to efficiently transfer electrons to the anode of an MFC with high faradaic efficiency, they became the model organisms for investigating extracellular electron transfer process in microbial electrochemical technologies. From the extensive studies on these electroactive model organisms, two general mechanisms for extracellular electron transfer were identified: direct electron transfer and indirect electron transfer (Fig. 13.1A).

In direct electron transfer, the bacteria establish a direct contact with insoluble substrates or electrodes via outer membrane *c*-type cytochromes (*c*-Cyts) or through cellular appendages such as pili or nanowires that were shown to be also associated with *c*-Cyts. In indirect electron transfer, bacteria use exogenous or endogenous soluble redox-active compounds as electron shuttles to mediate electron transfer between surface-exposed proteins and insoluble electron acceptors (Fig. 13.1B) [6].

Both mechanisms are linked to the presence of redox-active multiheme *c*-Cyts that are key players in extracellular electron transfer process [7]. Multiheme *c*-Cyts are involved in the electron transport from the inner membrane, through the periplasmic space, toward the outer membrane and finally to the extracellular terminal electron acceptors. Note that the potential of c-type cytochrome can span a large range of almost 1 V from ca. −0.4 V versus standard hydrogen electrode (SHE) for acetate-oxidizing microbial bioanodes [8] to almost +0.6 V versus SHE in *Acidithiobacillus ferrooxidans*, a candidate for microbial cathodes [9].

Thus, multiheme *c*-Cyts seem remarkably suitable to ensure efficient downstream relay and storage of electrons for both reductive and oxidative processes. In fact, 111 genes coding for *c*-Cyts are found in *G. sulfurreducens* PCA and 72 genes code for proteins with more than two heme groups [10]. From the 42 genes that code

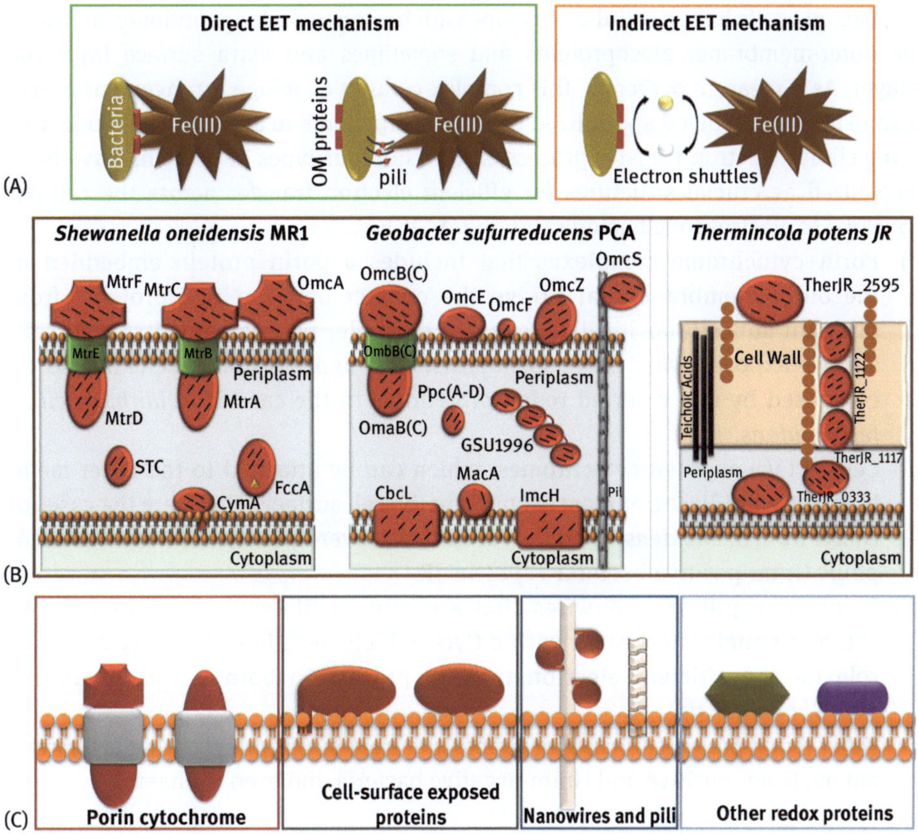

Fig. 13.1: General mechanisms for bacteria-electrode interaction. (A) Extracellular electron transfer mechanisms; (B) schematic representation of the extracellular electron transfer pathways in Gram-negative (Shewanella and Geobacter) and Gram-positive bacteria (Thermincola potens JR; from Costa [13]). (C) Representation of the four types of cell-surface proteins (from Costa et al. [14]).

for *c*-Cyts in *S. oneidensis* MR1, 33 have more than one heme group and are known to participate in the various respiratory pathways of this bacterium, which is linked to its wide variety of terminal electron acceptors [11].

Among the various proteins involved in the extracellular electron transfer process of electroactive bacteria, the cell-surface-exposed cytochromes are those that play a crucial role in this process being responsible for both direct and indirect electron transfers [12].

Due to their subcellular localization, at the bacteria–terminal acceptor interface, these proteins are the last physiological redox relay between the microbial metabolism and the solid metal oxide surfaces or the electrodes.

Bacteria cells have a cellular envelope which often includes peptidoglycan walls, the outer-membrane, glycoproteins and sometimes and extra surface layer the s-layer. As a mean to overcome this complex cellular envelope arrangement microorganisms have evolved specialized cellular components to ensure the efficiency of extracellular electron transfer processes. To date, four types of proteins have been recognized as crucial structures for efficient electron transfer across the cellular envelope of electroactive bacteria [14] (Fig. 13.1C):

(i) Porin–cytochrome complexes that includes a porin protein embedded in the outer membrane that bridge the contact between two proteins from different subcellular localizations, for example, periplasmic MtrA and cell-surface-exposed MtrC in *S. oneidensis*. This type of complex can also be composed by one inserted redox protein, as in the case of *Acidithiobacillus ferrooxidans*.

(ii) Cell-surface-exposed cytochromes, which can be attached to the outer membrane by a lipidic tag or loosely bound to the cell surface. Those are the cases of OmcA from *S. oneidensis* and the recently discovered TherJR2595 from thermophilic Gram-positive bacteria *T. potens* JR.

(iii) Conductive pili or nanowires that are protein filaments anchored to the cell and usually decorated with *c*-Cyts, which were found to play a crucial role for the efficient electron transfer process in both *G. sulfurreducens* and *S. oneidensis*.

(iv) Other redox proteins, including copper and iron–sulfur proteins that are spread among Gram-positive and Gram-negative bacteria and even Archaea.

Besides being a transversal feature to most microorganisms that bares an outer membrane, electroactivity has now being observed in some Gram-positive bacteria, Archaea, microalgae, and even fungi [2, 15]. The outer-membrane electroactivity of these bacteria is the main component of the MFC and related technologies – not only at the bioanode where organic compounds may be oxidized, but also at the less well-characterized and understood biocathode where inorganic compounds may be reduced.

13.3 Principle of MFCs for waste water treatment

An MFC is composed of at least one microbial electrode, either a bioanode or a biocathode, or both biocathode and bioanode. Most conventional MFCs consist of a biotic anode and an abiotic cathode in two separated compartments filled with electrolyte medium, separated by ion-exchange membrane or salt bridge (Fig. 13.2). The electrochemically active biofilm grows on the anode and it is

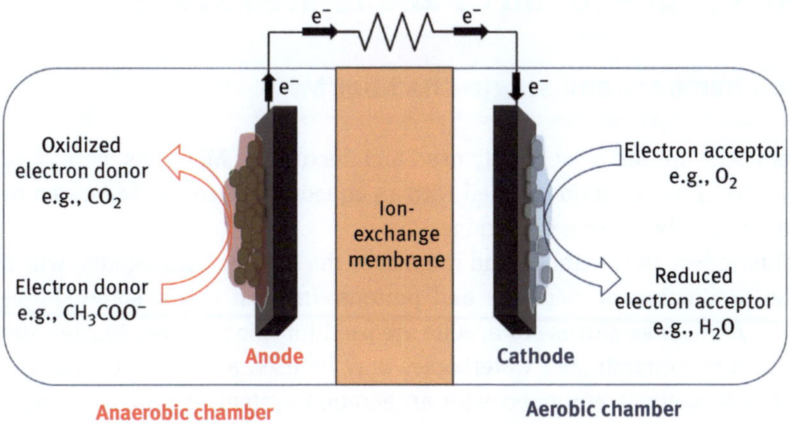

Fig. 13.2: A schematic diagram of a two-chamber MFC with anaerobic anodic chamber and aerobic cathodic chamber. In this case, the anode and cathode are both colonized by an electroactive biofilm (biotic), but the cathode can also be abiotic.

able to convert various biodegradable organic substances into electrons, CO_2, and protons [16].

The electrons are then transported to the cathode via the external circuit and ions are transferred through the ion-exchange membrane. Simultaneously, in the cathodic chamber, a reduction process occurs. The final electron acceptor can be oxygen – either dissolved in water or taken directly from air if an air cathode is used. In those cases, the cathodic compartments are aerobic, while the anodic chamber should be devoid of any electron acceptors like oxygen to ensure the transport of the electrons from bacteria to the anode. Dioxygen is not the only electron acceptor used in a MFC; ferrocyanide and nitrate, for example, can be used as a cathode electron acceptor.

The main advantages of using MFC instead of typically used activated sludge or trickling filters are as follows: the production of a useful product or electricity recovery during the treatment, the lack of energy-consuming aeration (in air-cathode MFC), the lower bacterial biomass production compared to conventional systems, and also potentially odor control [16]. In addition, the reactions at both the cathode and the anode can be used for water treatment: anodic oxidation of organic compounds found in the wastewater from various sources mostly in their reduced form and cathodic reduction of pollutants such as nitrogen compounds or metals. Several configurations have been proposed in the literature and are described in the following section.

13.4 Cell designs for MFCs and related systems

13.4.1 Two-chambers and single-chamber MFC

Two-chambers cells are the most widely used architecture for MFCs. As mentioned earlier, it consists of two containers filled with an aqueous electrolyte, separated by an ion-exchange membrane or a salt bridge.

The biofilm grows on the anode and it oxidizes the organic compounds, which results in the production of electrons and protons. In most cases, some simple organic chemicals such as glucose or acetate are used for this purpose. Wastewater taken directly from contaminated waterbodies may be used as well. The cathodic compartment is sometimes equipped with an aeration system to provide enough dissolved oxygen [17].

Typically, ion-exchange membranes are used to separate the anodic and cathodic compartments, which allow the transport of ions between compartments and prevent the transport of other species. The key parameters for those membranes are as follows: resistance, price, and proton-exchange capabilities because significant difference between pH of the two chambers leads to lower efficiency. In addition, the membrane must not be permeable for other compounds, especially for oxygen as its presence in anodic chamber would limit the energy production.

The membrane should also have antimicrobial or antiadhesion properties to prevent (or delay) an adverse biofilm forming on the anodic side of its surface [18, 19].

The most commonly used cation-exchange membranes are Nafion® [20] and Ultrex® [21]. Anion-exchange membranes can be also used in MFCs. In this case, some compounds present in anode chamber, such as carbonates or phosphates (used as buffers), act like proton carriers [22]. It has been proved that the use of anion-exchange membranes leads to the smaller difference of pH on the two sides of membrane. However, anion-exchange membranes might be less selective and allow for the transfer of electron donor compounds, for example, acetate [23].

Because of their complex structure, the scaling-up and practical application of two-chambers MFCs is difficult to perform. Therefore, single-chamber MFCs have been proposed. They usually consist of only the anodic chamber, which is linked to the porous cathode exposed to the air, and so oxygen gas may be captured and used for the reduction. With this system, there is no need for energy-costly aeration. The cathode can be separated from the electrolyte by a cation-exchange membrane.

The most typical single-chamber MFC reactor, designed by Park and Zeikus, is in the shape of cuboid, with a cathode on one of the walls [24]. Another example from

Liu et al. is a cylindrical reactor with a single air cathode (also of cylindrical shape, with air flow inside) and eight graphite rod anodes placed around [25].

13.4.2 Sediment MFC and microbial electrochemical snorkel

Sediment MFCs, also known as benthic MFCs, are membraneless MFCs consisting of an anode buried in the sediment and a cathode placed above in overlying water, usually connected by the external circuit attached to data-measuring apparatus.

The electricity generation is based on the natural potential difference between the anoxic sediment and aerobic water above. In this system, the biofilm on the anode is oxidizing organic matter in sediment and the produced electrons are passed to the anode, and so the current is generated. The electrons are then transferred to the cathode and used to reduce dissolved oxygen [26]. Moreover, numerous configurations have been proposed, such as using multiple anodes instead of one or multiple anode–cathode pairs [27].

As mentioned earlier, MFCs might be used for both energy generation and wastewater treatment. If the main aim is to produce energy, one must choose the optimum current and resistance values to achieve the highest power. However, if the role of MFC is only water purification, it is more beneficial to reach the highest current. This may be achieved by using a short-circuited system. In such a case, voltage is limited, and power is decreased. The rate of organic compounds oxidation can however be more efficient.

In 2011, Bergel and coworkers introduced the concept of the microbial electrochemical snorkel [28]. The idea was to introduce the conductive material – a snorkel electrode – to sediment MFC and allow the development of biofilms at the sediment level and the water level. In the anaerobic zone, the microbial biofilm oxidizes the organic matter and transfers the electrons to the snorkel.

The transport of these electrons to the aerobic zone is then facilitated by the snorkel so they can be used by cathodic biofilms for the reduction process. Similarly for dual-chamber MFC, the electron acceptor is usually oxygen. The whole concept is shown in Fig. 13.3.

Both the microbial electrochemical snorkel and the short-circuited MFC showed better performance in removing chemical oxygen demand than an MFC with applied resistance. Microbial electrochemical snorkel can be used for bioremediation of crude oil-contaminated marine sediments [29-31] and was also proposed for nitrate removal [32]; (*Cf.* Section 13.6.2). A simple technology consisting of carbon felt anode in the sediment linked to an iron rod in the water column significantly increased the rate of denitrification. The following section will focus on electrodes and possible electrode modifications to improve MFCs.

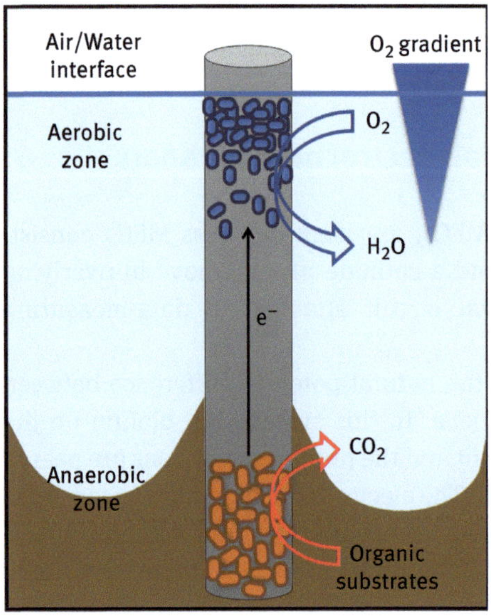

Fig. 13.3: A schematic representation of microbial electrochemical snorkel (from Erable et al. [28]). Near the surface, dioxygen is reduced by an electroactive biofilm (in blue); an abiotic cathodic system may also be used.

13.5 Electrode materials, surface modifications, and biocomposites

Generally, electrodes used in fuel cells should have good electrical conductivity, chemical and mechanical stability as well as anticorrosion properties, and low cost. Electrodes that are meant to be covered with the biofilm must additionally have features such as high surface roughness, strong biocompatibility, and efficient electron transfer between bacteria and electrode surface. It is possible to modulate the composition of the electrode to reduce the energy of activation or improve the biofilm formation onto the electrode. Modulation of the electrode by modifications or the use of composite is also possible.

13.5.1 Electrode materials

In MFCs at least one electrode reaction is catalyzed by microorganisms that are supported onto electrodes acting as an electron donor (cathode) or an electron acceptor (anode). Hence, the electrode materials and its surface properties are the key factors in the performance and cost of MFCs [33, 34].

The most common cathode materials are based on carbon (graphite, carbon cloth, carbon paper, etc.) [35]. Improving cathode performance involves modification of the electrode with a catalyst that reduces the cathodic reaction activation energy and increases the reaction rate.

The often chosen catalyst is platinum because it offers high catalytic performance; however, its availability is limited [36], it is expensive (value in March 2018 > 2,000 € kg^{-1}), its surface can be easily poisoned in the natural conditions, and its production results in a strong environmental impact [37]. This is why efforts have been made to lower its amount or to use non-Pt catalysts such as metal oxides (manganese, rutile oxides, and PbO_2) and metal complexes (phthalocyanine and porphyrin derivatives) [35].

The microbial catalysis mechanism is currently more often studied at the anode than at the cathode [38, 39]. Therefore, several anode material modification techniques have been investigated to improve the bacteria–electrode interfacial interactions and, thus, to increase MFC performances. Carbon-based materials (graphite, carbon cloth, carbon paper, carbon felt, reticulated vitreous carbon, and glassy carbon) are the most used anodes in the MFC studies, although the interest for metal (stainless steel, platinum, gold, titanium, copper nickel, and silver) and composite (carbon + metal) materials is increasing [34, 35, 38].

13.5.2 Surface modification

Current and power densities can be considerably enhanced by anode surface modifications, which can be classified into surface functionalization and thin film coating [31]. More precisely, modification techniques include surface treatments with physical or chemical methods, immobilization of conductive and electroactive species, and depositions of metals or metal seeds [30].

The modification of the electrode materials' surface roughness and porosity by thermal treatment enhances the biofilm density and development, probably because of the addition of hydrophilic functional groups and/or positive charges on the electrode surface depending on the used gas atmosphere (nitrogen, oxygen, and ammonia) [38, 40-42].

As recently demonstrated, what really impacts the attachment or current production of mixed culture anodic biofilms on carbon are rather the changes in surface chemistry than those in surface roughness (topography) [42]. The introduction of functional groups containing nitrogen and oxygen aims to improve cell attachment and biofilm growth on electrode surface [33, 41, 43]. Indeed, surface charge, surface hydrophobicity (wettability), and surface affinity are key factors that directly affect the anodic biofilm formation, its community composition, and thus the current generation in MFCs [44].

Positively charged and hydrophilic surfaces are more selective to microbes (e.g., *Geobacter sulfurreducens*) and more conductive for electroactive biofilm formation [39, 41, 45]. In the case of electrode surface functionalization with a self-assembled monolayer, the redox behavior of *Geobacter sulfurreducens* is strongly dependent on the terminal group that directly interacts with outer-membrane *c*-type cytochrome redox proteins [41, 44].

For example, the carboxyl-anchoring group provides biocompatible conditions for the outer-membrane *c*-type cytochrome of *Geobacter sulfurreducens*, which facilitate the extracellular electron transfer at the microorganism/electrode interface[44]. Other anode modification methodologies including oxidation of the electrode surface by using acid, electrochemical methods, UV/O_3 or plasma treatment, incorporation of quinone entities or redox/conductive polymers, and the use of carbon nanotubes have been successful in increasing the current and power densities of MFCs [33, 39, 43, 46].

An attractive chemical surface modification method for the anode materials functionalization is the grafting by reduction of aryl diazonium salts [36]. Phenylboronic acid (targeting sugars on the outer membrane) grafted on MFCs graphite anodes by electrochemical reduction of aryl diazonium salts allows a faster connection of electroactive biofilms on the electrodes and leads to higher MFC performances [47]. A recent study showed that MFCs with pyridine-functionalized graphite anodes also exhibit faster development and improved performances [48].

In all cases of electrode surface modification, the increase of MFC current and power is due to a combination of several factors including an enhanced electrical conductivity, an increased surface area, and an increased biocompatibility of the electrode surface [49]. Hence, understanding the correlation between electrode performances and its surface chemistry is crucial for controlling the electrode materials functionalization and then for continuing to improve the MFC performances, especially for water treatment.

13.5.3 Biocomposite electrodes

A biocomposite electrode, also called electroactive artificial biofilm, is a recent theme of research. The principle is to associate a material that is not produced by the microorganism with bacteria in a biofilm-like architecture. These exogenous elements can be polymeric materials, carbon nanomaterials, or proteins that may participate to the immobilization of bacteria and to extracellular electron transfer reactions [50].

Three different strategies are proposed in the literature for the implementation of these biocomposite electrodes:

(i) The first strategy involves the growth of bacteria on the electrode surface and the deposition of the polymer, either silica [51] or polypyrrole [52], on a biofilm to improve extracellular electrode transfer [52] or to provide a standardized basis for MFCs [51].
(ii) The second strategy involves the growth of bacteria in the presence of carbon nanomaterials [53, 54] or mineral nanoparticles [55, 56]. For example, Yong et al.

took advantage of the reduction of graphene oxide by *Shewanella oneidensis* to produce a biocomposite system allowing high current density either at the anode or at the cathode.

(iii) The last strategy involves either the preparation of the biocomposite from planktonic bacteria by the immobilization of bacteria in a polymeric matrix (silica [57, 58] or redox polymers [58, 59]) or the self-assembly of planktonic bacteria with carbon nanotubes and cytochromes [60, 61].

Biocomposite electrodes allow a rapid analysis of the electroactivity of selected or engineered bacterial strains, for fundamental purposes, for example, for evaluating the effect of gene deletion on extracellular electron transfer [61]. They also allow to find an application in ready-to-use bioelectrodes and biosensors [57].

13.6 Microbial biocathode-catalyzed reaction of relevance to water treatment

The anode of an MFC can oxidize organic compounds. Electrons are then delivered through the external circuit to the cathode and to the micro-organisms of cathodic biofilms. Bioanodes are used to purify water from the organic matter and cathodic microorganisms can treat inorganic contaminants. Herein, we summarize recent research in biocathodes for oxygen reduction, nitrate removal, metal recovery, and hydrogen electrosynthesis.

13.6.1 Oxygen microbial biocathode

To generate a high potential difference, it is necessary to reduce a strong oxidant. Dioxygen is attractive for that purpose, because it is a strong oxidizer and is naturally available in large quantities. From the potential values of the bioanodic oxidation of the organic compounds such as acetate ($E^{0'} = -0.38$ V vs SHE at pH 7) and the cathodic reduction of oxygen ($E^{0'} = +0.81$ V at pH 7), one can conclude that it is theoretically possible to generate a potential difference of 1.2 V while purifying the wastewater of the organic matter (oxidation of acetate to carbon dioxide).

However, the kinetics of the reduction of molecular oxygen to water (eq. (13.1)) is slow on carbon electrodes; its use in MFC technology is therefore dependent on a catalyst deposited on the electrode to accelerate the reaction rate.

$$O_2 + 2H_2O + 4e^- \rightarrow 4OH^- \, (E^{0'} = +0.81\,\text{V vs SHE at pH 7}) \quad (13.1)$$

There are a few possible choices for a good catalyst for the oxygen reduction. One of them is platinum, which is highly active and is often used a catalyst. There are some disadvantages of using it though and they have been discussed in Section 13.4.1.

Microorganisms have developed effective oxidoreductases (enzymes) able to reduce oxygen and these enzymes can be extracted and used directly as reduction catalysts on the cathode. However, these oxidoreductases are very sensitive to inhibition and denaturation, they often require complex chemical operations to be immobilized on electrodes, and their lifespan generally does not exceed a few days under operation.

Microbial catalysis is more suitable for long-term production and the microorganisms have various mechanisms for reducing oxygen. Unlike platinum, they are inexpensive, readily available and in addition they are capable of self-regeneration thanks to cell division. Biofilms that connect directly or indirectly to the cathode surface must be capable of using solid inorganic compounds as electron donors. These features may explain why the majority of biocathode studies have focused on reducing dioxygen.

The first example of an MFC capable of reducing dioxygen to water consisted of an abiotic anode and a stainless steel plate as the cathode [37]. A marine aerobic biofilm covered the cathode and could catalyze the reduction of oxygen. A number of subsequent reports highlighted a variety of *inocula* (e.g., marine water, wastewater, sediments, and pure cultures) and methods used to establish a cathodic biofilm that implies a diversity of communities and mechanisms.

Different methods have been described for designing an aerobic biocathode:

(i) Developing a biocathode on electrodes left at open circuit or connected to an anode with high resistance (>1 kΩ) [37, 62–66] and relying on the natural ability of microorganism to exchange electrons and to develop on solids.
(ii) By polarizing the electrode at a potential lower than the open circuit potential. Currently this method is widely used with *inocula* from sewage sludge, wastewater, industrial/agricultural wastes, seawater, sediment, and so on [67–75].
(iii) Polarity reversal of already established bioanodes. An acetate-fed bioanode can be transformed into aerobic cathodic biofilm after consecutive exhaustion of acetate and supply of dioxygen. The biomass can be developed during anodic phase and the cathodic phase allows a better control of pH, which helps retaining the MFC performance during long-term operation [76, 77].

The cathode is usually the electrode that limits the overall performance of many MFCs and represents a significant portion of the total costs [78]. To improve aerobic biocathodes, it is necessary to develop the interactions between air (oxygen and gas diffusion), water (proton diffusion to biofilm and species enabling development of a cathodic biofilm), and solid (material of electrode permitting the establishment of a biofilm and power recovery) [78–83].

Studies on these three interactions with the biofilm led to the production of different aerobic biocathode devices [84]. These different *inocula*, methods, and devices can explain why a wide range of microbial groups has been isolated from the cathode biofilms. These biocathodes have various onset potentials (E_{onset}) for the oxygen reduction ranging from −0.05 V to +0.6 vs SHE. The highest onset potential found for oxygen reduction by a biocathode is around +0.6 V [65]. The biofilm was obtained from sediment as the source of bacteria and the biocathode was connected to an abiotic anode oxidizing ferrocyanide.

A wide range of bacterial species belonging to the *Alphaproteobacteria, Betaproteobacteria, Gammaproteobacteria, Bacteroidetes*, and other less well-known taxonomic classes have been identified as dominant in mixed-community aerobic biocathodes [85]. The high diversity of bacteria isolated on biocathodes does not allow the identification of model organisms. Without model organisms, in-depth studies of the metabolism and mechanisms of electronic transfer taking place in oxygen biocathode are hardly feasible.

When a pure culture of bacteria is used for the establishment of a cathodic biofilm [86], the performances obtained are much lower than those of mixed biofilms, suggesting that the reduction of dioxygen is linked to metabolic and/or physiologic associations between several microbes in mixed biofilms.

Although associations of microbial metabolisms are essential in the catalysis of the reduction of oxygen, they are not yet fully understood. However, the mechanisms of electron transfer from cathode to bacteria are probably very similar to the mechanisms of electron transfer from bacteria to an anode, which was explained in the first section.

Apart from oxygen reduction, a great value of MFC is the possibility given to the microorganisms to use many other electron acceptors including nitrogenous species such as nitrate or metallic species, as discussed in the following sections.

13.6.2 Nitrate microbial biocathode

Since synthetic nitrogen fertilizers are commonly used in agriculture, the rivers and groundwater are often contaminated with nitrates. This may cause eutrophication and decreases the water quality. According to European Union regulations, the water intended for human consumption may contain max. 50 mg L^{-1} of nitrate and 0.5 mg L^{-1} of nitrite, and satisfy the following condition: [nitrate]/50 + [nitrite]/3 ≤ 1 mg L^{-1} [87]. However, water with >25 mg L^{-1} should already undergo purification [88].

Nitrate removal with the use of MFC is an interesting idea because of its low cost, no need for power input, possibly high efficiency, and energy-producing ability. Nitrate can be reduced at the cathode in the anaerobic environment, as shown in the following reactions:

$$NO_3^- + 2e^- + 2H^+ \rightarrow NO_2^- + H_2O (E^{0'} = +0.433 V \text{ vs SHE at pH 7}) \quad (13.2)$$

$$NO_2^- + e^- + 2H^+ \rightarrow NO + H_2O(E^{0'} = +0.350V \text{ vs SHE at pH } 7) \quad (13.3)$$

$$NO + e^- + H^+ \rightarrow \tfrac{1}{2}N_2O + \tfrac{1}{2}H_2O(E^{0'} = +1.175V \text{ vs SHE at pH } 7) \quad (13.4)$$

$$\tfrac{1}{2}N_2O + e^- + H^+ \rightarrow \tfrac{1}{2}N_2O + \tfrac{1}{2}H_2O(E^{0'} = +1.355V \text{ vs SHE at pH } 7) \quad (13.5)$$

In total, 2 moles of nitrate and 10 moles of electrons per 1 mole of gas nitrogen (N_2) are used:

$$2NO_3^- + 10e^- + 12H^+ \rightarrow N_2 + 6H_2O \quad (13.6)$$

The idea of using a biocathode for denitrification has been first proposed in 1966 [89]. The power production of denitrifying MFC is lower than in the systems where oxygen is reduced; however, such reactor allows for simultaneous carbon and nitrogen removal. In addition, since in denitrifying MFCs both cathodic and anodic chambers are anaerobic, there is no risk of oxygen leaking to the anodic compartment [90, 91].

The denitrification on the biocathode has been also performed in a single-chamber MFC [90]. Some of the denitrifying bacteria have been recognized and identified as belonging to *Proteobacteria, Actinobacteria,* or *Bacteroidetes* phyla [92, 93]. The experiments with the process of incubation of the denitrifying bacteria on the cathode, prior to the denitrification, have also been conducted. This might be done either by operating the MFC without the addition of nitrate while the acclimatization is done and introducing nitrate afterward [92] or by placing the cathode in the real environment before using it in bioelectrochemical systems [94].

13.6.3 Metal recovery using microbial electrochemistry

Metals are among the most important raw material in our society. The use of metals is essential in many branches of industry, especially in the high-tech industry. In addition, it is noticeable that the market of some metals is in a monopolistic state, with some countries controlling almost all the global extraction and production of these specific metals. Today, the most revealing example of this situation is the case of China, which controls almost half of arsenic production, two-third of the germanium production, and the quasi totality of the lanthanides production [95]. In this context, the ability to recover those metals from wastes is slowly becoming more and more attractive.

Mining industrial water can be used as a source of a raw metal. The use of microbial-electrochemical technologies appears promising for many reasons like the possibility to process complex successions of enzymatic reactions (like oxidation of acetate into carbon dioxide) through bacterial metabolism.

The microbial electrometallurgy is based on the exploitation of microorganisms metabolism at the anode to provide electrons from organic matter and the reduction of metal ions present in the liquid medium at the cathode.

An advantage of the microbial electrometallurgy compared to other methods to recover raw metals from waters is that this system is efficient even with very low concentrations of metal ions in the liquid medium.

One specificity of the reduction of metal ions is that their reduction potentials are very diverse, for example, from $E^0 = -0.76$ V vs SHE for zinc (eq. (13.7)) to $E^0 = +1.50$ V vs SHE for gold (eq. (13.8)) and can be even higher or lower in case the ion is coupled with other elements in solution:

$$Zn^{2+} + 2e^- \rightarrow Zn_{(s)} \left(E^0 = -0.76V \text{ vs SHE}\right) \quad (13.7)$$

$$Au^{3+} + 3e^- \rightarrow Au_{(s)} \left(E^0 = +1.50V \text{ vs SHE}\right) \quad (13.8)$$

Hence, the type of microbial reduction will take a configuration based on the reduction potential of the metal. For instance, for metals ions with high reduction potential like copper or silver [96, 97], an MFC configuration is used, meaning no external power source is required and that energy from the metal reduction reaction can even be recovered.

On the other hand, for metals with low reduction potentials like cadmium or zinc [98], a microbial electrolysis cell (MEC) configuration is used with the need of a power source to allow the reduction (Fig. 13.4). It is also noticeable that sometimes intermediaries are needed between the cathode and the bacteria itself for metal reduction, such as hydrogen peroxide that act as an electron shuttle (Fig. 13.4) [99].

The recovery of metals using microbial electrometallurgy is an interesting field of research, especially because of the possibility to be applied on the medium with very low concentrations of metals [99]. Currently, this method is studied for common metals, like copper or zinc, and highly profitable metals, like uranium or chromium, but is also promising for other metals recovery like lanthanides that represent high economical stakes [95].

Electroactive microbial technologies not only allow to treat organic and inorganic pollutants from water but also enable metal recovery. One of the advantages to these technologies is to produce energy by treating water. The production of energy can occur by the conversion of the chemical energy of substrates into electrical energy or by the production of hydrogen. Hydrogen production with microbial electrochemical systems will be discussed next.

13.6.4 Hydrogen production through biocatalysis in a microbial electrolysis cell

In a political and social context where developed societies are trying to greatly reduce any form of waste and increase recycling, the use of wastewaters to produce useful elements has become an important research field. Nowadays, those wastewaters can mainly be used to produce two different products: methane and hydrogen.

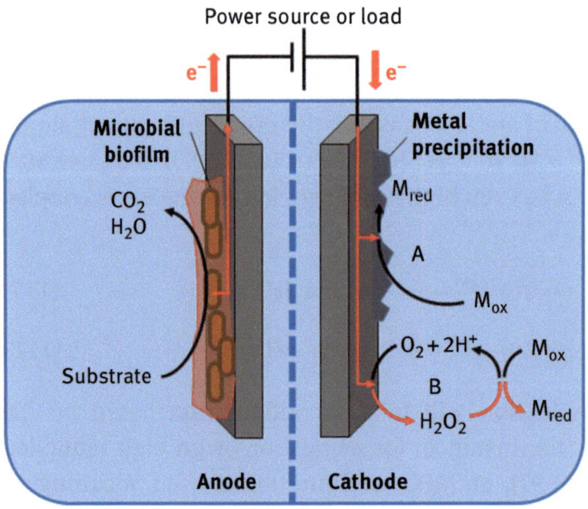

Fig. 13.4: Representation of metal recovery system using microbial bioelectrochemistry. The reduction potential of the metal recovered will determine whether this system is an MFC (with a load) or an MEC (with a power source).

Methane is a biogas that can be used for heat or electricity generation, making it an interesting output product from wastewater treatment. Methane is produced through the anaerobic metabolism of some microorganisms.

However, it appears that hydrogen is actually more valuable than methane when both molecules are produced from a specific quantity of wastewater [100]. For example, hydrogen is used as a reactive molecule not only in the petrochemical fields industry but also in the food industry for fat saturation.

There are other ways to produce hydrogen from wastewaters apart from MEC. An example is dark fermentation [101], a process based on anaerobic fermentation of some bacteria like *Clostridiacea*. Known for many years and already applied in the production of hydrogen, this process has shown its limit because of a low hydrogen yields caused by thermodynamically unfavorable reactions [102, 103]. For this reason, the MEC technology is promising for producing hydrogen gas from wastewaters.

The production of hydrogen through an MEC is based on electroactive bacterial community fixed on the anode that will oxidize organic water and generate CO_2 or HCO_3^- or others catabolic products (depending on the metabolic pathway of the bacteria), electrons, and protons. Electrons and protons combine at the cathode to produce dihydrogen (H_2) [103].

It is important to note that this reaction needs an external applied voltage. Indeed, the potentials at which protons can be reduced are $E^0 = 0$ V vs SHE at pH = 0 and $E^{0\prime} = -0.414$ V vs SHE at pH 7. In the case of acetate, the thermodynamic oxidation potential is around $E^{0\prime} = -0.300$ V vs SHE at pH 7 [16].

The reduction potential of protons is more negative than the acetate oxidation at the same pH, making the reaction thermodynamically unfavorable. To overcome this, an external potential can be permanently applied to allow the oxidation of acetate and the reduction of hydrogen to occur by electrolysis at low-energy input [104].

When acetate is used as organic matter, the reactions at the two electrodes are as follows:

Anode:

$$CH_3COO^- + 2H_2O \rightarrow HCO_3^- + 8e^- + 8H^+ (E^{0'} = -0.300V \text{ vs SHE at pH 7}) \quad (13.9)$$

Cathode:

$$8e^- + 8H^+ \rightarrow 4H_2 (E^0 = 0V \text{ vs SHE and } E^{0'} = -0.414V \text{ vs SHE at pH 7}) \quad (13.10)$$

There is more than one configuration that have been tested to produce hydrogen through MEC. The most common type of configuration is the two-chambers MEC with the anodic and catholic cells separated by a membrane (commonly a proton-exchange membrane) and with the cathode and the anode connected to a power supply (Fig. 13.5A).

Nonetheless, there are many other configurations that have been tested like a single-chamber cell-producing hydrogen without any membrane between the anode and the cathode.

The most interesting configuration for hydrogen production from wastewater is the continuous flow configuration. In this configuration, a gas-phase cathode is used, where hydrogen is produced. The absence of proton-exchange membrane

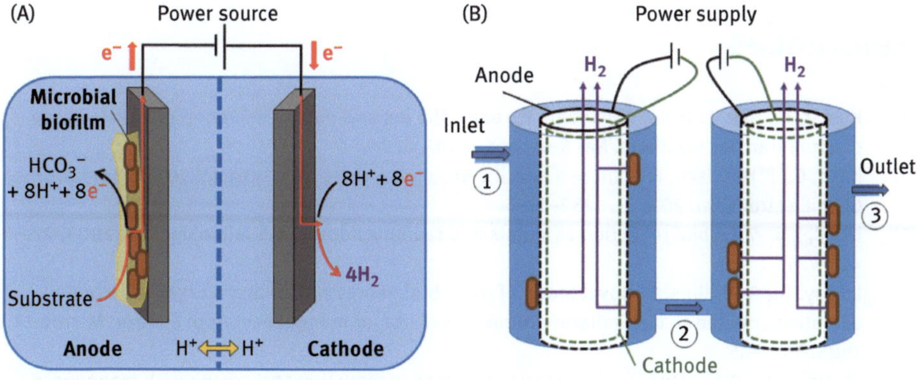

Fig. 13.5: Schematic representations of a common two chambers type configuration MEC to produce hydrogen (A) and a continuous flow configuration MEC (B). From Kadier et al. [105].

allows to decrease the distance between the two electrodes and hence the internal resistance.

Thanks to electroactive biofilms onto electrode, it is possible to treat different water streams while producing high added-value compounds, particularly hydrogen or metals [106]. This method has the main advantage of being more environmental friendly than the conventional ones.

13.7 Conclusion

The interest for MFC and all related microbial electrochemical technologies comes from the many potentials that they offer, not only for treating waste water but also for recovering energy, metals, and hydrogen. Water treatment is achieved by removing organics in the waste at the bioanode, and removal of nitrogen species or metal ions at the biocathode. Research that developed in this field during the last decades has permitted the identification of many electroactive strains. One can be optimistic for the future developments in this field of electromicrobiology. While the limitations of MFCs have been now clearly identified (low and variable conductivity of the wastewater and relatively low current densities), the possibility offered by the technology are engaging researchers in exploring new cell configurations and applications while progressing toward a better fundamental understanding of the complex interaction of microbes with electrodes for their optimal use. In this context and as discussed in the chapter, the challenge of mastering and understanding biocathodes, or the demonstration of efficient electrochemical snorkels for water treatment, represent an example of current challenges being tackled by the multidisciplinary scientific community interested in microbial electrochemistry.

References

[1] Kumar, A., Hsu, L.-H.-H., Kavanagh, P., et al. The ins and outs of microorganism–electrode electron transfer reactions. Nat Rev Chem. 2017, 1, 24.

[2] Koch, C., & Harnisch, F. Is there a specific ecological niche for electroactive microorganisms? ChemElectroChem. 2016, 3, 1282–1295.

[3] Lloyd, J. R. Microbial reduction of metals and radionuclides. FEMS Microbiol Rev. 2003, 27, 411–425.

[4] Lovley, D., & Phillips, E. Novel mode of microbial energy metabolism: organic carbon oxidation coupled to dissimilatory reduction of iron or manganese. Appl Environ Microbiol. 1988, 54, 1–15.

[5] Myers, C. R., & Nealson, K. H. Bacterial manganese reduction and growth with manganese oxide as the sole electron acceptor. Science. 1988, 240, 1319–1321.

[6] Nevin, K. P., & Lovley, D. R. Mechanisms for Fe(III) oxide reduction in sedimentary environments. Geomicrobiol J. 2002, 19, 141–159.

[7] Gralnick, J. A., & Newman, D. K. Extracellular respiration. Mol Microbiol. 2007, 65, 1–11.

[8] Kracke, F., Vassilev, I., & Krömer, J. O. Microbial electron transport and energy conservation – The foundation for optimizing bioelectrochemical systems. Front Microbiol. 2015, 6, 1–18.

[9] Yarzábal, A., Brasseur, G., Ratouchniak, J., et al. The high-molecular-weight cytochrome c Cyc2 of Acidithiobacillus ferrooxidans is an outer membrane protein. J Bacteriol. 2002, 184, 313–317.

[10] Methé, B. A., Nelson, K. E., Eisen, J. A., et al. Genome of Geobacter sulfurreducens: metal reduction in subsurface environments. Science (80-). 2003, 302, 1967–1969.

[11] Shi, L., Chen, B., Wang, Z., et al. Isolation of a high-affinity functional protein complex between OmcA and MtrC: two outer membrane decaheme c-type cytochromes of Shewanella oneidensis MR-1. J Bacteriol. 2006, 188, 4705–4714.

[12] Richter, O. M. H., & Ludwig, B. Electron transfer and energy transduction in the terminal part of the respiratory chain – lessons from bacterial model systems. Biochim Biophys Acta – Bioenerg. 2009,1787,626–634.

[13] Costa, N. L. Gram positive bacteria do it differently? – Probing the molecular bases for the efficient extracellular electron transfer performed by Thermincola potens JR. PhD Thesis, ITQB Universidade Nova de Lisboa, 2017.

[14] Costa, N. L., Clarke, T. A., Philipp, L. A., Gescher, J., Louro, R. O., & Paquete, C. M. Electron transfer process in microbial electrochemical technologies: the role of cell-surface exposed conductive proteins. Bioresour Technol. 2018,255,308–317.

[15] Yilmazel, Y. D., Zhu, X., Kim, K. Y., Holmes, D. E., & Logan, B. E. Electrical current generation in microbial electrolysis cells by hyperthermophilic archaea Ferroglobus placidus and Geoglobus ahangari. Bioelectrochemistry. 2018, 119, 142–149.

[16] Logan, B.E., Microbial Fuel Cells. John Wiley & Sons, Inc., New York, 2008.

[17] Logan, B. E., & Regan, J. M. Electricity-producing bacterial communities in microbial fuel cells. Trends Microbiol. 2006, 14, 512–518.

[18] Mook, W. T., Aroua, M. K. T., Chakrabarti, M. H., Noor, I. M., Irfan, M. F., & Low, C. T. J. A review on the effect of bio-electrodes on denitrification and organic matter removal processes in bio-electrochemical systems. J Ind Eng Chem. 2013, 19, 1–13.

[19] Leong, J. X., Daud, W. R. W., Ghasemi, M., Liew, K. B., & Ismail, M. Ion exchange membranes as separators in microbial fuel cells for bioenergy conversion: a comprehensive review. Renew Sustain Energy Rev. 2013, 28, 575–587.

[20] Pous, N., Puig, S., Dolors Balaguer, M., & Colprim, J. Cathode potential and anode electron donor evaluation for a suitable treatment of nitrate-contaminated groundwater in bioelectrochemical systems. Chem Eng J. 2015, 263, 151–159.

[21] Samrat, N., Kaza, K. R., Ruggeri, B., & Tommasi, T. Denitrification of water in a microbial fuel cell (MFC) using seawater bacteria. Journal of Cleaner Production, 2018, 178, 449–456.

[22] Varcoe, J. R., Atanassov, P., Dekel, D. R., et al. Anion-exchange membranes in electrochemical energy systems. Energy Environ Sci. 2014, 7, 3135–3191.

[23] Pandit, S., Ghosh, S., Ghangrekar, M. M., & Das, D. Performance of an anion exchange membrane in association with cathodic parameters in a dual chamber microbial fuel cell. Int J Hydrogen Energy. 2012, 37, 9383–9392.

[24] Park, D. H., & Zeikus, J. G. Improved fuel cell and electrode designs for producing electricity from microbial degradation. Biotechnol Bioeng. 2003, 81, 348–355.

[25] Liu, H., Ramnarayanan, R., & Logan, B. E. Production of electricity during wastewater treatment using a single chamber microbial fuel cell. Environ Sci Technol. 2004, 38, 2281–2285.

[26] Zabihallahpoor, A., Rahimnejad, M., & Talebnia, F. Sediment microbial fuel cells as a new source of renewable and sustainable energy: present status and future prospects. RSC Adv. 2015, 5, 94171–94183.

[27] Ewing, T.; Thi, P.; Babauta, J. T.; Trong, N.; Heo, D.; Beyenal, H. Scale-up of Sediment Microbial Fuel Cells. J. Power Sources, 2014, 272, 311–319.

[28] Erable, B., Etcheverry, L., & Bergel, A. From microbial fuel cell (MFC) to microbial electrochemical snorkel (MES): maximizing chemical oxygen demand (COD) removal from wastewater. Biofouling. 2011, 27, 319–326.

[29] Bellagamba, M., Cruz Viggi, C., Ademollo, N., Rossetti, S., & Aulenta, F. Electrolysis-driven bioremediation of crude oil-contaminated marine sediments. N Biotechnol. 2017, 38, 84–90.

[30] Viggi, C. C., Matturro, B., Frascadore, E., et al. Bridging spatially segregated redox zones with a microbial electrochemical snorkel triggers biogeochemical cycles in oil-contaminated River Tyne (UK) sediments. Water Res. 2017, 127, 11–21.

[31] Cruz Viggi, C., Presta, E., Bellagamba, M., et al. The "Oil-Spill Snorkel": an innovative bioelectrochemical approach to accelerate hydrocarbons biodegradation in marine sediments. Front Microbiol. 2015, 6, 881.

[32] Yang, Q., Zhao, H., & Liang, H. H. Denitrification of overlying water by microbial electrochemical snorkel. Bioresour Technol. 2015, 197, 512–514.

[33] Cornejo, J. A., Lopez, C., Babanova, S., et al. Surface modification for enhanced biofilm formation and electron transport in Shewanella anodes. J Electrochem Soc. 2015, 162, H597–5603.

[34] Guo, K., Prévoteau, A., Patil, S. A., & Rabaey, K. Materials and their surface modification for use as anode in microbial bioelectrochemical systems. 2017.

[35] Guo, K., Prévoteau, A., Patil, S. A., & Rabaey, K. Engineering electrodes for microbial electrocatalysis. Curr Opin Biotechnol. 2015, 33, 149–156.

[36] Jones, N. Platinum pollution issue gets measured. Nature News Blog. 2009. doi:10.1038/news.2009.375

[37] Bergel, A., Féron, D., & Mollica, A. Catalysis of oxygen reduction in PEM fuel cell by seawater biofilm. Electrochem commun. 2005, 7, 900–904.

[38] Zhou, M., Chi, M., Luo, J., He, H., & Jin, T. An overview of electrode materials in microbial fuel cells. J Power Sources. 2011, 196, 4427–4435.

[39] Picot, M., Lapinsonnière, L., Rothballer, M., & Barrière, F. Graphite anode surface modification with controlled reduction of specific aryl diazonium salts for improved microbial fuel cells power output. Biosens Bioelectron. 2011, 28, 181–188.

[40] Cheng, S., & Logan, B. E. Ammonia treatment of carbon cloth anodes to enhance power generation of microbial fuel cells. Electrochem commun. 2007, 9, 492–496.

[41] Santoro, C., Babanova, S., Artyushkova, K., et al. Influence of anode surface chemistry on microbial fuel cell operation. Bioelectrochemistry. 2015, 106, 141–149.

[42] Pierra, M., Golozar, M., Zhang, X., et al. Growth and current production of mixed culture anodic biofilms remain unaffected by sub-microscale surface roughness. Bioelectrochemistry. 2018, 122, 213–220.

[43] Saito, T., Mehanna, M., Wang, X., et al. Effect of nitrogen addition on the performance of microbial fuel cell anodes. Bioresour Technol. 2011, 102, 395–398.

[44] Kuzume, A., Zhumaev, U., Li, J., et al. An in situ surface electrochemistry approach towards whole-cell studies: the structure and reactivity of a Geobacter sulfurreducens submonolayer on electrified metal/electrolyte interfaces. Phys Chem Chem Phys. 2014, 16, 22229–22236.

[45] Guo, K., Freguia, S., Dennis, P. G., et al. Effects of surface charge and hydrophobicity on anodic biofilm formation, community composition, and current generation in bioelectrochemical systems. Environ Sci Technol. 2013, 47, 7563–7570.

[46] Flexer, V., Marque, M., Donose, B. C., Virdis, B., & Keller, J. Plasma treatment of electrodes significantly enhances the development of anodic electrochemically active biofilms. Electrochim Acta. 2013, 108, 566–574.

[47] Lapinsonnière, L., Picot, M., Poriel, C., & Barrière, F. Phenylboronic acid modified anodes promote faster biofilm adhesion and increase microbial fuel cell performances. Electroanalysis. 2013, 25, 601–605.

[48] Smida, H., Lebègue, E., Bergamini, J. F., Barrière, F., & Lagrost, C. Reductive electrografting of in situ produced diazopyridinium cations: tailoring the interface between carbon electrodes and electroactive bacterial films. Bioelectrochemistry. 2018, 120, 157–165.

[49] Scott, K., Rimbu, G. A., Katuri, K. P., Prasad, K. K., & Head, I. M. Application of modified carbon anodes in microbial fuel cells. Process Saf Environ Prot. 2007, 85, 481–488.

[50] Pinck, S., Jorand, F., & Etienne, M. Electrochemistry of biofilms [Internet]. In: Wandelt, K., (Ed.) Encyclopedia of Interfacial Chemistry: Surface Science and Electrochemistry, 7, 182–189.

[51] Luckarift, H. R., Sizemore, S. R., Roy, J., et al. Standardized microbial fuel cell anodes of silica-immobilized Shewanella oneidensis. Chem Commun. 2010, 46, 6048.

[52] Yu YY, Chen HL, Yong YC, Kim DH, Song H.Conductive artificial biofilm dramatically enhances bioelectricity production in Shewanella-inoculated microbial fuel cells. Chem Commun 2011,47,12825–7.

[53] Yong YC, Yu YY, Zhang X, Song H. Highly active bidirectional electron transfer by a self-assembled electroactive reduced-graphene-oxide-hybridized biofilm. Angew Chemie – Int Ed 2014,53,4480–3.

[54] Zhao C, Wu J, Ding Y, et al. Hybrid Conducting Biofilm with Built-in Bacteria for High-Performance Microbial Fuel Cells. ChemElectroChem 2015,2,654–8.

[55] Kato S, Hashimoto K, Watanabe K. Microbial interspecies electron transfer via electric currents through conductive minerals. Proc Natl Acad Sci 2012,109,10042–6.

[56] Nakamura R, Kai F, Okamoto A, Newton GJ, Hashimoto K. Self-constructed electrically conductive bacterial networks. Angew Chemie - Int Ed 2009,48,508–11.

[57] Estevez-Canales M, Pinto D, Coradin T, Laberty-Robert C, Esteve-Núñez A. Silica immobilization of Geobacter sulfurreducens for constructing ready-to-use artificial bioelectrodes. Microb Biotechnol 2018,11,39–49

[58] Ghach W, Etienne M, Urbanova V, Jorand FPA, Walcarius A. Sol–gel based 'artificial' biofilm from Pseudomonas fluorescens using bovine heart cytochrome c as electron mediator. Electrochem commun 2014,38,71–4.

[59] Vostiar, I., Ferapontova, E., & Electrical, G. L. "Wiring" of viable Gluconobacter oxydans cells with a flexible osmium-redox polyelectrolyte. Electrochem commun. 2004, 6, 621–626.

[60] Pinck, S., Etienne, M., Dossot, M., & Jorand, F. P. A. A rapid and simple protocol to prepare a living biocomposite that mimics electroactive biofilms. Bioelectrochemistry. 2017, 118, 131–138.

[61] Pinck, S., Xu, M., Clement, R., Lojou, E., Jorand, F., & Etienne, M. Influence of cytochrome charge and potential on the cathodic current of electroactive artificial biofilms. Bioelectrochemistry. 2018, 124, 185–194.

[62] Faimali, M., Chelossi, E., Garaventa, F., Corrà, C., Greco, G., & Mollica, A. Evolution of oxygen reduction current and biofilm on stainless steels cathodically polarised in natural aerated seawater. Electrochim Acta. 2008, 54, 148–153.

[63] Dumas, C., Mollica, A., Féron, D., Basseguy, R., Etcheverry, L., & Bergel, A. Checking graphite and stainless anodes with an experimental model of marine microbial fuel cell. Bioresour Technol. 2008, 99, 8887–8894.

[64] Commault, A. S., Laczka, O., Siboni, N., et al. Electricity and biomass production in a bacteria-Chlorella based microbial fuel cell treating wastewater. J Power Sources. 2017, 356, 299–309.

[65] Rothballer, M., Picot, M., Sieper, T., et al. Monophyletic group of unclassified γ-Proteobacteria dominates in mixed culture biofilm of high-performing oxygen reducing biocathode. Bioelectrochemistry. 2015, 106, 167–176.

[66] Cournet, A., Bergé, M., Roques, C., Bergel, A., & Délia, M.-L. Electrochemical reduction of oxygen catalyzed by Pseudomonas aeruginosa. Electrochim Acta. 2010, 55, 4902–4908.
[67] Clauwaert, P., van der Ha, D., Boon, N., et al. Open air biocathode enables effective electricity generation with microbial fuel cells. Environ Sci Technol. 2007, 41, 7564–7569.
[68] Cha, J., Choi, S., Yu, H., Kim, H., & Kim, C. Directly applicable microbial fuel cells in aeration tank for wastewater treatment. Bioelectrochemistry. 2010, 78, 72–79.
[69] Rabaey, K., Read, S. T., Clauwaert, P., et al. Cathodic oxygen reduction catalyzed by bacteria in microbial fuel cells. ISME J. 2008, 2, 519–527.
[70] Aldrovandi, A., Marsili, E., Stante, L., Paganin, P., Tabacchioni, S., & Giordano, A. Sustainable power production in a membrane-less and mediator-less synthetic wastewater microbial fuel cell. Bioresour Technol. 2009, 100, 3252–3260.
[71] Renslow, R., Donovan, C., Shim, M., et al. Oxygen reduction kinetics on graphite cathodes in sediment microbial fuel cells. Phys Chem Chem Phys. 2011, 13, 21573.
[72] Chen, Z., Huang, Y. C., Liang, J. H., Zhao, F., & Zhu, Y. G. A novel sediment microbial fuel cell with a biocathode in the rice rhizosphere. Bioresour Technol. 2012, 108, 55–59.
[73] Zhang, G., Zhao, Q., Jiao, Y., Wang, K., & Lee D-J, R. N. Biocathode microbial fuel cell for efficient electricity recovery from dairy manure. Biosens Bioelectron. 2012, 31, 537–543.
[74] Strycharz-Glaven, S. M., Glaven, R. H., Wang, Z., Zhou, J., Vora, G. J., & Tender, L. M. Electrochemical investigation of a microbial solar cell reveals a nonphotosynthetic biocathode catalyst. Appl Environ Microbiol 2013,79,3933–3942.
[75] Reimers, C. E., Girguis, P., Stecher, H. A., Tender, L. M., Ryckelynck, N., & Whaling, P. Microbial fuel cell energy from an ocean cold seep. Geobiology. 2006, 4, 123–136.
[76] Cheng, S., Liu, H., & Logan, B. E. Increased performance of single-chamber microbial fuel cells using an improved cathode structure. Electrochem commun. 2006, 8, 489–494.
[77] Li, W., Sun, J., Hu, Y., Zhang, Y., Deng, F., & Chen, J. Simultaneous pH self-neutralization and bioelectricity generation in a dual bioelectrode microbial fuel cell under periodic reversion of polarity. J Power Sources. 2014, 268, 287–293.
[78] Logan, B. E. Scaling up microbial fuel cells and other bioelectrochemical systems. Appl Biochem Biotechnol. 2010, 85, 1665–1671.
[79] Behera, S. D., Kumari, U., Shankar, R., & Mondal, P. Performance analysis of a double-chambered microbial fuel cell employing a low-cost sulfonated polystyrene proton exchange membrane. Ionics (Kiel). 2018, 24, 3573–3590.
[80] Khalid, S., Alvi, F., Fatima, M., et al. Dye degradation and electricity generation using microbial fuel cell with graphene oxide modified anode. Mater Lett. 2018, 220, 272–276.
[81] Estrada-Arriaga, E. B., Hernández-Romano, J., García-Sánchez, L., et al. Domestic wastewater treatment and power generation in continuous flow air-cathode stacked microbial fuel cell: effect of series and parallel configuration. J Environ Manage. 2018, 214, 232–241.
[82] Fan, Y., Hu, H., & Liu, H. Enhanced Coulombic efficiency and power density of air-cathode microbial fuel cells with an improved cell configuration. J Power Sources. 2007,171,348–354.
[83] Wei, J., Liang, P., & Huang, X. Recent progress in electrodes for microbial fuel cells. Bioresour Technol. 2011, 102, 9335–9344.
[84] Rimboud, M., Barakat, M., Bergel, A., & Erable, B. Different methods used to form oxygen reducing biocathodes lead to different biomass quantities, bacterial communities, and electrochemical kinetics. Bioelectrochemistry. 2017, 116, 24–32.
[85] Milner, E. M., Popescu, D., Curtis, T., Head, I. M., Scott, K., & Yu, E. H. Microbial fuel cells with highly active aerobic biocathodes. J Power Sources. 2016, 324, 8–16.
[86] Debuy, S., Pecastaings, S., Bergel, A., & Erable, B. Oxygen-reducing biocathodes designed with pure cultures of microbial strains isolated from seawater biofilms. Int Biodeterior Biodegradation. 2015, 103, 16–22.

[87] The Council of the European Union. Council Directive 98/83/EC of 3 November 1998 on the quality of water intended for human consumption. Off J Eur Communities. 1998, L330, 32–54.
[88] Piren-Seine, P. Synthèses Pratiques agricoles et nitrates dans les milieux aquatiques. 2011,
[89] Guilbault, G. G. Symposium on bioelectrochemistry of microorganisms. 3. Electrochemical analysis of enzymatic reactions. Bacteriol Rev. 1966, 30, 94–100.
[90] Lefebvre, O., Al-Mamun, A., & Ng, H. Y. A microbial fuel cell equipped with a biocathode for organic removal and denitrification. Water Sci Technol. 2008, 58, 881–885.
[91] Zhang, G., Zhang, H., Zhang, C., et al. Simultaneous nitrogen and carbon removal in a single chamber microbial fuel cell with a rotating biocathode. Process Biochem. 2013, 48, 893–900.
[92] Zhang, Y., & Angelidaki, I. Bioelectrode-based approach for enhancing nitrate and nitrite removal and electricity generation from eutrophic lakes. Water Res. 2012, 46, 6445–6453.
[93] Zhu, G., Chen, G., Yu, R., Li, H., & Wang, C. Enhanced simultaneous nitrification/denitrification in the biocathode of a microbial fuel cell fed with cyanobacteria solution. Process Biochem. 2016, 51, 80–88.
[94] Wang, Y., Hu, J., Wang, L., et al. Acclimated sediment microbial fuel cells from a eutrophic lake for the in situ denitrification process. RSC Adv. 2016, 6, 80079–80085.
[95] Price, J. G. The world is changing. Soc Econ Geol. 2010, 82, 12–14
[96] Ter, H. A., Liu, F., Weijden, R., van Der, Weijma, J., Cjn, B., & Hamelers, H. V. M. Copper recovery combined with electricity production in a microbial fuel cell. Environ Sci Technol. 2010, 44, 4376–4381.
[97] Choi, C., & Cui, Y. Recovery of silver from wastewater coupled with power generation using a microbial fuel cell. Bioresour Technol. 2012, 107, 522–525.
[98] Modin, O., Wang, X., Wu, X., Rauch, S., & Fedje, K. K. Bioelectrochemical recovery of Cu, Pb, Cd, and Zn from dilute solutions. J Hazard Mater. 2012, 235–236, 291–297.
[99] Dominguez-Benetton, X., Varia, J. C., Pozo, G., et al. Metal recovery by microbial electro-metallurgy. Prog Mater Sci. 2018, 94, 435–461.
[100] Logan, B. E., Peer reviewed: extracting hydrogen and electricity from renewable resources. Environ Sci Technol. 2004, 38, 160–167.
[101] Angenent, L. T., Karim, K., Al-Dahhan, M. H., Wrenn, B. A., & Domíguez-Espinosa, R. Production of bioenergy and biochemicals from industrial and agricultural wastewater. Trends Biotechnol. 2004, 22, 477–485.
[102] Drake, H. L., Küsel, K., & Matthies, C. Acetogenic prokaryotes. In: The Prokaryotes: prokaryotic Physiology and Biochemistry. Springer Berlin Heidelberg 2013, p. 3–60.
[103] Rozendal, R. A. Hydrogen production through biocatalyzed electrolysis. PhD Thesis, Wageningen University & Research E-depot, Netherlands 2007.
[104] Tartakovsky, B., Manuel, M., Wang, H., & Guiot, S. High rate membrane-less microbial electrolysis cell for continuous hydrogen production. Int J Hydrogen Energy. 2009, 34, 672–677.
[105] Kadier, A., Simayi, Y., Abdeshahian, P., Azman, N., Chandrasekhar, K., & Kalil, M. A comprehensive review of microbial electrolysis cells (MEC) reactor designs and configurations for sustainable hydrogen gas production. Alexandria Eng J 2016, 55, 427–443.
[106] Luo H, Liu G, Zhang R, Bai Y, Fu S, Hou Y. Heavy metal recovery combined with H2 production from artificial acid mine drainage using the microbial electrolysis cell. J Hazard Mater 2014, 270, 153–9.

Index

https://doi.org/10.1515/9783110570526-014